Physical Activity and Health Guidelines

Recommendations for Various Ages, Fitness Levels, and Conditions from 57 Authoritative Sources

Riva L. Rahl, MD
Cooper Clinic

Human Kinetics

Library of Congress Cataloging-in-Publication Data

Rahl, Riva L., 1973-
  Physical activity and health guidelines : recommendations for various ages,
fitness levels, and conditions from 57 authoritative sources / Riva L. Rahl.
     p. ; cm.
  Includes bibliographical references and index.
  ISBN-13: 978-0-7360-7943-3 (print)
  ISBN-10: 0-7360-7943-2 (print)
  1. Exercise. 2. Physical fitness. 3. Health. I. Title.
  [DNLM: 1. Exercise--Guideline. 2. Physical Fitness--Guideline. 3. Exercise
Therapy--Guideline. 4. Health Behavior--Guideline. QT 255 R147p 2010]
  QP301.R25 2010
  612.7'6--dc22

                                                           2009053208

ISBN-10: 0-7360-7943-2 (print)
ISBN-13: 978-0-7360-7943-3 (print)

The Web addresses cited in this text were current as of November 30, 2009, unless otherwise noted.

Acquisitions Editor: Michael S. Bahrke, PhD; **Developmental Editors:** Judy Park and Jillian Evans; **Assistant Editor:** Dena P. Mumm; **Copyeditor:** Jocelyn Engman; **Indexer:** Bobbi Swanson; **Permission Managers:** Dalene Reeder and Martha Gullo; **Graphic Designer:** Nancy Rasmus; **Graphic Artist:** Kathleen Boudreau-Fuoss; **Cover Designer:** Keith Blomberg; **Photographer (cover):** © Human Kinetics; **Photographer (interior):** © Human Kinetics unless otherwise noted. Part I photo collage (l to r): iStockphoto/Catherine Yeulet; Bananastock; Jon Feingersh/Blend Images/Getty Images. Part II photo collage (l to r): Eyewire; Brand X Pictures; Photodisc/Getty Images. Part III photo collage (l to r): Bananastock; Blend Stock; iStockphoto/Lisa F. Young. Part IV photo collage (l to r): Dmitry Ersler/fotolia.com; Bill Crump/Brand X Pictures; © Elke Dennis-Fotolia.com; **Photo Asset Manager:** Laura Fitch; **Photo Production Manager:** Jason Allen; **Art Manager:** Kelly Hendren; **Associate Art Manager:** Alan L. Wilborn; **Illustrator:** Alan L. Wilborn; **Printer:** Sheridan Books

Printed in the United States of America   10 9 8 7 6 5 4 3 2 1

The paper in this book is certified under a sustainable forestry program.

Human Kinetics
Web site: www.HumanKinetics.com

United States: Human Kinetics
P.O. Box 5076
Champaign, IL 61825-5076
800-747-4457
e-mail: humank@hkusa.com

Canada: Human Kinetics
475 Devonshire Road Unit 100
Windsor, ON N8Y 2L5
800-465-7301 (in Canada only)
e-mail: info@hkcanada.com

Europe: Human Kinetics
107 Bradford Road
Stanningley
Leeds LS28 6AT, United Kingdom
+44 (0) 113 255 5665
e-mail: hk@hkeurope.com

Australia: Human Kinetics
57A Price Avenue
Lower Mitcham, South Australia 5062
08 8372 0999
e-mail: info@hkaustralia.com

New Zealand: Human Kinetics
P.O. Box 80
Torrens Park, South Australia 5062
0800 222 062
e-mail: info@hknewzealand.com

E4682

To my family.
You all inspire and amaze me.

CONTENTS

FOREWORD   viii
PREFACE   ix
ACKNOWLEDGMENTS   xi

PART I   **General Health and Fitness Guidelines**   **1**

1 Exploring the Relationship Between Physical Activity and Health   **3**

History and Development of Physical Activity Guidelines 4 •
Basic Concepts and Definitions 5 • Changes and Benefits Resulting
From Physical Activity 7 • Risks of Physical Activity 11 • Patterns in
Physical Activity Participation 12 • Summary 14

2 Exercise Guidelines for Physical Fitness and Health   **15**

Most Recent National Guidelines in the United States 16 • Significant
Historical Guidelines 21 • Specialized Guidelines 30 • State and
International Guidelines 36 • Summary 41

3 Guidelines for Personal Exercise Programs   **43**

Basic Definitions and the FITT Principle 43 • Components of
Physical Fitness 47 • Guidelines for Individualized Exercise
Prescription 48 • Summary 57

PART II   **Physical Activity Guidelines by Population**   **59**

4 Infants and Toddlers   **61**

Guidelines From the American Academy of Pediatrics 61 • Guidelines
From Other Organizations 63 • Guidelines for Prevention of Obesity and
Chronic Disease 67 • Physical Activity as Play and Nutritional Goals for
Children 69 • Summary 69

5 School-Aged Children   **71**

Basic Facts About Physical Activity and Health in Children 71 •
Establishment of the Earliest Guidelines for Children 74 • Landmark
Guidelines 76 • School and Community Health Guidelines 81 •
Guidelines Outside of the School Environment 87 • International
and State Guidelines 90 • Guidelines Focused on Specific Health
Goals 94 • Summary 101

6 Pregnant and Postpartum Women   103

Encouraging Pregnant Women to Be Physically Active 103 • Benefits and Risks of Physical Activity for Pregnant Women 104 • Primary Concerns of Exercise and Physical Activity During Pregnancy 106 • Guidelines for General Exercise Prescription 108 • Important National and International Guidelines 109 • Benefits and Risks of Physical Activity During the Postpartum Period 115 • Summary 119

7 Older Adults   121

Benefits of Physical Activity for Older Adults 121 • Major National Guidelines 124 • Other National Guidelines 128 • International and State Guidelines 132 • Guidelines for Adults Who Are Frail or at Risk for Falling 134 • Contraindications to Exercise 141 • Summary 142

PART III   Physical Activity Guidelines by Disease States   143

8 Cancer Prevention and Optimal Cardiometabolic Health   145

How Exercise Reduces the Risk of Cancer 146 • General Cancer Prevention Guidelines 147 • Guidelines for Specific Types of Cancer 150 • Metabolic Syndrome 150 • Type 2 Diabetes 151 • Summary 154

9 Cancer   155

Benefits of Physical Activity 155 • American Cancer Society Guidelines 157 • Other Notable Guidelines 158 • Guidelines for Specific Cancers 163 • Side Effects of Cancer Affecting Physical Activity 164 • Summary 165

10 Hypertension and Cardiovascular Disease   167

Benefits of Exercise for Hypertension 167 • American College of Sports Medicine Guidelines 169 • International Guidelines 171 • Guidelines for Coronary Artery Disease 173 • Summary 180

11 Arthritis and Osteoporosis   181

Rheumatoid Arthritis 181 • Osteoarthritis 183 • General Arthritis Guidelines 185 • Osteoarthritis Guidelines 187 • Osteoporosis 193 • Guidelines for Prevention of Osteoporosis 195 • Guidelines for Individuals With Osteoporosis 201 • Summary 203

12 Diabetes   205

Benefits of Physical Activity for Preventing and Managing Diabetes 206 • Potential Concerns Regarding Physical Activity 207 • Guidelines for Individuals With Type 2 Diabetes 209 • Guidelines for Individuals With Type 1 Diabetes 213 • Summary 218

13 Neuromuscular Disorders219

Benefits of Physical Activity 219 • General Recommendations From *Physical Activity Guidelines for Americans* 220 • Guidelines for Cerebral Palsy 220 • Guidelines for Parkinson's Disease 224 • Guidelines for Muscular Dystrophy 226 • Guidelines for Multiple Sclerosis 231 • Guidelines for Spinal Cord Injury and Disability 233 • Guidelines for Stroke and Brain Injury 235 • Summary 238

14 Asthma239

Exercise-Induced Asthma 239 • General Guidelines for People With Asthma 240 • Guidelines for Children With Asthma 244 • Summary 246

PART IV Guidelines for Exercise Testing and Beyond247

15 Exercise Testing249

Standards for Exercise Testing in Adults 249 • Benefits of Exercise Testing for Children 252 • Summary 253

16 Cardiac Exercise Testing and Prescription255

Candidates for Testing 255 • Utility of Information Acquired From Exercise Testing 257 • Protocols for Exercise Testing 258 • Exercise Testing Procedures 259 • Other Methods of Cardiac Testing 260 • Summary 261

17 Diet and Weight Management263

Basic Facts About Body Weight 263 • Dietary Guidelines From National Organizations 265 • International Dietary Guidelines 273 • Hydration, Energy, and Supplementation During Activity 274 • Dietary Guidelines for Special Populations 278 • Dietary Guidelines for Various Diseases 285 • Summary 292

18 Exercise Equipment and Facilities293

Aerobic Exercise Machines 293 • Weights 297 • Tools for Monitoring Physical Activity 298 • Exercising at Home 299 • Exercising in a Workout Facility 301 • Summary 305

APPENDIX A   307
APPENDIX B   313
REFERENCES   319
INDEX   347
ABOUT THE AUTHOR   355

FOREWORD

Physical Activity and Health Guidelines: Recommendations for Various Ages, Fitness Levels, and Conditions from 57 Authoritative Sources by Riva Rahl, MD, is the first compilation and interpretation of guidelines for recommending physical activity in both clinical and public health settings. Publication of such guidelines by various professional organizations and government agencies worldwide was initiated in the 1970s and has been stimulated, in part, by the large volume of new scientific data available on the complex relations between profiles of habitual physical activity and the numerous health outcomes that occur throughout the life span. In addition to the extensive review of the recommendations that have been made for youth, adults, older adults, and patients with various diseases or disabilities, Dr. Rahl provides commentary on the implications for the implementation of the recommendations. This review is a resource for health and exercise professionals who need to locate specific recommendations or want to compare one set of recommendations with others. Also, it is a unique resource for students interested in understanding the evolution of the development of physical activity and health guidelines over the past four decades.

Physical Activity and Health Guidelines is organized into four parts and 18 chapters, which allow easy access to specific recommendations or related information. Part I, General Health and Fitness Guidelines, includes a historical background of physical activity guidelines and a conceptual framework for prescribing physical activity to improve health. Part II, Physical Activity Guidelines by Population, provides an overview of recommendations for generally healthy persons throughout the life span, including a particularly informative chapter on guidelines and advice regarding exercise by pregnant and postpartum women. Part III, Physical Activity Guidelines by Disease States, reviews guidelines for the prevention of major chronic diseases with specific chapters focused on patients with cancer, cardiovascular disease and hypertension, arthritis and osteoporosis, diabetes, asthma, and neuromuscular disorders. Having all of these recommendations for specific patient populations in one location should prove especially valuable to physicians and other clinicians who provide exercise recommendations to a diverse population of patients. Part IV, Guidelines for Exercise Testing and Beyond, addresses issues aligned to exercise recommendations, including a chapter on the rationale and procedures for exercise testing in adults and children, a chapter reviewing nutrition guidelines especially when they have been incorporated with physical activity guidelines, and a final chapter on guidelines related to logistical issues involved in the implementation of a health-oriented program of physical activity.

As a practicing physician at the Cooper Clinic and medical director for the Cooper Aerobics Center in Dallas, Dr. Rahl has integrated a high-quality review of existing exercise guidelines with her substantial experience in implementing those guidelines in a clinical setting. This combination of expertise and experience has resulted in a valuable addition to the professional exercise literature.

William L. Haskell
March 2010

PREFACE

This book is designed for anyone interested in exploring the recommendations for exercise and health across all age groups and common disease states. Many people will benefit from this text, including the general population interested in fitness and health, students studying toward fitness and health certifications, and professionals setting up fitness and health programs for companies or civic organizations.

Although an abundance of information is available on physical activity and exercise recommendations for better health, this book is the first single source dedicated to compiling the many different recommendations on physical activity. This book presents a comprehensive overview of the current guidelines from the appropriate national and international organizations as well as provides helpful information for implementing these guidelines. It also includes a historical record of some of the earliest guidelines and covers significant milestones up to and including the recently published *Physical Activity Guidelines for Americans (PAGA)*. *PAGA* is notable in that it contains guidelines for many different subsets of the American population and is based on the most comprehensive review to date of the scientific literature on physical activity and health. More than 200 health professionals met over a span of 2 years to discuss and create these guidelines, which were published by the U.S. Department of Health and Human Services. This book refers to the *PAGA* guidelines as well as to many others as it details the physical activity recommendations for individuals beginning in infancy all the way through older adulthood, and for people with chronic disease or disability ranging from diabetes, asthma, osteoporosis, and osteoarthritis to cerebral palsy, stroke injury, and others. In addition, it addresses national guidelines geared toward the prevention of common diseases such as cancer, coronary artery disease, and the metabolic syndrome.

This text also explains the core components used to develop an exercise program, with attention to national guidelines, so that readers can be prepared to implement an appropriate exercise program, whether for an individual, a corporation, or a community group. Information on how physical activity recommendations can help people meet weight management guidelines is also included, as there is a close association between obesity and a sedentary lifestyle. An entire chapter is dedicated to the national dietary guidelines that have been published alongside physical activity guidelines; these are also organized by age group and disease state where available. Guidelines for cardiac and other exercise testing are included, as many health organizations utilize this information not only in implementing and evaluating physical activity programs but also in assessing the safety of recommending a program for certain individuals.

Part I of this book covers the history of physical activity guidelines as well as some of the science behind the relationship between physical activity and health. Chapter 2 describes the general guidelines for all Americans. A review of the components of a personal exercise program, provided in chapter 3, completes the first section. Part II covers physical activity guidelines by population, including infants, toddlers, school-aged children and adolescents, pregnant women, and adults who are older and even frail. The chapters in part III are arranged by disease state, and they cover physical activity guidelines for individuals as they relate to chronic health conditions such as hypertension, coronary artery disease, diabetes, cancer, arthritis, osteoporosis, asthma, cerebral palsy, and other disabling conditions. Part IV, the final part of the book, contains guidelines for exercise and cardiac testing, national nutritional guidelines as they relate to physical activity, and guidelines concerning the implementation of a physical activity program. The latter includes advice for setting up a home gym, selecting activity and health equipment, joining a fitness facility, and selecting appropriate exercise videos and other associated devices. At the end of this book is an appendix containing Web resources to help the interested reader look up more details on the various guidelines presented throughout the text.

Each set of physical activity and exercise guidelines is presented in a template that names the date issued (or most recently updated), the issuing organization, the target population, and the location of the source from which the guidelines are taken. Wherever possible, Web links to the original sources are provided, so that the reader can easily find the most recent version available. Many guidelines follow the principles of the physical activity prescription, dividing the physical activity recommendations into those for aerobic exercise, resistance training, and flexibility training. These are presented in table format for easy viewing. When provided, the frequency, intensity, duration, and type of activity for the three components of the physical activity prescription are included in these tables. Some guidelines do not address these specifics of the physical activity prescription, while others simply do not recommend that the target population participate in one component or another. Other organizations emphasize solely the aerobic element and do not comment on strength or flexibility. For incomplete guidelines, the template contains whatever information is available and designates which information is not specified. For example, many recommendations specify all the details for aerobic exercise but simply mention that flexibility training should be considered as part of a warm-up or cool-down. In these instances, the flexibility row within the table

does not include much detail. Other guidelines are somewhat vague, even when it comes to the prescription for aerobic exercise, and these guidelines are listed within the chapter text.

This book is focused on guidelines that apply to Americans and therefore most of the guidelines referenced herein are issued by American entities. Some notable guidelines, however, are established by important international bodies (such as the International Osteoporosis Foundation or the World Cancer Research Fund) and prescribe recommendations that apply just as well to inhabitants of the United States as they do to people from around the world. Thus this text includes relevant international guidelines. In a few cases, guidelines issued by other national entities (such as the Canadian Hypertension Society or the Society of Obstetricians and Gynaecologists of Canada) are included; these serve to reiterate the similarities they share with American guidelines and to highlight important differences that might prompt discussion. In other cases, foreign guidelines might simply offer more detail than the corresponding American recommendations.

Having gained an understanding of the various physical activity recommendations available today, readers may find they require access to the knowledge, equipment, and adjunctive tools needed to follow and advance these recommendations. The appendixes of this text provide these resources, divided by topic, and include links to key organizations, statements, Web sites, and other tools related to physical activity and health.

Nowhere else is there a written work that pulls together all this information into a single source. After finishing this book, readers will have a complete understanding of the needs of any individual or group beginning an exercise program as well as have the pertinent background information and adjuncts needed to design an exercise program and put it into action.

ACKNOWLEDGMENTS

Thank you to my husband, Brian, for your patience and support throughout this work. You accepted my aspiration for this ambitious project and helped out in so many ways. Your understanding and encouragement for more than a year mean a lot to me. I appreciate your support of my running, my career, and our family. To my wonderful boys, Evan and Reagan, thank you for your constant love, growth, and joy. Thanks also to my mother, Suzie; my father, Dick; and of course my sister, Marisa, for providing breaks so I could keep working on the book. Dad is my original inspiration for lifestyle physical activity, having walked 3 miles each way to work every day. Thank you also to John Baer not only for being my constant running partner but also for introducing me to Dr. Cooper several years ago.

This book came about after a conversation with Dr. Michele Kettles, who suggested I talk to Mike Bahrke at Human Kinetics, who brought this project to my attention. I appreciate the interest and curiosity that my friends, family, and patients have shown as I worked through writing this book. I am encouraging each and every one to be physically active!

I also want to thank my team at the Cooper Clinic—Claudia Hilton, Jeanne Finseth, David Stewart, and Marie Williams-Albright—for allowing me to focus on my patients and give them the best experience possible. Lastly, I am grateful for all the contributions of Dr. Kenneth Cooper to the fields of preventive medicine, fitness, and aerobics; without all your work and perseverance, I would not have the opportunity to do what I love to do every single day.

PART I

General Health and Fitness Guidelines

Chapter 1 introduces the history of physical activity and health guidelines. It then details some of the science behind the relationship between physical activity and health and presents the benefits and risks of regular physical activity. Basic concepts as they relate to physical activity are defined. Many of these concepts are repeated in the various guidelines included throughout the text as well as in the physical activity and health literature.

Chapter 2 introduces the general physical activity guidelines for adults, presenting both current and historical guidelines from national organizations in the United States, international, and state organizations. Because guidelines vary considerably—in both the recommendations and how to achieve the recommendations—the different purposes of each of the guidelines are discussed. Guidelines from the American College of Sports Medicine are presented alongside those of the American Heart Association, U.S. Department of Health and Human Services, and U.S. Office of the Surgeon General.

Chapter 3 introduces the personal exercise program. While national physical activity guidelines make broad suggestions for overall improved health, the personal exercise prescription individualizes exercise by dividing physical activity into three components: aerobic exercise, strength training, and flexibility exercise. Then the four variables of the FITT principle—frequency, intensity, time, and type—are delineated for each of these components in order to provide a specific plan. Chapter 3 also includes physiology concepts to help illustrate the guidelines.

Exploring the Relationship Between Physical Activity and Health

Physical inactivity is a major public health problem, and an estimated 200,000 deaths occurring in the United States each year are related to a sedentary lifestyle (Powell and Blair 1994). Recent data incorporating poor diet along with physical inactivity suggest that the two together are responsible for more than 400,000 premature deaths each year, accounting for 1 in 6 deaths in the United States (Mokdad and others 2004)! The causes of these deaths include, but certainly are not limited to, cardiovascular disease, type 2 diabetes, and cancers of the colon and breast.

Both physical activity and exercise (which are differentiated later in this chapter) play a central role in maintaining good health. They are a vital part of preventive medicine because they affect present and future health tremendously at a relatively low cost compared with other interventions. Because physical activity is so important, many public health organizations have emphasized the value of participating in physical activity and so have established many sets of guidelines. These guidelines include not only activity guidelines

set by the Centers for Disease Control and Prevention (CDC), the American Heart Association (AHA), the U.S. Office of the Surgeon General, the U.S. Department of Health and Human Services (USDHHS), and the American College of Sports Medicine (ACSM) but also the Dietary Reference Intakes (DRIs) and recommendations in the *Dietary Guidelines for Americans* provided by the Institute of Medicine (IOM). In the United States, the first national physical activity guidelines for the general population were published by the ACSM in 1976. Since that time, many updates have been made not only to these general recommendations but also to guidelines subsequently created for youths, pregnant women, older adults, and other special populations—all of which are included in this book. For specific disease states, national or other respected bodies also periodically issue guidelines, and these are included in the book as well.

There is a large body of evidence supporting the relationship between physical activity and health; less evidence supports the efficacy of the specific activity recommendations, but they are under study. It is important to differentiate guidelines developed for general health maintenance and disease prevention from those created with more specific goals in mind. For example, the growing obesity epidemic has led to guidelines focused solely on weight loss or prevention of weight gain. Although the recommended quality and quantity of physical activity differ somewhat depending on which guidelines are used, the objectives of *all* of the various guidelines are to improve health through enhanced metabolic, structural, and physiological parameters. As the volume or intensity of activity increases, more significant health improvements can be achieved; there is a dose–response relationship between physical activity and health. The minimal recommendations originally set by the Office of the Surgeon General and more recently by the Physical Activity Guidelines Advisory Committee may enable a degree of improved health while being minimal enough to encourage the high number of sedentary individuals in the United States to begin at least *some* exercise. Other guidelines that recommend more intense and longer durations of exercise may reduce the risk for chronic disease, prevent weight gain, and maintain weight loss in people who were previously obese. Other recommendations focus on improving functional capacity or avoiding disability associated with specific disease states. As research continues to demonstrate the relationship between physical activity and health outcomes, individuals may be able to select physical activity models appropriate for their specific health risks, desired physiological benefits, and personal preferences.

History and Development of Physical Activity Guidelines

The ACSM was founded in 1954 by 11 physicians, physiologists, and educators. Around the same time, the President's Council on Youth Fitness was formed under Dwight D. Eisenhower in response to the finding that American youths were relatively unfit when compared with their European counterparts. Both groups endeavored to promote fitness for Americans. The ACSM was the first to develop physical activity guidelines that were widely disseminated in the United

States. These initial guidelines focused on promoting fitness instead of health for two reasons: (1) There was an accumulation of knowledge regarding fitness benefits rather than general health benefits and (2) fitness for sport participation was and continues to be one of the primary areas of focus of the organization. The ACSM published the first edition of *ACSM's Guidelines for Exercise Testing and Prescription* in 1976, and, now in its eighth edition, this publication remains one of the most widely referenced guidebooks of its kind. The ACSM was also the source of the first position stand on physical activity and health for the general public, which was issued in 1978. A position stand is created by a group of leaders within an organization to reflect their thoughts on a specific topic. The initial position stand of the ACSM established the relationship between physical activity and good health and suggested a concise guide for activity within the framework of the statement. Earlier in the 1970s, the AHA issued a handbook on the use of endurance exercise training and testing for the diagnosis and prevention of heart disease. However, until the AHA (along with other groups, including the ACSM) identified physical inactivity as a risk factor for coronary artery disease (CAD) in 1992 and the Office of the Surgeon General issued the landmark *Physical Activity and Health: A Report of the Surgeon General* in 1996, the ACSM's guidelines served as the primary source of information for fitness and activity recommendations in the United States.

The ACSM has since published position stands on several areas of focus, including cardiorespiratory fitness, muscle fitness, flexibility, body composition, disease prevention, hypertension, and osteoporosis. These position stands are widely regarded as national guidelines because of the wealth of scientific knowledge accumulated to develop them.

Motivated by the deteriorating fitness and growing waistlines of Americans that began a few decades ago, the U.S. federal government became involved in recommending physical activity for health. In 1979, it issued the first *Healthy People* report, recommending endurance exercise for health promotion and disease prevention; the most recent version, *Healthy People 2010*, was issued in 2000 and put forth goals for individuals and groups to achieve improved health through many different parameters. This latest report is expected to be supplanted by *Healthy People 2020*. The government has also been involved in recommendations through the publication of various reports by the Office of the Surgeon General, the *Dietary Guidelines for Americans*, and reports by the USDHHS.

Despite the government involvement, the ACSM guidelines continue to be the health and fitness industry's standard template for physical activity and exercise prescription for both healthy populations and special populations, particularly when the goal is improved fitness. Other guidelines such as the *Physical Activity Guidelines for Americans (PAGA)* recently published by the USDHHS (USDHHS 2008), the newer joint AHA and ACSM guidelines (Haskell and others 2007), and the 2005 *Dietary Guidelines for Americans* (USDHHS 2005) are considered appropriate for achieving general health or reducing cardiovascular and cancer risk. In September of 2002, the IOM—a private organization—made headlines by issuing new physical activity guidelines calling for 60 or even 90 min of daily activity. Many people assumed these recommendations were intended to replace those referenced in the 1996 report issued by the Office of the Surgeon General. The IOM recommendations, however, were designed to address the burgeoning obesity epidemic; the report suggested that 60 min of daily, moderate-intensity physical activity helps prevent weight gain and accomplishes additional weight-independent health benefits. As these guidelines were part of a focus on nutrition and weight management, their goal was maintaining a healthy weight; the difference in focus between this report and others reflects the importance of interpreting various guidelines in the context of their stated health goals.

Over time, there have been shifts back and forth between general and specific guidelines. Earlier guidelines for promoting physical activity are somewhat vague. As the idea of regular physical activity for everyone has become more accepted in the United States, guidelines have become more detailed. For example, in 1996 the Office of the Surgeon General recommended 30 min of moderate-intensity exercise on most if not all days of the week. By the 2007 update to the joint ACSM and AHA guidelines, the recommendations were detailed such that individuals were given options of different intensities, durations, frequencies, or combinations of those. The most current national guidelines, *PAGA*, actually return to a simpler recommendation, simply delineating a total volume of weekly activity and giving examples to illustrate a number of ways to achieve the suggested volume. Additionally, the recent guidelines provide more details for resistance exercise, flexibility, balance, and other important fitness components.

Because many healthy young adults may view vigorous exercise negatively, newer guidelines have emphasized that a lower dose of activity—in both intensity and duration—is acceptable and provides health benefits. A greater dose, of course, confers more and larger health benefits but may not be practical or realistic for everyone. This shift to more modest recommendations reflects the strategy to promote activity that provides the greatest benefit for the greatest number of people. Motivating the multitude of sedentary Americans to commit to a basic but critical amount of activity has enormously positive economic and health implications.

Basic Concepts and Definitions

Physical activity is defined as any bodily movement produced by skeletal muscles that results in caloric expenditure (Caspersen 1985). Physical activity is further qualified as lifestyle, light, moderate, or vigorous. These classifications help determine what quantity of activity is recommended. There is no consensus on the exact definition of these various intensity levels. Some guidelines define intensity levels while others just suggest that activity be performed at a particular level. Subsequent sections of this chapter present examples of subjective or objective measures of these classifications and illustrate the various differences in these classifications.

Physical fitness is the body's ability to perform specific tasks or activities for a prolonged duration without experiencing undue physical stress or fatigue. The scope of physical fitness includes multiple parameters such as cardiorespiratory endurance, skeletal muscular endurance, skeletal muscular strength, skeletal muscular power, speed, flexibility, agility, balance, reaction time, and body composition. Many of these parameters are defined in the following sections of this chapter. Physical fitness allows people to complete daily activities more easily, protects against coronary heart disease, and augments caloric expenditure.

Physical Activity Versus Exercise

Although the terms *physical activity* and *exercise* are often used interchangeably, *exercise* differs slightly from *physical activity* in concept and has been defined as planned, structured, and repetitive bodily movement done to improve or maintain physical fitness. For this text, the terms *physical activity* and *exercise* are used under different circumstances, although the intent is not to make a specific distinction between the two. Instead, the various guidelines and recommendations use one term or another, and this text attempts to replicate the various guidelines and recommendations as closely as possible. Although this book is titled *Physical Activity and Health Guidelines*, it could just as easily be titled *Exercise and Health Guidelines*. One interesting observation is that public health guidelines almost always use *physical activity* while guidelines for specific disease states frequently use *exercise*. Sedentary individuals may find it less intimidating to contemplate beginning a physical activity program as opposed to an exercise program.

Components of Exercise Programs

The major components of most exercise programs are aerobic, muscle-strengthening, and flexibility exercises. Some programs also include neuromuscular training, balance exercises, and functional exercises, depending on the goals desired. Most physical activity programs are based on the FITT principle, which refers to the *f*requency, *i*ntensity, *t*ime, and *t*ype of exercise.

Aerobic exercise, or exercise that improves the efficiency of aerobic energy production and improves cardiorespiratory endurance, usually consists of training that utilizes large muscles in continuous and repetitive or rhythmic motion. Arms and legs can be used either together or separately. Aerobic exercise provides many health benefits, most notably improved cardiorespiratory fitness, which in turn decreases all-cause mortality as evidenced by data from the Harvard Alumni Study (Paffenbarger and others 1986) and the Aerobics Center Longitudinal Study (Blair and others 1989).

Resistance training, or exercise that strengthens muscle, is used to improve muscular strength and muscular power. The former is defined as the maximum ability of a muscle to develop force or tension, while the latter is defined as the maximum ability of a muscle to apply a force or tension at a given velocity. Muscular endurance occurs with training and is the ability of a muscle or group of muscles to sustain repeated contractions against a resistance for an extended duration. Resistance training in general may consist of exercises utilizing free weights, weight machines, elastic bands, body weight, or other implements (such as medicine balls). Resistance training can be isokinetic, isometric, or isotonic. Table 1.1 defines these terms.

Flexibility is the intrinsic property of body tissues that determines the range of motion achievable without injury at a joint or group of joints (Holt and others 1996). There are two types of flexibility. Static flexibility is a linear or angular measurement of the actual limits of motion in a joint or complex of joints—it is a clinical measure-

TABLE 1.1 Types of Resistance Training

Type	Definition
Isokinetic	Muscle fibers shorten at a constant speed so that the tension developed remains maximal over the full range of joint motion.
Isometric	The ends of a contracting muscle are fixed so that contraction increases tension at a constant overall length (there is no joint movement).
Isotonic	Muscle fibers shorten with varying tension as the result of a constant load.

Definitions from Ehrman, Clinical Exercise Physiology, p. 125, 2003.

ment defining the range of motion at the joint. Dynamic flexibility is a measure of stiffness and accounts for the resistance to stretch or the rate of increase in tension that occurs in a relaxed muscle as it is stretched through the entire range of motion.

General Principles of Training

Periodization, a method of alternating training loads to produce peak performance for a specific competitive event, can be applied to either aerobic training or resistance training. Variables that are manipulated in periodization include the intensity and volume of training; individuals cycle between higher-volume, lower-intensity periods and lower-volume, higher-intensity periods. For example, a swimmer may swim a longer distance at a slower speed for 3 wk followed by 3 wk of speed sessions with a lower volume. This strategy stresses different types of muscle and different motor memory pathways. Specificity is a concept in exercise physiology that states that the benefits of physical activity are specific to the body system being stressed. That is, a person who runs sprints frequently is likely to become better and faster at sprinting but not necessarily faster at swimming or upper-body resistance training. Overload is the physical stress placed on the body when it engages in physical activity that is greater in amount or intensity than usual. When overloaded, the body can adapt by improving efficiency, strength, or capacity. The overload principle makes way for progression, which is the concept of gradually increasing a load after the body adapts to a certain overload by improving its level of fitness. Progression is an important concept in many physical activity guidelines—beginning exercisers who are very unfit can gradually overload their body and subsequently progress in order to meet recommendations, fulfill guidelines, and eventually improve their health.

Changes and Benefits Resulting From Physical Activity

The following sections describe the various benefits a person can expect to achieve by engaging in physical activity. There is a dose–response relationship between physical activity and many of these benefits, which include not only reduced all-cause mortality but also improved quality of life due to outcomes such as less joint pain, better body composition and muscle mass, and improved energy, sleep, and mood. Many chronic illnesses can be prevented through regular physical activity, and people who already have chronic illnesses may use physical activity to improve their condition.

The Dose–Response Relationship

Evidence suggests there is a dose–response relationship between physical activity or exercise and health outcomes (Paffenbarger and others 1986; Blair and others 1989; Blair and others 1995; Laukkanen and others 2001). Figure 1.1 demonstrates this relationship and illustrates that people with lower baseline fitness can achieve greater health benefits with a given increase in physical activity. One study looking at physical fitness (as measured by a maximal treadmill stress test) and risk of all-cause and cause-specific mortality found that all-cause mortality rates declined across physical fitness quintiles from 64.0 per 10,000 person y in adult men who were the least

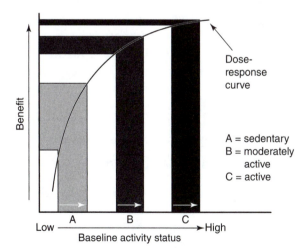

Figure 1.1 The dose–response relationship of activity and health. A higher dose of physical activity produces greater benefits; people with lower baseline fitness can achieve greater health benefits with a given increase in physical activity.

fit to 18.6 per 10,000 person y in adult men who were the most fit! Corresponding values among women were 39.5 and 8.5 per 10,000 person y (Blair and others 1989). Another study found that physical activity improves life expectancy; data from the Framingham Heart Study show that moderate and high levels as opposed to low levels of physical activity increase life expectancy for men aged 50 y by 1.3 and 3.7 y. Corresponding values in women are 1.5 and 3.5 y (Franco and others 2005). Yet another study found a higher prevalence of cardiovascular risk factors among individuals with low fitness as measured by a submaximal stress test. This study included both adults and adolescents, and thus its findings suggest that regular physical activity can significantly affect long-term health, starting at an age when cardiovascular risks are thought to be rare (Carnethon and others 2005). In a 1993 study, it was found that 14% of all deaths in the United States could be attributed to activity patterns and diet (McGinnis and Foege 1993). By 2000, this number increased to 16.6% (Mokdad and others 2004). Another study suggested that a sedentary lifestyle is responsible for 23% of deaths due to major chronic disease (Hahn and others 1998).

There is still much to be learned about the optimal dose of physical activity. What is the minimum amount of exercise needed for health maintenance? Is there a different threshold of activity necessary to prevent heart disease versus diabetes versus cancer? Is there a further risk reduction attained by higher levels of activity, whether higher in frequency, intensity, or duration? Ongoing and future research is likely to address and answer these questions.

Physiological Adaptations

Musculoskeletal adaptations to aerobic physical activity include an increase in slow-twitch muscle fibers. The number of muscle capillaries also increases, leading to improved energy efficiency. Adaptations to resistance training include increased size and recruitment of muscle fibers as well as greater strength and cross-sectional area of ligaments and tendons. Glycogen storage improves, although glycogen is spared as efficient muscles utilize fat as an energy source. These improvements augment maximal oxygen uptake and lactate threshold. Cardiac output increases through improvements in stroke volume; resting and submaximal exercise heart rates decrease. Peripheral resistance decreases due to enhanced capillary capacity for blood flow. Individuals with high blood pressure as well as those with normal blood pressure all experience decreased arterial blood pressures at rest, submaximal exercise, and peak exercise. Physical activity increases circulating levels of testosterone, growth hormone, and catecholamines, while serum levels of insulin decrease.

Aerobic activity induces beneficial changes in lipoproteins, decreasing the levels of very low-density lipoprotein (VLDL) and increasing the levels of high-density lipoprotein (HDL); resistance training diminishes low-density lipoprotein (LDL). This last change is perhaps one of the more important effects of resistance training with respect to CAD and stroke risk, since LDL is one of the factors most strongly correlated to CAD and stroke. Exercise may promote the production of atheroprotective cytokines (Smith 2001), and in people who already have CAD, exercise training improves cardiac autonomic function (Iellamo and others 2000). There is evidence that regular, moderate-intensity physical activity also enhances the function of the immune system. Three randomized studies found that regular physical activity reduces the number of sick days taken due to colds and other viral illnesses (Nieman and others 1990, 1993, 1998). Epidemiological data also confirm this finding; adults who exercise regularly experience fewer upper respiratory tract infections, with the amount of energy expended during moderate physical activity inversely related to respiratory tract infection symptomatology (Matthews 2002; Kostka and others 2000). There may be a J-shaped relationship between exercise and immune function, however, and a single, high-intensity episode of heavy exercise such as running a marathon results in a marked decline in immune cellular function.

Health Benefits of Regular Physical Activity

Regular physical activity provides a multitude of benefits. Thus individuals have much to gain by becoming and remaining physically active. From a public health standpoint, the gains may be even larger. In 1994, Powell and Blair estimated that deaths in the United States due to colon cancer, coronary heart disease, and diabetes could be reduced by 5% to 6% if there were small increases in physical activity practices among Americans. This estimate corresponds to 30,000 to 35,000 deaths each year, or 1% to 1.5% of the overall mortality (Powell and Blair 1994)! Even small changes cause significant gains in health; these gains are detailed in the following sections.

Weight Control

Diet and exercise are the foundation for maintaining a health body weight. Compared with diet alone, diet combined with physical activity is associated with a greater reduction in body fat and greater preservation of lean body mass. Even without caloric restriction, exercise reduces body fat, although overall body weight may not change. As being overweight or obese is associated with an increased risk for a multitude of diseases, the ramifications of preventing weight gain are considerable. Current estimates of adults who are overweight worldwide number 1.6 billion, and the World Health Organization (WHO) estimates that this number will rise to 2.3 billion by 2015 (WHO 2006). Even though many individuals exercise to improve or maintain weight secondary to concerns about external appearance, they gain much by doing so. Because of the close relationship between physical activity and diet, many guidelines focusing on weight control or even diet explicitly address physical activity. To omit this component would undermine the significance of the interplay among diet, physical activity, and health.

Reduced Risk of Cardiovascular Disease and Type 2 Diabetes

CAD is the number one cause of death in the United States, and multiple studies have shown a strong inverse relationship between regular physical activity and the risk of CAD, cardiac events, and death. Physical activity influences several of the most important CAD risk factors, including the following:

- Hypertension (improves exercise and resting blood pressures)
- Type 2 diabetes (reduces the risk of getting diabetes and improves glycemic control in diabetic patients)
- Tobacco use (facilitates smoking cessation)
- Dyslipidemia (increases HDL and lowers LDL and VLDL)

It is likely that physical activity significantly reduces CAD through a combination of improved individual risk factors and other factors that are less quantifiable. Long-term aerobic exercise improves arterial blood pressure, body composition, and thrombotic hemostatic factors. As inflammation is now known to play a fundamental role in atherogenesis and vascular plaque rupture, these latter two effects are critical; improvements in body composition lower circulating inflammatory markers, thus making thrombotic myocardial infarction and stroke less likely.

Type 2 diabetes now affects more than 20 million Americans. Physical activity not only prevents or delays the onset of type 2 diabetes but also is a powerful tool in reducing complications from the disease once it is present. Exercise improves glycemic control and insulin sensitivity. Chapters 8 and 12 detail the available information and guidelines on exercise for people with diabetes.

Cancer Prevention and Treatment

Exercise provides modest protection against breast, colon, and prostate cancer (Rockhill and others 1999; McTiernan and others 2003; Bardia and others 2006; Dallal and others 2007). In people who already have cancer, physical activity improves measures of quality of life, including fatigue experienced both during and after cancer treatment (Courneya and others 2000). Chapter 8 details guidelines for cancer prevention while chapter 9 details the benefits and guidelines of exercise for people with cancer.

Improved Functional Status and Cognition

Physical activity helps to ensure functional independence throughout life. While aerobic activity is the cornerstone of improving cardiorespiratory fitness, exercises increasing strength, endurance, and flexibility provide continued physical function with increasing age. Muscle strength naturally declines with age, and this loss of strength is related closely to losses in physical function (Brown and others 1995). Age-related loss of strength may be attenuated by strengthening exercises, which may provide enough muscular strength and endurance to allow older adults to continue daily activities and functioning. This ability makes the difference between living independently and living with disability. In addition to preserving functional independence, muscular strengthening and flexibility exercises improve balance and reduce the risk for falls. Lastly, improved joint flexibility has a positive effect on the ability to walk, stoop, rise from sitting, and— an extremely important measure of independence for many older adults—drive. In the past it was assumed that improved flexibility reduces the risk of injury, but this assumption has not been proven. Review of the available literature shows that only normal levels of static flexibility are necessary for a low risk of injury—people with very high or very low static flexibility probably possess an increased risk for injury (Knudson and others 2000).

After the age of 30, many bodily processes decline at an estimated rate of 2% per year; exercise can attenuate this decline to only 0.5% per year (Pyron 2003). The benefit of activity continues to have an ever-increasing effect on older individuals who continue to exercise. Whether started in the earlier or the later decades of life, physical activity may also prevent or delay cognitive decline in the elderly (Abbott and others 2004; Barnes and others 2003; Weuve and others 2004; Yaffe and others 2001). In one prospective study among adults older than 65 y, people who exercised at least 3 times per week had a significantly reduced incidence of Alzheimer's disease compared with those who exercised less frequently over the course of the 6 y study (Larson and others 2006). Another study looked at adults who already had subjective memory impairment; after just 6 mo of a physical activity program, participants demonstrated a modest improvement in cognition throughout an 18 mo follow-up (Lautenschlager and others 2008).

Increased Psychological Well-Being and Other Benefits

Exercise reduces stress, anxiety, and depression. Although the relationship between exercise and mood is not completely understood, evidence suggests that exercise raises the levels of mood-enhancing neurotransmitters such as endorphins in the brain. Also, by increasing body temperature, physical activity may have a calming effect (MFMER 2007). Additionally, sleep quality is improved in people who exercise routinely; this superior quality of sleep alone may boost mood and the ability to cope with any given situation. For postpartum mothers, a regular exercise program is recommended to reduce the incidence of postpartum depression and anxiety.

Regular exercise also plays a role for individuals who are attempting to quit smoking. When combined with a cognitive-behavioral smoking cessation program, vigorous exercise modestly facilitates short- and long-term smoking cessation in women (Marcus and others 1999). In people attempting to quit smoking, vigorous exercise can also improve exercise capacity and delay subsequent weight gain.

Effect on Health Care Costs

Annual medical costs in the United States now account for 16% of the gross domestic product, and it is estimated that this percentage will continue to rise. In 2009, total health care spending is expected to reach $2.5 trillion U.S., and this number is expected to increase to $4.4 trillion U.S. by 2018 (Siska 2009)! Direct health care costs incurred each year related to physical inactivity are $77 billion U.S. (Taylor and Whitt-Glover 2007). Another estimate taking into account both sedentary lifestyle and obesity puts the annual burden at $90 billion U.S. (Manson and others 2004). In addition to contributing to chronic disease, impaired physical functioning, and reduced quality of life, inactivity is responsible for an estimated 300,000 premature deaths each year in the United States. The economic benefits of regular physical activity may be substantial for many reasons, including the reduction of acute and chronic illnesses. One study found that physically active individuals had lower annual direct medical costs than inactive individuals had, at a difference of $330 U.S. per person (Pratt and others 2000). Physical activity as a part of cardiac rehabilitation plays a significant economic role as well; the cost effectiveness of cardiac rehabilitation is estimated to be $4,950 U.S. per year of life saved (Ades and others 1997).

Direct medical costs are only a part of the picture; physical activity may also reduce absenteeism or disability expenses. On a corporate scale, encouraging physical activity through employee wellness or workplace programs may improve health, reduce health care costs, and substantially improve corporate economy. In one 14 y study, the fittest employees experienced one-eighth as many injuries as the least fit employees experienced, while the unfit employees incurred twice the amount of injury cost (Cady 1985). Another study comparing physically active and sedentary employees at Mesa Petroleum showed that the physically active employees spent $217 U.S. per person less on medical claims and had 21 h per person less of sick time each year (Gettman 1986). In 1996, Shephard reported a 0% increase in medical costs for a company with a fitness program and a 35% increase in medical costs for a company with no fitness program.

Physical activity is one of the key components of attaining and maintaining a healthy weight. It has been demonstrated that when compared with their leaner counterparts, men who are overweight cost their employers $170 U.S. per year more in medical expenses while women who are overweight cost a staggering $495 U.S. more (Businessweek 2008). A CDC study found that physically active people as a group have lower direct medical costs than inactive people have each year (Pratt and others 2000). The same study estimated that if the inactive Americans—who account for more than 88 million people over the age of 15—partook in regular moderate physical activity, the national direct medical costs could be reduced by $76.6 million U.S. (in 2000 dollars) annually (Pratt and others 2000). It has been shown that in addition to improving productivity, participation in a corporate fitness or wellness program improves job satisfaction (Opatz 1994). This not only helps the corporate bottom line but also has positive ramifications on a larger scale.

Risks of Physical Activity

There are risks involved with physical activity, although in general the benefits certainly out-

weigh the risks. By far, the most common risk is musculoskeletal injury. The risk for this type of injury is closely correlated with overuse. Additionally, a major change in exercise frequency, intensity, or duration can increase the risk for musculoskeletal damage. Possible musculoskeletal injuries include the following:

- Acute muscle strains or tears
- Inflammation
- Chronic muscle or joint strains
- Stress fractures
- Traumatic fractures
- Nerve palsies
- Tendonitis
- Bursitis

Other, less-common health risks include arrhythmias, sudden cardiac death, postactivity myocardial infarction, infection, rhabdomyolysis, bronchoconstriction, and dehydration. For women, the female athlete triad (eating disorder, amenorrhea, and osteoporosis) can negate some of the positive effects of habitual activity. In the female athlete triad, disordered eating and chronic energy deficits disrupt the hypothalamic-pituitary axis. This in turn interrupts menstrual cycles and reduces bone density.

Most of the risks associated with physical activity can be attenuated through sensible actions and planning. People who are at a higher risk for a particular condition or injury may modify their activity accordingly. Gradually increasing activity intensity, frequency, and duration is the most sensible way to prevent activity-related injury. For example, older individuals may need a year of gradual progression to ultimately attain the desired activity level. Increasing frequency or duration before increasing intensity is recommended; this concept appears in multiple different guidelines, particularly guidelines for older adults or people with chronic disease. Obtaining medical clearance for exercise is addressed later in this chapter and also in chapters 15 and 16.

Although habitual physical activity reduces the long-term risk for CAD events in individuals who have not been active, vigorous activity may transiently increase the risk for sudden cardiac death. Structural heart disease in younger people and CAD in older people are the most common causes of sudden cardiac death. Acute coronary thrombosis may be provoked by vigorous-intensity activity, especially in individuals who are not used to that level of intensity. Acute coronary thrombosis is caused in part by increased platelet activation, which is related to the relative intensity of exercise—that is, at a given *absolute* intensity the *relative* intensity of a specific workload may be higher for individuals who are unaccustomed to regular physical activity. Acute coronary thrombosis may cause sudden cardiac death by occluding coronary arteries. Proinflammatory molecules generally are higher in people who are habitually inactive, and coronary thrombosis accelerates the inflammatory cascade and ultimately can lead to acute obstruction and sudden cardiac death.

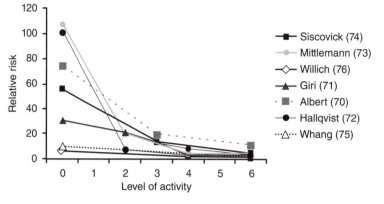

Figure 1.2 Risk of sudden adverse cardiac event by level of activity.

Figure G10.4, PAGAC report. From Physical Activity Guidelines Advisory Committee. *Physical Activity Guidelines Advisory Committee Report, 2008.* Washington DC: U.S. Department of Health and Human Services, 2008.

The absolute risk of sudden cardiac death (whether ultimately fatal or nonfatal) depends on the population being studied; in general it is lower for younger individuals. Estimates of annual sudden deaths due to activity range from 1 in 769,000 among female athletes in high school and college (lowest) to 1 in 15,260 among joggers (highest; Thompson and others 1982 and Van Camp and others 1995). When looking at risk from a workout perspective, estimates suggest that 1 death occurs per 2.57 million workouts (Franklin and others 2000). More common than sudden cardiac death is acute myocardial infarction (heart attack), which is 6.75 times more likely to occur during exercise (Siscovick and others 1991). Because of the increased myocardial oxygen demand and reduced coronary perfusion that occur during physical activity, the risk of acute myocardial infarction during or soon after vigorous exertion is between 6 and 64 times higher than the risk of infarction without exertion (Cobb and Weaver 1986). Overall, however, this rate is 50 times higher for individuals who are habitually inactive than it is for their most active counterparts (Mittleman and others 1993). Therefore, the risk of experiencing a heart attack or cardiovascular event while exercising is greater than the risk of such an event is when not exercising, but the long-term risk associated with habitually not exercising significantly outweighs the transient risk occurring during physical activity, as manifested by the fact that sedentary individuals have a higher likelihood of experiencing a cardiovascular event while either exercising or not exercising. Studies have shown that the more active the individual is, the lower the risk for an acute cardiovascular event per minute of activity. A graph summarizing this finding is presented in figure 1.2.

Patterns in Physical Activity Participation

Only about 23% of adults in the United States report that they participate in regular, vigorous physical activity for 20 min on at least 3 d of the week. Only 15% of adults report participating in any physical activity for 30 min on at least 5 d of the week. Some more specific demographics are as follows:

- Women are more likely than men are to report no participation in leisure-time physical activity.
- African Americans and Hispanics are less active than Caucasians are.
- Wealthier individuals are more physically active than are people who are less affluent.
- Older adults are less active than younger adults are. In fact, the time spent in physi-

Obtaining Clearance for a Physical Activity Program

Overall, the benefits of quality-adjusted life expectancy obtained through the development of health-related fitness warrant participation in regular physical activity, but in light of the concerns and risks inherent to an exercise program, it is necessary to consider readiness and medical clearance for physical activity. Different populations may have different needs and concerns for clearance; when appropriate, these needs are addressed in the various chapters of this book. The ACSM and AHA do not currently endorse routine preexercise clearance for asymptomatic individuals who plan to undertake a program that fulfills the minimum recommended physical activity levels. People who are symptomatic or who have medical issues causing concern warrant at least a discussion with a physician. At that point, the decision can be made for the person to pursue graded exercise testing or other evaluation before embarking on the physical activity program.

The Physical Activity Readiness Questionnaire (PAR-Q) was developed in the 1970s to screen for participant readiness for the Canadian Home Fitness Test. This self-administered screening questionnaire has been used widely in Canada and abroad and in the 1990s was endorsed by the ACSM for healthy adults. Since that time, the PAR-Q has been revised in order to increase specificity without sacrificing sensitivity. Thus the newer version, while still preventing high-risk individuals from unsafe exercise, permits exercise among those who may have previously been screened out unnecessarily. The current PAR-Q is available in appendix B.

---

cal activity peaks during adolescence and declines with age; the typical elderly person is almost entirely sedentary (USDHHS 1996).

- Americans in the north-central and western states tend to be more active than the Americans in the northeastern and southern states.

On the individual level, an important factor affecting physical activity participation is social influence. People who have active social influences tend to be more active themselves. For children the main source of influence is their parents. Adolescents focus almost entirely on their peers and tend to exercise if they identify with a peer group that values, supports, and participates in physical activity. For adults, the social support stems from friends, coworkers, and significant others.

There are also trends in the types of physical activity people choose to do. Walking has been the most popular form of physical activity in the United States since 1990. The National Sporting Goods Association (NSGA) reported that in 2007 89.8 million Americans walked for exercise. Other popular forms of physical activity are exercising with equipment (such as a stair stepper, stationary recumbent bicycle, or elliptical trainer), swimming, and cycling (NSGA 2008).

When asked why they do not exercise, most individuals say that they do not have enough time for physical activity. Although time seems like an obvious barrier for many adults, this same complaint is frequently encountered in children as well (Sallis 1994). Other reasons include pain or disability, perceived risk for pain or disability, and lack of access to exercise facilities or a safe environment in which to exercise. One study found that adults were more likely to be active if they had a number of exercise facilities located within a short distance of their homes (Sallis and others 1990). In both adults and adolescents, the number and strength of perceived barriers to activity are consistently related to physical activity levels (Sallis and others 1992; Tappe and others 1989). Both children and adults prefer activities with lower levels of exertion; those who participate in vigorous- or moderate-intensity activity tend to drop out at a higher frequency. This is one of the reasons why the more recent guidelines emphasize lower-intensity activity. Lack of enjoyment of activity is another major obstacle for both adults and children (Stucky-Ropp and DiLorenzo 1993; Garcia and King 1991).

SUMMARY

Physical activity plays an important role in health, and there is a dose–response relationship between physical activity and health. Although physical activity carries short-term risks, such as sudden cardiac death or musculoskeletal injuries, these generally diminish as a person becomes more fit and accustomed to exercise. The benefits of activity are numerous and include a reduction in the incidence of multiple chronic disease processes, an effective method of managing many chronic illnesses, and financial benefits in terms of fewer missed days of work and reduced workplace and insurance costs. Despite these clear advantages, participation rates for physical activity are considerably low. Individuals face physical and psychological hurdles to initiating and maintaining a regular physical activity program.

Exercise Guidelines for Physical Fitness and Health

The first physical activity guidelines were developed in the 1950s, shortly after the inception of the ACSM. Before the 1970s, there was little evidence supporting the benefits of physical activity. As a result there were few official guidelines specifying recommended amounts of physical activity. The ACSM's original guidelines focused on physical fitness and fitness promotion as opposed to general health and disease prevention. Since that time, the body of evidence linking physical activity to improved health has grown, and there now exist more formal physical activity recommendations. Major organizations that have created and distributed guidelines for physical activity include private associations, governmental bodies, and professional groups (see sidebar). As each of these organizations has a different mission and different priorities, the purpose behind the different physical activity recommendations varies. For example, in 1992 the AHA identified physical inactivity as a distinct risk factor for CAD, and so the AHA guidelines focus on using moderate-intensity physical activity as a method to reduce the risk for heart disease.

The National Association for Sport and Physical Education (NASPE) has issued activity guidelines for infants, toddlers, and children in order to promote motor development, reduce childhood obesity, and encourage a pattern of lifelong physical activity that reduces the future risk of chronic disease. In 1996, the U.S. Office of the Surgeon General issued guidelines for sedentary adults; these were presented in light of the fact that more than 50% of the U.S. population does not meet physical activity guidelines (USDHHS 1996) and that 23.7% of the population does not participate in any leisure-time activity whatsoever (Haskell and others 2007). Despite the fact that over the past few years engagement in regular physical activity has increased among U.S. men and women, more than half of each sex still does not participate in regular physical activity (Kruger and others 2007).

The guidelines presented in this chapter represent a collection of general guidelines focused on improving fitness, muscular strength, flexibility, or overall health. Some are included for historical significance, while others are included to emphasize

Organizations With Established Physical Activity Guidelines

- American Academy of Family Physicians
- American Academy of Pediatrics
- American Alliance for Health, Physical Education, Recreation and Dance
- American Cancer Society
- American College of Obstetricians and Gynecologists
- American College of Rheumatology
- American College of Sports Medicine
- American Council on Exercise
- American Diabetes Association
- American Geriatrics Society
- American Heart Association
- American Medical Association
- Australian Government Department of Health and Ageing
- Canadian Society for Exercise Physiology
- Centers for Disease Control and Prevention
- Health Canada
- Institute of Medicine
- International Osteoporosis Foundation
- National Academy of Sciences
- National Association for Sport and Physical Education
- National Coalition for Promoting Physical Activity
- National Institutes of Health
- National Osteoporosis Foundation
- Office of the Surgeon General
- President's Council on Physical Fitness and Sports
- Royal College of Obstetricians and Gynaecologists
- U.S. Department of Health and Human Services and U.S. Department of Agriculture

the differences in recommendations that appear when the intended health goal changes. Overall these guidelines establish a fitness foundation for all Americans and serve as a starting point for the other, more specific guidelines that are presented in subsequent chapters of this book, such as the guidelines for pregnant women or older adults. Some of the more specific guidelines are issued by the same organizations issuing the more general guidelines, while other sets of specific guidelines are created by other entities.

Most Recent National Guidelines in the United States

This section discusses the recent guidelines issued by notable national entities in the United States; namely, the most recent ACSM and AHA guidelines and the most recent U.S. government recommendations. The first nationally recognized physical activity guidelines in the United

States were issued by the ACSM. These have been updated several times, and the most recent ACSM guidelines were endorsed by the AHA and were released in 2007 with the purpose of improving and maintaining health. Previous ACSM guidelines were issued with other entities or have stood alone. In 2008, the U.S. government, through the USDHHS, released the *Physical Activity Guidelines for Americans (PAGA). PAGA* provided the first comprehensive national guidelines containing recommendations for not only the general adult population but also youths, older adults, and various segments of the population with chronic disease. These guidelines are not only the most recent but also the most scientifically rigorous in terms of data collection, evaluation, and discussion by the large panel of experts who met to develop these guidelines.

American College of Sports Medicine and American Heart Association Guidelines

The original ACSM guidelines for physical fitness have been updated several times. These guidelines are perhaps the most rigorous of the national recommendations, as the goal is to improve physical and cardiorespiratory fitness rather than simply prevent disease or weight gain. Initially issued in 1978 with a target audience of athletes, the ACSM recommendations were most recently updated and distributed in 2007. The 2007 guidelines were supported by the AHA and endeavored to clarify the types and amount of physical activity necessary to improve and maintain health. Although the previous set of guidelines, which were copublished in 1995 with the CDC, recommended that "every U.S. adult should accumulate 30 minutes or more of moderate-intensity physical activity on most, preferably all, days of the week" (Pate and others 1995, p. 402), it became apparent that many individuals misinterpreted the recommendations or were not compliant. Some people mistakenly believed that their light daily activities met the guidelines, while others felt that only vigorous activity could improve health (Haskell and others 2007). Additionally, since 1995 the body of evidence supporting the link between physical activity and improved health has been

increased significantly, and in-depth data have been provided on this link.

The 2007 ACSM and AHA guidelines apply to healthy adults between 18 and 65 y. They also target individuals with chronic medical problems that do not impair physical activity (such as hearing impairment); guidelines for pregnant women and adults older than 65 y are addressed elsewhere in this book. In these guidelines, the purpose of physical activity is to promote and maintain health and reduce the risk of chronic disease and premature mortality. The following three sections detail the ACSM and AHA recommendations on aerobic activity, muscle-strengthening exercises, and clearance for physical activity.

Aerobic Activity

Healthy adults should strive to complete 30 min of moderate-intensity activity 5 d/wk or 20 min of vigorous-intensity activity 3 d/wk. The intensity of physical activity may be estimated by using METs. A MET is a metabolic equivalent and 1.0 MET is the amount of energy expended by an individual who is sitting quietly. For the ACSM and AHA guidelines, moderate-intensity physical activity is regarded as 3.0 to 6.0 METs, while vigorous-intensity physical activity is more than 6.0 METs. Appendix B lists common activities and their MET equivalents. A combination of moderate- and vigorous-intensity activities can be used to meet the recommendations. This makes sense when using the MET calculation: An individual may ballroom dance (4.5 METs) for 30 min on 5 d/wk for a total of 675 MET-min or could jog (10.0 METs) for 20 min on 3 d/wk for a total of 600 MET-min.

The newer guidelines allow some flexibility and variety in accumulating the recommended amounts of activity. A minimum of 450 to 750 MET-min/wk is advised, with the lower end suggested for individuals who are just starting a physical activity program. As their fitness improves, these individuals should progress toward the higher end of the recommended range. Activity may be accumulated in exercise bouts as short as 10 min; this recommendation may enable people to meet their activity goal by achieving short sessions of physical activity throughout the day and thus may reduce noncompliance due to lack of time or energy. Lastly, these guidelines differentiate aerobic activity (of moderate or vigorous intensity) from routine activities of daily living (ADLs), which are generally of light intensity. Casual walking, self-care, and light office activities do not qualify as moderate intensity and thus do not count toward the recommended accumulation of activity—even if performed for prolonged durations.

Muscular Strength and Endurance

Because of the mounting evidence that muscular strength and endurance benefit health, the ACSM and AHA guidelines now include a specific recommendation in this area. Before now, only the ACSM guidelines for cardiorespiratory fitness included recommendations for resistance training—the guidelines for general health benefits did not. In 1998, the ACSM issued guidelines specifically for improving muscular strength and flexibility; these are presented later in this chapter (see guideline 2.9).

It is now recommended that resistance training using the major muscle groups be performed on at least 2 or more nonconsecutive days per week. Each session should include 8 to 10 exercises and

---

GUIDELINE 2.1

Title: "Physical Activity and Public Health: Updated Recommendation for Adults"

Organization: ACSM and AHA

Year published: 2007

Purpose: To update and clarify the 1995 recommendations on the types and amounts of physical activity needed by healthy adults to improve and maintain health

Location: *Medicine & Science in Sports & Exercise* 39(8): 1423-1434

Population: Healthy adults aged 18 to 65 y

GUIDELINES

Type	Frequency	Intensity	Duration
Aerobic exercise	3-5 d/wk, depending on intensity	Moderate to vigorous intensity, depending on frequency; can be quantified by METs	A minimum of 20-30 min/d, depending on the frequency and intensity; can be accumulated in bouts lasting 10 min or longer
Muscular strengthening and endurance exercise	2 d/wk	Heavy enough to cause substantial fatigue by the last repetition	1 set of 8-12 repetitions for 8-10 exercises using the major muscles

Other considerations include the following:

- Combinations of moderate- and vigorous-intensity activity can be performed to meet this recommendation. For example, a person can meet the recommendation by walking briskly for 30 min 2 times during the week and then jogging for 20 min on 2 other days.

- These moderate- or vigorous-intensity activities are in addition to the light-intensity activities frequently performed during daily life (e.g., self-care tasks, washing dishes, using light tools at a desk) or activities of very short duration (e.g., taking out the trash, walking from the parking lot to a store or office).

- Because of the dose–response relationship between physical activity and health, individuals who want to improve their personal fitness, further reduce their risk for chronic health conditions, and prevent unhealthy weight gain should exceed the minimum recommended amounts of activity.

Adapted, by permission, Haskell WL et al., 2007, Physical Activity and Public Health: Updated Recommendations for Adults. Med Sci Sports Exerc, 39(8), 1423-1434.

8 to 12 repetitions of each exercise. Weights that cause substantial fatigue after 8 to 12 repetitions are appropriate.

Preactivity Screening and Clearance

There is significant controversy regarding the usefulness of undergoing exercise testing before starting a regular physical activity program. The ACSM and AHA 2007 guidelines state that asymptomatic men and women who plan to be physically active at the minimum levels of moderate-intensity activity set forth in the guidelines do not need to consult with a physician or health care provider before beginning their exercise program unless they have specific questions or concerns. People with symptoms suggestive of cardiovascular disease (CVD) or diabetes should consult a medical provider who can determine whether exercise testing before an activity program is appropriate.

Physical Activity Guidelines for Americans

In 2007, USDHHS Secretary Mike Leavitt appointed the Physical Activity Guidelines Advisory Committee (PAGAC) in order to provide a scientific review of all the data on physical activity and health. An earlier 2006 workshop held jointly by the IOM and the USDHHS had determined that significant scientific advances justified physical activity guidelines separate from those published previously as part of the *Dietary Guidelines for Americans*. Although the 1996 *Physical Activity and Health: A Report of the Surgeon General* brought together a large body of information on the two topics, since the publishing of the report more studies, information, and other reports had been made available. Additionally, new information regarding different age groups and populations with certain diseases had been published in the ensuing decade. For this reason, the PAGAC met, discussed the new data, and set out to provide the most comprehensive collection to date of information connecting physical activity and health. The result, an extensive report detailing the scope and strength of the evidence linking physical activity and health, was issued in June 2008 and made available for public comments. In October of 2008, the committee issued *PAGA*—guidelines

not only for American adults but also for American youths, pregnant women, older adults, and individuals at risk for or diagnosed with an array of medical problems.

The *PAGA* recommendations are similar to the 2007 ACSM and AHA guidelines in that they acknowledge that higher-intensity activity provides greater benefits and thus allows for a shorter duration or frequency of exercise. One difference between *PAGA* and the ACSM and AHA guidelines, however, is the definition of *moderate intensity* and *vigorous intensity*. Whereas the ACSM and AHA guidelines consider 3.0 to 6.0 METs to be moderate intensity, *PAGA* specifically notes that the committee felt that 3.0 to 5.9 METs was more consistent with moderate intensity and that 6.0 METs and higher would be considered vigorous intensity. What this means is that activities of 6.0 METs previously considered to be moderate intensity are now considered to be vigorous intensity by the *PAGA* guidelines. Table 2.1 presents a sample of activities expending 6.0 METs.

Also, the *PAGA* guidelines attempted to make the recommended activity level more straightforward. Instead of giving various combinations of activities needed to meet goals, the new guidelines stated a total weekly duration of activity. One reason for doing so is that the evidence doesn't suggest any difference in health benefits for people who achieve a similar volume of activity in different ways. For example, conclusions from the literature have been based on a total number of MET-minutes per week and have not distinguished between someone performing moderate-intensity activity for 40 min 3 times a week and someone performing 30 min 4 times a week.

Lastly, although the 2007 ACSM and AHA guidelines simply refer to the dose–response relationship between physical activity and health, the 2008 *PAGA* report emphasizes this feature in numerous places. *PAGA* also details the amounts of activity associated with expected health benefits. The minimum recommendation of performing 150 min of moderate-intensity activity each week falls into the category of providing "substantial" health benefits. Table 2.2 summarizes the *PAGA* information on health benefits associated with various levels of activity.

TABLE 2.1 Activities Classified as 6.0 METs (Vigorous-Intensity Activities)

Type	Description
Conditioning	Lifting weights (free weight, Nautilus, or Universal), vigorous effort
Conditioning	Taking a Jazzercise class
Conditioning	Teaching an aerobic exercise class
Fishing and hunting	Fishing in a stream in waders
Home activities	Moving furniture or household items, carrying boxes
Lawn and garden	Mowing the lawn, walking, using a hand mower
Occupation	Shoveling, light (less than 10 lb or 4.5 kg in 1 min)
Running	Jogging and walking in combination (jogging component is less than 10 min)
Sports	Boxing, punching a bag
Sports	Playing tennis, doubles
Walking	Walking 3.5 mph (5.6 kph), uphill
Water activities	Water-skiing
Water activities	Swimming, leisurely, not lap swimming
Winter activities	Skiing, downhill, moderate effort

Data from Ainsworth BE et al., 2000, Compendium of physical activities: An update of activity codes and MET intensities. Med Sci Sports Exerc. 32:S498-S516.

TABLE 2.2 Expected Health Benefits Associated With Various Levels of Physical Activity

Physical activity level	Range of moderate-intensity minutes performed each week	Summary of overall health benefits	Comment
Inactive	No activity beyond baseline	None	Being inactive is unhealthy.
Low	Activity beyond baseline but fewer than 150 min/wk	Some	Activity at the high end of this range provides additional and more extensive health benefits than activity at the low end of the range provides.
Medium	150-300 min/wk	Substantial	Activity at the high end of this range provides additional and more extensive health benefits than activity at the low end of this range provides.
High	More than 300 min/wk	Additional	Current science does not allow researchers to identify an upper limit of activity above which there are no additional health benefits.

Reprinted from 2008 Physical Activity Guidelines for Americans, U.S. Department of Health and Human Services, www.health.gov/paguidelines, Accessed 10/10/2009.

GUIDELINE 2.2

Title: *Physical Activity Guidelines for Americans*

Organization: USDHHS

Year published: 2008

Purpose: To lower the risk for all-cause mortality, coronary heart disease, stroke, hypertension, and type 2 diabetes

Location: www.health.gov/paguidelines

Population: Adults

GUIDELINES

- Adults should get at least 150 min/wk of moderate-intensity aerobic activity, 75 min/wk of vigorous-intensity aerobic activity, or an equivalent combination of moderate- and vigorous-intensity activity.

- Aerobic activity should be performed in episodes lasting at least 10 min and preferably should be spread throughout the week.

- For additional and more extensive health benefits, adults should increase their aerobic physical activity to 300 min (5 h) of moderate-intensity aerobic activity, 150 min of vigorous-intensity aerobic activity, or an equivalent combination of moderate- and vigorous-intensity activity each week. Additional health benefits are gained by engaging in physical activity beyond this amount.

- Adults should also engage in muscle-strengthening activities on 2 or more days a week, as muscle-strengthening activities provide additional health benefits. These activities should be moderate or high intensity and should involve all the major muscle groups. For each exercise, perform 1 to 3 sets of 8 to 12 repetitions.

Reprinted from 2008 Physical Activity Guidelines for Americans, U.S. Department of Health and Human Services, October 2008, Accessed from www.health.gov/paguidelines 10/10/2009.

Significant Historical Guidelines

While the ACSM and AHA and the *PAGA* guidelines represent the most recent comprehensive guidelines published in the United States, they also reflect the changes and improvements of the recommendations that preceded them. Understanding the importance of specific details in the current guidelines requires an appreciation of the historical significance of the previous guidelines as well as of the circumstances and data present at the time these past guidelines were written. This section summarizes landmark reports, guidelines, and recommendations on physical activity that have been issued in the United States.

Original American College of Sports Medicine Recommendations

The original ACSM recommendations were issued in 1978, 2 y after the first edition of *ACSM's Guidelines for Exercise Testing and Prescription*. In 1990, the 1978 guidelines were updated and included resistance training and a suggestion that accumulating activity is acceptable. The intensity of exercise heart rate as a percentage of heart rate reserve (HRR) was lowered slightly, and the quantifying of intensity using a percentage of maximum heart rate (HRmax) was added. This latter option may seem more straightforward to the many people who are not aware of their HRR.

GUIDELINE 2.3

Title: "The Recommended Quantity and Quality of Exercise for Developing and Maintaining Cardio-respiratory and Muscular Fitness in Healthy Adults"

Organization: ACSM

Year published: 1978

Purpose: To develop and maintain cardiorespiratory fitness

Location: *Medicine & Science in Sports & Exercise* 10: vii-x

Population: Healthy adults

GUIDELINES

Component	Frequency	Intensity	Duration	Type
Cardiorespiratory fitness	3-5 times per week	60%-90% of HRR or 50%-85% of $\dot{V}O_2$max	15-60 min	Endurance activities

GUIDELINE 2.4

Title: "The Recommended Quantity and Quality of Exercise for Developing and Maintaining Cardio-respiratory and Muscular Fitness in Healthy Adults"

Organization: ACSM

Year published: 1990

Purpose: To develop and maintain cardiorespiratory and muscular fitness

Location: *Medicine & Science in Sports & Exercise* 22: 265-274

Population: Healthy adults

GUIDELINES

Component	Frequency	Intensity	Duration	Type
Cardiorespiratory fitness	3-5 times per week	60%-90% of HRmax or 50%-85% of HRR or $\dot{V}O_2$max	20-60 min performed continuously or accumulated in bouts of at least 10 min; 20 min or more for higher intensity and 30 min or more for lower intensity	Endurance activities
Resistance training	2 times per week	1 set of 8-12 repetitions to near fatigue	8-10 exercises	Dynamic exercises using the major muscle groups

Special consideration:
- There are potential health benefits of regular exercise performed more frequently and for longer durations but at lower intensities than the recommendations outlined in this table.

Original American Heart Association Recommendations

In 1992, the AHA for the first time included physical inactivity as a major risk factor for coronary heart disease (CHD), along with cigarette smoking, hypertension, and hypercholesterolemia. As a result, the 1992 statement on the benefits of exercise recognized the importance of exercise in both individuals with CHD and individuals wishing to prevent CHD as opposed to only individuals with CHD. The statement was somewhat generalized and vague with respect to the details of exercise frequency, intensity, and duration; the significance of the pronouncement was in the attention it brought to the value of a physically active lifestyle in preventing CHD. The statement was updated in 2003 with the following changes: (1) The frequency, intensity, and duration of the recommended exercise were in line

---

GUIDELINE 2.5

Title: "Statement on Exercise. Benefits and Recommendations for Physical Activity Programs for All Americans" and "Exercise and Physical Activity in the Prevention and Treatment of Atherosclerotic Cardiovascular Disease" (update)

Organization: AHA Committee on Exercise, Rehabilitation, and Prevention, Council on Clinical Cardiology

Year published: 1992 (updated in 2003)

Purpose: For health promotion

Location: *Circulation* 86(1): 340-344 and *Circulation* 107(24): 3109-3116 (update)

Population: Healthy adults

GUIDELINES

Component	Frequency	Intensity	Duration	Type
Aerobic exercise (1992 paper)	3-4 times per week	At least 50% of exercise capacity (the point of maximal ventilatory oxygen uptake or the highest achievable work intensity)	30-60 min	Dynamic exercises working the large muscles (walking, hiking, stair-climbing, jogging, bicycling, rowing)
Aerobic exercise (2003 update)	Most, preferably all, days of the week	Moderate intensity (40%-60% of $\dot{V}O_2max$ or absolute intensity of 4-6 METs)	30 min or more	A variety of physical activities

Special considerations include the following:

- Low-intensity activities performed daily may have some long-term health benefits and may lower CVD risk.
- Low-intensity leisure activities such as walking, playing golf, or playing ping-pong are recommended for the elderly.
- Lower-intensity activities such as walking have a lower risk of sudden cardiac death, falls, and joint injuries than higher-intensity activities have.
- The 2003 update states that only healthy men >45 y and women >55 y considering initiating a vigorous exercise program should be considered for routine exercise stress testing.
- Health professionals should personally involve themselves in an active lifestyle.

with the recommendations made in the Surgeon General's 1996 report (see guideline 2.8); (2) resistance exercise was also suggested, as recommended by the AHA 2000 guidelines endorsed by the ACSM (see guideline 2.13); and (3) health professionals were urged to engage in a healthy and active lifestyle to serve as role models for their patients and to advocate for public health policy!

Centers for Disease Control and Prevention and American College of Sports Medicine Public Health Recommendations

In 1993 the CDC sought to develop a statement with the ACSM in order to recommend appropriate physical activity for the promotion of good health. The report came out in 1995 and was developed by a group of five scientists chosen by the CDC and ACSM. This committee added 15 other experts, including epidemiologists, exercise physiologists, public health professionals, and health psychologists, who all helped to discuss the research and create the paper. The recommendations, titled "Physical Activity and Public Health," were published in the *Journal of the American Medical Association* and released to the public. As opposed to other position stands previously issued by the ACSM, which focused on exercise training for fitness and made the attendant recommendation of vigorous-intensity activity, this statement—which took 2 y to develop—promoted the concept of moderate-intensity activity for general health. Another first was the suggestion that activity could be accumulated in bouts of as little as 8 to 10 min. Although this model had been mentioned in the 1990 ACSM position statement, it was not widely promoted until this 1995 report. This concept was based on a limited number of studies comparing accumulated activity to continuous exercise. Subsequently, this model has been promoted in nearly every statement making physical activity recommendations; it appears to be correct.

GUIDELINE 2.6

Title: "Physical Activity and Public Health"

Organization: CDC and ACSM

Year published: 1995

Purpose: For health promotion and disease prevention

Location: *Journal of the American Medical Association* 273(5): 402-407

Population: Healthy adults

GUIDELINES

Component	Frequency	Intensity	Duration	Type
Aerobic activity	Most, preferably all, days of the week	Moderate	30 min accumulated in bouts of 8-10 min	Endurance activities

Special considerations include the following:

- Muscular strength and flexibility training is "not to be overlooked" but not specifically addressed.
- Higher-intensity activity may confer greater health benefits.
- People who are already physically active should consider increasing their activity for further benefits.
- Sedentary individuals should gradually incorporate increased activity, building up to 30 min daily.
- Men older than 40 y and women older than 50 y who plan a vigorous program (60% or greater of $\dot{V}O_2max$) or who have a chronic disease or risk factors for a chronic disease should consult their physician before beginning.

National Institutes of Health

At the end of 1995, the National Institutes of Health (NIH) issued a consensus statement regarding physical activity and cardiovascular health. It was a result of a 3-day conference convened to provide both physicians and the general public with a responsible assessment of the relationship between physical activity and cardiovascular health. A 13-member panel presented information, experts and conference members discussed the information, the statement was revised, and finally the statement was produced. The conclusions were similar to those released by the Office of the Surgeon General the next year: All Americans should engage in regular physical activity for at least 30 min on most days of the week. One important point addressed in this consensus statement, also addressed in the Surgeon General's report and frequently overlooked by individuals, the media, and physicians themselves, is that additional health benefits may be derived by performing activity of greater duration or intensity. Another point that has been endorsed in various iterations by other entities is that certain individuals should undergo a medical evaluation before initiating a vigorous exercise program.

GUIDELINE 2.7

Title: "Physical Activity and Cardiovascular Health"

Organization: NIH Consensus Development Panel on Physical Activity and Cardiovascular Health

Year published: 1995 (online) and 1996 (in *JAMA)*

Purpose: To reduce the risk for CVD through physical activity and to define the benefits and risks of different types of physical activity for people with CVD

Location: www.consensus.nih.gov/1995/1995ActivityCardivascularHealth101html.htm and *Journal of the American Medical Association* 276: 241-246

Population: Children and adults as well as people with CVD

GUIDELINES

- All Americans should engage in regular physical activity at a level appropriate to their capacities, needs, and interests. All children and adults should accumulate at least 30 min of moderate-intensity physical activity on most, preferably all, days of the week. People who currently meet these standards may derive additional health and fitness benefits by becoming more physically active or including more vigorous activity.

- Cardiac rehabilitation programs that combine physical activity with reduction in other risk factors should be more widely applied to patients with known CVD.

- Individuals with CVD and men older than 40 y or women older than 50 y with multiple cardiovascular risk factors should undergo a medical evaluation before embarking on a vigorous exercise program.

Societal Recommendations

- Programs should be developed to help health care providers communicate to patients the importance of regular physical activity.

- Increased community support of regular physical activity, with environmental and policy changes at schools, workplaces, community centers, and other sites, is needed.

- A coordinated national campaign involving a consortium of collaborating health organizations should be initiated to encourage regular physical activity.

From NIH Consensus Development Program, downloaded on 10/10/2009 from www.consensus.nih.gov/1995/1995ActivityCardivascularHealth101html.htm

Office of the Surgeon General

Physical Activity and Health: A Report of the Surgeon General was published in 1996. Perhaps one of the recommendations most widely known among Americans, the "30 minutes of moderate physical activity on most, if not all, days of the week" was advertised and publicized to a great degree. Although the report—which was 278 pages long—included a number of other details about the current state of American physical activity and health as well as made other recommendations to improve these, most individuals took home the 30 min suggestion and ignored the other information. One major purpose of this report was to encourage previously sedentary individuals to initiate a physical activity program. The idea was that some physical activity—any physical activity—is better than none, and this report was meant to appeal to individuals who found the previously published ACSM guidelines overwhelming. Many individuals felt that since they were not athletic or trying to improve their sport performance, they did not necessarily have any reason to exercise or be physically active. A major point of the Surgeon General's report was to support the notion that physical activity is good for almost everyone and that general health benefits are gained even at minimal levels of activity—levels that do not necessarily produce significant improvements in cardiorespiratory fitness, such as those obtained by following the ACSM guidelines. The report's recommendation for 2 sessions of strength training each week seemed almost an afterthought—many mentions of the guidelines do not address this aspect, although the report does specify sets, repetitions, and number of exercises.

GUIDELINE 2.8

Title: *Physical Activity and Health: A Report of the Surgeon General*

Organization: Office of the Surgeon General

Year published: 1996

Purpose: To improve health and to reduce the risk of premature mortality in general and the risk of CHD, hypertension, colon cancer, and diabetes in particular

Location: www.cdc.gov/nccdphp/sgr/sgr.htm

Population: People of all ages

GUIDELINES

- People of all ages should perform 30 min of moderate-intensity physical activity on most, if not all, days of the week.

- Most people can obtain greater health benefits by engaging in physical activity of more vigorous intensity or of longer duration.

- Previously sedentary people who begin a physical activity program should start with short sessions (5-10 min) of physical activity and gradually build up to the desired level of activity.

- Adults who have or are at risk for chronic health problems such as heart disease, diabetes, or obesity should consult a physician before beginning a new program of physical activity. Men older than 40 y and women older than 50 y who plan to begin a new program of vigorous activity should consult a physician to be sure they do not have heart disease or other health problems.

- Adults should supplement cardiorespiratory endurance activity with strength-developing exercises at least 2 times each week. At least 8 to 10 strength-developing exercises working the major muscle groups of the legs, trunk, arms, and shoulders should be performed at each session, and 1 or 2 sets of 8 to 12 repetitions of each exercise should be performed.

Reprinted, from U.S. Department of Health and Human Services, Physical Activity and Health: A Report of the Surgeon General, Atlanta, GA: U.S. Department of Health and Human Services, Centers for Disease Control and Prevention, National Center for Chronic Disease Prevention and Health Promotion, 1996.

American College of Sports Medicine Position Statements on Fitness and Health Benefits and Weight Maintenance

The ACSM updated its original 1978 position statement on fitness and the health benefits of exercise in 1990 and again in 1998. The 1998 position statement reflected the significant amount of scientific evidence that had become available in the time since the original position statement, which made an important distinction between physical activity for overall health and physical activity for fitness. Because varying amounts of physical activity are necessary to derive health benefits with respect to conditions such as hypertension, diabetes, osteoporosis, obesity, and CAD, the ACSM then issued specific recommendations in independent position statements; these are addressed in the relevant chapters throughout this book. The 1998 ACSM position stand was designed for the middle to the higher end of the exercise and physical activity spectrum, although it notes that its recommendations are adaptable to a broad cross section of the healthy adult population.

Guidelines for aerobic activity, muscular strength and endurance, body composition, and flexibility are provided. Because the emphasis is on improved fitness and not merely reduced risk of chronic disease, there is a wide range of appropriate levels of activity that correspond to active individuals all the way to elite athletes.

The ACSM has also issued guidelines regarding weight maintenance (see guideline 2.10). In November of 2001 the position stand "Appropriate Intervention Strategies for Weight Loss and Prevention of Weight Regain for Adults" was published as part of an effort to address the growing obesity problem in America. Other guidelines published by the ACSM specifically address adults who are overweight or obese, as defined by a body mass index (BMI) of 25.0 to 29.9 kg/m^2 or 30 kg/m^2 and greater, respectively. The guidelines contain some details about energy intake and pharmacological interventions, but the primary focus is on physical activity as it

---

GUIDELINE 2.9

Title: "Recommended Quantity and Quality of Exercise for Developing and Maintaining Cardiorespiratory and Muscular Fitness and Flexibility in Healthy Adults"

Organization: ACSM

Year published: 1998

Purpose: To develop and maintain cardiorespiratory fitness, body composition, muscular strength and endurance, and flexibility

Location: *Medicine & Science in Sports & Exercise* 30(6): 975-991

Population: Healthy adults

GUIDELINES

- Cardiorespiratory activity is recommended for 20 to 60 min 3 to 5 times per week. Any activity that uses large muscles continuously and is rhythmic and aerobic is considered appropriate. Intensity ranges from 55% to 90% of HRmax (55%-64% for unfit individuals and 65%-90% for all others) or from 40% to 85% of maximum oxygen consumption ($\dot{V}O_2$max) or HRR.

- Resistance training is recommended 2 times per week and should include 8 to 10 exercises performed for 8 to 12 repetitions to near fatigue (10-15 repetitions for older or more frail individuals). Dynamic exercises using the major muscle groups are appropriate.

- Flexibility training utilizing static and dynamic techniques is recommended at least 2 to 3 d/wk. The appropriate length of the stretch hold is that which is sufficient to develop and maintain range of motion.

GUIDELINE 2.10

Title: "Appropriate Intervention Strategies for Weight Loss and Prevention of Weight Regain for Adults"

Organization: ACSM

Year published: 2001

Purpose: To attain safe and effective weight loss and to prevent weight regain after weight loss

Location: *Medicine & Science in Sports & Exercise* 33(12): 2145-2156

Population: Adults who are overweight (BMI of 25-29.9 kg/m²) or obese (BMI of 30 kg/m² or greater)

GUIDELINES

- Individuals should perform moderate-intensity (55%-69% of HRmax) exercise most days of the week.
- Some vigorous (70% or more of HRmax) exercise is necessary in the long term.
- Individuals should first progress to a minimum of 150 min/wk and then ultimately progress to 200 to 300 min/wk. An alternative is to expend at least 2,000 kcal/wk.
- Aerobic endurance exercise should be the primary form of activity, although resistance training may be used as an adjunct to enhance functional capacity.
- Accumulation of exercise in intermittent bouts may reduce barriers to initiation of a physical activity program and may be used by individuals who need to start a program.
- Resistance exercise should be used to supplement the endurance exercise program and should focus on improving muscular strength and endurance.

relates to weight control. Clinical guidelines relating to weight loss had been published previously by the National Heart, Lung, and Blood Institute (NHLBI); however, the ACSM specifically addressed exercise interventions for weight control due to its position as a leading professional organization for exercise, physical activity research, and public health initiatives. The NHLBI guidelines are presented in the next section.

National Heart, Lung, and Blood Institute

In 1998, the NHLBI issued guidelines on treatment for overweight and obesity. These were the result of the first Expert Panel on the Identification, Evaluation, and Treatment of Overweight and Obesity in Adults, which met in 1995. At that point, the obese population in the United States was growing at a staggering rate, and nearly 100 million Americans were overweight or obese, with an associated annual

cost of $100 billion U.S. Using an evidence-based model, the panel, which comprised physicians, researchers, exercise physiologists, clinical nutrition specialists, psychologists, and epidemiologists, developed clinical practice guidelines for treating overweight and obesity, including dietary, behavioral, pharmacological, and surgical therapies. The physical activity guidelines were similar to those presented 2 y earlier in the Surgeon General's report; however, the NHLBI guidelines suggested that a longer duration of exercise might be necessary to achieve weight loss. Walking in particular was promoted as a physical activity because of its ease of initiation, safety, and long-term feasibility. Two other points stressed in the NHLBI guidelines were (1) patients should increase everyday activities such as lifestyle activities and (2) sedentary time should be limited. This latter point is emphasized in many of the guidelines for young people (which are presented in chapter 5). The NHLBI guidelines were endorsed by the National Cholesterol

GUIDELINE 2.11

Title: *Clinical Guidelines on the Identification, Evaluation, and Treatment of Overweight and Obesity in Adults: The Evidence Report*

Organization: NHLBI

Year published: 1998 (update currently in development)

Purpose: To prevent further weight gain, to reduce body weight, and to maintain a lower body weight over the long term

Location: www.nhlbi.nih.gov/guidelines/obesity

Population: Adults who are overweight (BMI of 25-29.9 kg/m²) or obese (BMI of 30 kg/m² or greater)

GUIDELINES

Frequency	Intensity	Duration	Type
Most days of the week	Combination of higher intensity and shorter duration or lower intensity and longer duration	Depends on intensity; start with 10 min 3 d/wk, progress to 30-45 min at least 5 d/wk	Walking or other safe, accessible activity

Special considerations include the following:

- All adults should set a long-term goal of accumulating at least 30 min or more of moderate-intensity physical activity on most, preferably all, days of the week.

- Individuals should plan and schedule physical activity 1 wk in advance, budget the time needed for activity, and document their physical activity by keeping a diary in which the duration and intensity of exercise are recorded.

- An appropriate level of activity is that which uses approximately 150 kcal of energy a day or 1,000 kcal/wk.

Education Program (NCEP), the National High Blood Pressure Education Program (NHBPEP), the North American Association for the Study of Obesity (NAASO), and the National Institute of Diabetes and Digestive and Kidney Diseases (NIDDK) National Task Force on the Prevention and Treatment of Obesity.

Healthy People 2010

Published in 2000, *Healthy People 2010* outlines a set of goals and health objectives for Americans to achieve in order to improve overall health. It was the third in a set of initiatives developed to establish better health among Americans through state and community plans. The objectives are based on scientific data and include measurable statistics in order to quantify improvement. *Healthy People 2010* built upon *Healthy People 2000* (published in 1990) and the 1979 Surgeon General's report pub-

lished a decade earlier. The two primary goals of *Healthy People 2010* are (1) to increase quality and years of healthy life and (2) to eliminate health disparities. The report includes multiple sections aiming to improve health through positive changes in the leading health indicators, which are as follows:

- Physical activity
- Overweight and obesity
- Tobacco use
- Substance abuse
- Responsible sexual behavior
- Mental health
- Injury and violence
- Environmental quality
- Immunization
- Access to health care

The progress in working toward these goals is broken up into 28 focus areas, and there are a total of 467 health objectives. These objectives compare values seen in the late 1990s with target levels to be achieved by the year 2010. For example, at the time of the report, only about 23% of adults in the United States reported participating in regular, vigorous physical activity involving the large muscles in dynamic movement for 20 min or longer 3 or more days a week. Only 15% of adults reported participating in physical activity for 30 min or longer on 5 or more days a week, and another 40% reported participating in no regular physical activity. The *Healthy People 2010* goals for these parameters include having 30% of the adult population participating in both the vigorous and moderate levels of activity and only 20% not participating in any regular physical activity.

Healthy People 2010 was developed by the Healthy People Consortium, which is an alliance of more than 350 national organizations and 250 state health, mental health, substance abuse, and environmental agencies reflecting the ideas and expertise of a diverse range of individuals concerned about the health of the United States. When published in 2000, *Healthy People 2010* reflected the new perspective on physical activity, fitness, and exercise brought to light in the

1990s—a shift in focus from intense vigorous exercise for physical fitness to a broader range of health-enhancing physical activities. In addition to being listed at the top of the 10 leading health indicators targeted by *Healthy People 2010*, physical activity also affects some of the other targeted indicators. The effect of physical activity on reducing overweight and obesity is obvious; evidence suggests that physical activity may also positively affect mental health, tobacco and substance abuse, and possibly injury and violence.

Data were collected for various parameters related to physical activity goals in 1990 and 2000. Although the proportion of adults participating in vigorous physical activity for at least 20 min on 3 or more days of the week was significantly improved from 1990 to 2000, the percentage of adults who performed moderate activity for 30 min on at least 5 d/wk was significantly lower in 2000 (see figure 2.1).

Specialized Guidelines

The next section contains guidelines specific to resistance training. While many guidelines address both cardiorespiratory (or aerobic) and muscle strengthening (or resistance) training,

---

GUIDELINE 2.12

Title: *Healthy People 2010*

Organization: USDHHS

Year published: 2000

Purpose: To improve health, fitness, and quality of life through daily physical activity

Location: www.healthypeople.gov

Population: Adults

GUIDELINES

- Engage regularly, preferably daily, in moderate physical activity lasting at least 30 min.
- Engage in vigorous physical activity that promotes the development and maintenance of cardiorespiratory fitness 3 or more days a week for 20 or more minutes per occasion.
- Perform physical activities that enhance and maintain muscular strength and endurance 2 or more days each week.
- Perform physical activities that enhance and maintain flexibility.

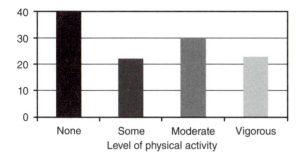

Figure 2.1 Percentage of Americans achieving moderate- or vigorous-intensity activity goals established in *Healthy People 2010*.

some focus on just one or the other. Also included are dietary guidelines, which often reference physical activity as part of maintaining a healthy balance.

Resistance Training

In general, muscle-strengthening or resistance training recommendations are included in overall physical activity guidelines, though the aerobic component of the guidelines is what is usually highlighted by the media. This emphasis on cardiorespiratory fitness was not always so. In the 1950s and 1960s, resistance training was utilized primarily as part of rehabilitation programs for people with orthopedic disabilities. By the 1970s, the widespread focus shifted to aerobic endurance activities, in part due to the popularity of Dr. Kenneth Cooper's *Aerobics* and Frank Shorter's 1972 Olympic marathon victory, but also because of the expanding body of scientific data confirming the positive health effects of aerobic activity. The original 1978 ACSM position statement on exercise guidelines for fitness omitted any guidelines regarding resistance training, primarily due to the fact that at that point in time there was very little evidence-based research quantifying the outcomes of resistance training. This lack of recommendations may have been misinterpreted as implying that resistance training is neither important nor necessary. The 1990 ACSM position statement did address resistance training, reflecting the new research documentation supporting its benefits such as bone health, improved basal metabolism, weight control, and low back health.

Resistance training provides results similar to those of aerobic conditioning but also has additional benefits. Cardiovascular benefits include decreased resting blood pressure as well as decreased exercise blood pressure and heart rate at a given workload. Resistance training can improve a person's lipid profile, primarily by reducing LDL, and can also improve glucose tolerance and hemoglobin A1c, particularly in people with diabetes. Body composition improves both through reduced fat mass and through maintained or increased lean body mass. Age-associated loss of bone mineral density is attenuated or prevented. The risk for injury is decreased. Daily activities are made easier through reduced demands on musculoskeletal, cardiovascular, and metabolic systems. Lastly, resistance training may reduce anxiety and depression and may enhance self-efficacy and overall psychological well-being.

In 2000, the AHA issued a position paper specifically addressing resistance exercise. Because of resistance exercise's numerous benefits not just for overall health but also for cardiovascular health, the Committee on Exercise, Rehabilitation, and Prevention of the Council on Clinical Cardiology felt resistance training deserved specific guidelines. Safety concerns and participation criteria were discussed, and guidelines were created for individuals with or without CVD. The guidelines reflect the committee's sentiment that endorsing a time-efficient program is the most favorable method for promoting the inclusion of resistance exercise in a physical activity program (or for promoting participation in any exercise program at all!). A training program performed twice a week with a single set of exercises may have an effect comparable to that of a theoretical program performed more often and with more sets; more individuals are apt to comply with the former and realize benefits, while the barrier for participation in the latter might prevent participation. In order to minimize the risk of injury and maximize health benefits, the guidelines recommend different combinations of intensity and number of repetitions for individuals of different ages or functional capacities. The guidelines for people without CVD are presented here, while the guidelines for people with CVD are presented in chapter 10.

GUIDELINE 2.13

Title: "Resistance Exercise in Individuals With and Without Cardiovascular Disease"

Organization: AHA Committee on Exercise, Rehabilitation, and Prevention, Council on Clinical Cardiology

Year published: 2000 (updated in 2007)

Purpose: To improve muscular strength and endurance, cardiovascular function, metabolism, coronary risk factors, and psychosocial well-being

Location: http://circ.ahajournals.org/cgi/content/full/101/7/828 (2000 publication) or *Circulation* 116: 572-584 (2007 update)

Population: Adults without CVD

GUIDELINES

Population	Frequency	Intensity	Duration
Individuals younger than 50-60 y	2-3 d/wk	To volitional fatigue (50% ± 10% of 1RM); 13-15 RPE	8-12 repetitions
Individuals older than 50-60 y	2-3 d/wk	To volitional fatigue (50% ± 10% of 1RM); 13-15 RPE	10-15 repetitions

Special considerations include the following:

- Each repetition should include a slow, controlled movement performed in a rhythmical manner (~2 s up and 4 s down), one full inspiration and expiration, and no breath holding.
- Individuals should alternate between upper-body and lower-body work to allow for adequate rest between exercises.
- Individuals should start with a lighter weight (30%-40% 1RM for upper body and 50%-60% 1RM for lower body) and increase resistance by 5% each training session after they can comfortably lift the weight for 12 to 15 repetitions.
- Individuals who cannot complete the minimum number of repetitions using good technique should reduce the weight.
- Exercises should involve the major muscle groups of the upper and lower extremities, such as the chest press, shoulder press, triceps extension, biceps curl, pull-down (upper back), back extension, abdominal crunch or curl-up, quadriceps extension or leg press, leg curl (hamstrings), and calf raise.

The National Strength and Conditioning Association (NSCA) has published specific guidelines for resistance training as well as an official position statement on the role of resistance training for youths (see chapter 5). Although the guidelines do not follow the FITT principle, they do address other aspects of resistance training, including overload, specificity, and periodization. These guidelines are focused on athletes although the stated goals are appropriate for a general population striving for improved health.

Dietary Guidelines From the Institute of Medicine and *Dietary Guidelines for Americans*

Both the 2002 IOM report on DRIs and the 2005 *Dietary Guidelines for Americans* include physical activity recommendations along with the macronutrient guidelines. Why or how did physical activity recommendations come to be a part of dietary recommendations? Both reports, which were created with help from the USDHHS (see sidebar), set out to address nutrient intake

GUIDELINE 2.14

Title: *Basic Guidelines for the Resistance Training of Athletes*

Organization: NSCA

Year published: 2007

Purpose: To enhance cardiovascular health, improve body composition, increase bone mineral density, reduce anxiety and depression, reduce injury risk, and increase muscle strength and endurance

Location: www.nsca-lift.org/Publications/Basic%20guidelines%20for%20the%20resistance%20training%20of%20athletes.pdf

Population: Athletes

GUIDELINES

- Progressive overload should be a fundamental characteristic of resistance training programs directed at the development of neuromuscular capabilities and athletic performance.
- Resistance training programs for athletic performance must adhere to the principle of training specificity in order to match the demands of the sport for which the athlete is training.
- Resistance training programs for athletic performance should be periodized in order to optimize muscular adaptations over long-term training. Periodized training also helps to reduce the potential for overtraining.
- Multiple-set periodized resistance training programs are superior to single-set nonperiodized programs for physical development during long-term training.
- Care must be taken when developing resistance training programs for younger and older athletes because the volume of exercise and the intensity may have to be altered to meet their recovery demands.

as part of a plan to guide Americans toward achieving a healthy weight and avoiding chronic disease attributable to overweight or obesity. In fact, the 2002 IOM report was the product of a panel that was charged with developing guidelines for "good nutrition throughout the life span and that may decrease the risk of chronic disease." In the process of quantifying rates and elements of energy expenditure, the IOM found that it became essential to address the physical activity needed to maintain a weight in the healthy BMI range. The resulting physical activity recommendations were based not only on the epidemiological data used to help create other guidelines but also on doubly labeled water studies.

The doubly labeled water analyses generated estimates for total energy expenditure (TEE) that were then translated into physical activity levels (PALs). PALs, which are a way of representing the average intensity of daily physical activity, are calculated by dividing TEE by basal energy expenditure (BEE). For example, an individual with a TEE of 2,450 kcal and a BEE of 1,400 kcal has a PAL of 1.75 (2,450 kcal / 1,400 kcal). The expert panel identified that sedentary individuals had a PAL of 1.4 or less, while most individuals maintaining a BMI in the healthy range had a PAL of more than 1.6. This level of activity corresponded with 60 min or more of moderate-intensity activity performed daily. Thus, the recommendation for an hour or more of daily activity in order to manage body weight and prevent gradual, unhealthy weight gain was based on a balance of energy expenditure and intake and drew from both epidemiological and tissue-level research. Chapter 17 provides more details on the DRIs.

The physical activity recommendations contained in the 2005 *Dietary Guidelines for Americans*

Physical Activity Guidelines as Part of Dietary Guidelines

In 2002, the IOM report on DRIs was published by the National Academy of Sciences. The preparation of the report was supported financially by the USDHHS, U.S. Department of Agriculture (USDA), U.S. Department of Defense (USDoD), and Health Canada.

In 2005, the sixth edition of the *Dietary Guidelines for Americans* was published by the USDA. These guidelines, published every 5 y, are prepared by the USDA and USDHHS, as required by Congress. In 1995, the first mention of the importance of a physically active lifestyle was made; in 2005, the report gave structured recommendations on physical activity for prevention of chronic disease as well as weight gain or regain.

Report	Most recent publication year (edition)	Publisher	Contributing entities	Basis of recommendations
IOM	2002 (6th in the DRI series)	National Academy of Sciences	USDHHS, USDA, USDoD, Health Canada	Cross-sectional data from doubly labeled water studies
Dietary Guidelines for Americans	2005 (6th)	USDA	USDA, USDHHS	2002 IOM report

were based on the belief that individuals of different weight classifications need differing amounts of activity to achieve and then maintain an ideal healthy weight. Individuals of a lower weight tend to need a lower volume of exercise to realize health benefits and reduce chronic disease risk, while individuals who are overweight or trying to maintain weight loss require more activity to expend more calories in order to achieve the same results. While many people thought that both the 2002 IOM report and 2005 *Dietary Guidelines for Americans* were at odds with the recommendation for 30 min of activity on most if not all days of the week made just a few years earlier in the 1996 Surgeon General's report, this is not the case. The 30 min of daily activity, while perhaps the most memorable aspect of the Surgeon General's report, is not the only part of the report. As noted earlier in this chapter, the Surgeon General's report—like many other reports that have come out in the past decade—addresses the dose–response relationship between activity and health and clearly states that 30 min is a minimum recommendation and that additional health benefits can be attained at higher intensities or longer durations of exercise. The physical activity recommendations in the 2002 IOM report and 2005 *Dietary Guidelines for Americans* are summarized here, while the nutritional guidelines contained in those reports are addressed in chapter 17.

The President's Challenge

In 2008, the President's Council on Physical Fitness and Sports (PCPFS) developed the Active Lifestyle program and issued criteria for earning the Presidential Active Lifestyle Award. The program, which aims to encourage more people to initiate a physical activity program and ultimately become lifelong exercisers, endorses the same guideline of 30 min of daily activity that is recommended by several other entities. The council requires people wishing to earn the reward to participate in 5 d of activity each week; this amount of exercise is felt to be an acceptable amount in terms of expanding the numbers of people involved in activity that may provide health benefits. While more activity is encouraged, the 5 d were chosen in order to allow people who may miss 1 or 2 d each week the chance to earn an award. The council gives separate criteria for young people; these are presented in chapter 5.

GUIDELINE 2.15

Title: *Dietary Reference Intakes for Energy, Carbohydrate, Fiber, Fat, Fatty Acids, Cholesterol, Protein, and Amino Acids*

Organization: IOM Food and Nutrition Board

Year published: 2002

Purpose: To promote health and vigor and to balance food energy intake with total energy expenditure

Location: www.nap.edu/books/0309085373/html

Population: Adults of normal weight

GUIDELINES

- Participate in an average of 60 min of daily moderate-intensity physical activity (walking or jogging at 3-4 mph or 5-6 kph) or shorter bouts of more vigorous exertion (jogging for 30 min at 5.5 mph or 8.9 kph).
- This amount of physical activity should correspond to a PAL that is greater than 1.6.

GUIDELINE 2.16

Title: *Dietary Guidelines for Americans*

Organization: USDHHS (jointly with the USDA)

Year published: 2005

Purpose: To promote health, psychological well-being, and a healthy body weight

Location: www.health.gov/dietaryguidelines/dga2005/report/

Population: Adults

GUIDELINES

- Engage in regular physical activity and reduce sedentary activities to promote health, psychological well-being, and a healthy body weight.
- To reduce the risk of chronic disease in adulthood, engage in at least 30 min of moderate-intensity physical activity, above usual activity, at work or at home on most days of the week.
- Most people can obtain greater health benefits by engaging in physical activity of more vigorous intensity or of longer duration.
- To help manage body weight and prevent gradual, unhealthy weight gain in adulthood, engage in approximately 60 min of moderate- to vigorous-intensity activity on most days of the week while not exceeding caloric intake requirements.
- To sustain weight loss in adulthood, participate in at least 60 to 90 min of daily moderate-intensity physical activity while not exceeding caloric intake requirements. Some people may need to consult with a health care provider before participating in this level of activity.
- Achieve physical fitness by including cardiovascular conditioning, stretching exercises for flexibility, and resistance exercises or calisthenics for muscle strength and endurance.
- The benefits of resistance exercises are seen in people who perform them on 2 or more days each week.
- Physical activity may include short bouts (e.g., 10 min bouts) of moderate-intensity activity.
- Vigorous activity provides greater benefits for physical fitness than moderate activity provides.

GUIDELINE 2.17

Title: *The President's Challenge*

Organization: PCPFS

Year published: 2008

Purpose: To encourage regular physical activity and reduce the risk of chronic disease, including heart disease, cancer, stroke, high blood pressure, and diabetes

Location: www.presidentschallenge.org

Population: Adults 18 y and older

GUIDELINES

- Adults should perform 30 min of moderate physical activity (an intensity equal to that of brisk walking) 5 d/wk.
- Exercising in blocks of 5 min or more is acceptable, although blocks of 10 min or more are ideal.

Pedometers

Using a pedometer is a convenient way for many individuals to quantify their physical activity. The press has promoted 10,000 steps a day as a way of achieving physical activity recommendations (Feury 2000; Hellmich 1999; Quittner 2000). The target of 10,000 steps is a pleasant round number, and encouraging people to achieve this target is one way of increasing the number of people who improve upon a sedentary lifestyle. The correlation between 10,000 steps and 30 min of moderate physical activity, however, is not definite. One study that supported the 10,000 steps as a way of achieving the recommendation of 30 min/d of moderate physical activity found that healthy adults who added 30 min/d of moderate physical activity to regular ADLs accumulated between 9,000 and 11,000 steps per day (Welk and others 2000). Another study found that 8,000 daily steps correspond to approximately 33 min of moderate physical activity (Tudor-Locke and others 2002), so a lower step count might fulfill the CDC and ACSM or Surgeon General activity guidelines. An authority on pedometers and step counting has suggested that a universal step goal is inappropriate due to the wide variation in physical activity levels found among different age groups in the American population (Tudor-Locke 2002).

State and International Guidelines

This section examines guidelines that have been issued by either state or international organizations. These guidelines are notable for various reasons, including overall scope or potential for influence, variance from other national guidelines, and contrast to U.S. guidelines. Most of these entities also have guidelines targeting specific groups, such as youths and older adults, and these are discussed where appropriate in the later parts of this book.

California Department of Health Services

Recently the California Department of Health Services (CDHS) issued physical activity guidelines for preschool through adulthood. The guidelines, developed as part of the California Obesity Prevention Initiative by the California

Center for Physical Activity, are notable for many reasons: (1) Many U.S. states do not issue guidelines separate from those issued for the nation, (2) California is often on the forefront of health promotion and health policy and many other states follow its lead, and (3) California's policies and action for youth activity in public education have been progressive, paving the way for information detailing the relationship between greater fitness and physical education activity and academic achievement. It doesn't hurt that California's governor and visible leader, Arnold Schwarzenegger, is a past president of the PCPFS as well as a prominent bodybuilder, actor, and former Mr. Olympia! The California guidelines are unique in that they are divided not by types of activity but by desired benefits of activity. Different goals of the guidelines are (1) overall general health benefits, (2) improved cardiovascular fitness and body composition, and (3) enhanced muscular strength and endurance with skeletal and flexibility benefits. These guidelines reflect the dose–response relationship of exercise and health and clearly delineate the higher volume of activity necessary to achieve greater health benefits—that is, improved cardiorespiratory fitness and body composition as opposed to minimal general health benefits.

Health Canada

Canadian guidelines also acknowledge the association between exercise intensity and exercise volume, calling for a longer duration of lower-intensity exercise or vice versa in order to achieve health benefits. Like the ACSM guidelines published the same year (1998), these Canadian guidelines include recommendations for both strength training and flexibility training.

GUIDELINE 2.18

Title: *The California Center for Physical Activity's Guidelines for Physical Activity Across the Lifespan*

Organization: CDHS

Year published: 2002

Purpose: For general health and more specific benefits (see table)

Location: www.caphysicalactivity.org/facts_recomm1.html

Population: Adults

GUIDELINES

Type of benefits	Frequency	Intensity	Duration
General health benefits	Most, if not all, days of the week	Moderate to vigorous	Bouts of at least 8-10 min, with a minimum daily accumulation of 30 min
Cardiorespiratory fitness, body composition, and additional health benefits	3-5 d/wk	Moderate to vigorous; moderate-intensity activity is recommended for adults not training for athletic competition	20-60 min depending on intensity; lower-intensity activity should last 30 min or more while higher-intensity activity may last 20-30 min
Muscular strength, muscular endurance, and skeletal and flexibility benefits	2-3 d/wk	8-12 repetitions of each exercise; 10-15 repetitions may be more appropriate for older or more frail individuals	The time needed to complete 8-10 exercises that condition the major muscle groups as well as flexibility exercises to stretch the major muscle groups

From California Obesity Prevention Initiative Health Systems Workgroup, California Department of Health Services, California Department of Health Services Physical Activity Guidelines for Children, Youth, and Adults. 2002. www.caphysicalactivity.com Accessed 10/10/2009.

Australian Government Department of Health and Ageing

Australian guidelines were issued in 1999 and call for four steps to better health. The first three steps cover the minimum amount of health-enhancing activity. These include using physical activity as an opportunity to improve health, incorporating as much activity as possible into the daily routine, and accumulating at least 30 min (accrued in sessions lasting at least 10-15 min) of moderate-intensity activity on most, preferably all, days. The fourth step is an option for people wishing to achieve greater health and fitness benefits: Individuals may add regular, vigorous activity on some or all days of the week.

International Association for the Study of Obesity

In 2002, a group of experts met in Bangkok to develop consensus guidelines regarding physical activity requirements to prevent primary weight gain or weight regain in formerly obese or overweight individuals. While the panel agreed that the existing guidelines of 30 min of moderate-intensity daily activity is adequate to reduce the risk for many chronic diseases, it noted that this level is insufficient to prevent obesity. It recommended that healthy weight be achieved with either a combination of moderate intensity and duration of activity or an overall PAL of 1.7. Additionally, the panel stressed the importance of limiting sedentary behavior in order to achieve weight goals.

World Health Organization

In 1998, the WHO recommended that men and women should achieve a PAL of 1.75. Then, as part of the Fifty-Seventh World Health Assembly held in 2004, the WHO issued the *Global Strategy on Diet, Physical Activity and Health*. In this publication, dietary recommendations were made alongside those for physical activity. These guidelines were based on the fact that

GUIDELINE 2.19

Title: *Canada's Physical Activity Guide to Healthy Active Living*

Organization: Public Health Agency of Canada and the Canadian Society for Exercise Physiology

Year published: 1998

Purpose: To improve fitness and achieve health-related outcomes such as decreasing the risk of premature death from chronic disease

Location: www.phac-aspc.gc.ca/pau-uap/paguide/back1e.html

Population: Adults

GUIDELINES

Type	Frequency	Intensity and Duration
Aerobic exercise	4 d/wk or more	Light effort for 60 min, moderate effort for 30-60 min, or vigorous effort for 20-30 min
Strength exercises	2-4 d/wk	Not specified
Flexibility exercises	4-7 d/wk	Gentle stretching

Special considerations include the following:
- Light effort includes light walking and easy gardening.
- Moderate effort includes brisk walking, biking, swimming, water aerobics, and raking leaves.
- Vigorous effort includes aerobics, jogging, hockey, fast swimming, fast dancing, and basketball.
- Accumulating activity in bouts lasting at least 10 min is acceptable.

GUIDELINE 2.20

Title: *National Physical Activity Guidelines for Australians*

Organization: Australian Government Department of Health and Ageing

Year published: 1999

Purpose: To enhance health (extra activity may provide extra protection against heart disease)

Location: www.health.gov.au/internet/main/publishing.nsf/Content/phd-physical-activity-adults-pdf-cnt.htm/$File/adults_phys.pdf

Population: Adults

GUIDELINES

- Complete at least 30 min of moderate-intensity physical activity on most, preferably all, days.
- Activity may be accumulated by combining a few shorter bouts of 10 to 15 min.
- If possible, also perform regular, vigorous activity for extra health and fitness. This should be done 3 to 4 d/wk for a minimum of 30 min.

GUIDELINE 2.21

Title: "How Much Physical Activity Is Enough to Prevent Unhealthy Weight Gain? Outcome of the IASO 1st Stock Conference and Consensus Statement"

Organization: International Association for the Study of Obesity

Year published: 2003

Purpose: To prevent weight gain and the transition to overweight or obesity

Location: *Obesity Reviews* 4(2): 101-114

Population: Adults

GUIDELINES

- To prevent the transition to overweight or obesity, perform at least 45 to 60 min of moderate-intensity activity per day or aim for a PAL of 1.7 (see p. 44 for definition).
- Reduce sedentary behavior by incorporating more incidental and leisure-time activity into the daily routine.
- To prevent weight regain after eliminating former obesity, perform 60 to 90 min of moderate-intensity activity or lesser amounts of vigorous-intensity activity per day.

(as described in the *World Health Report 2002)* the following six risk factors contribute to the bulk of the burden caused by noncommunicable diseases worldwide:

- High blood pressure
- High blood cholesterol
- Inadequate intake of fruits and vegetables
- Overweight or obesity
- Physical inactivity
- Tobacco use

Five of these risk factors are very closely related to physical activity and diet. Goals of the WHO global strategy include reducing disease related to these risk factors through public policy, increased awareness, and scientific evaluation and intervention. The strategy makes physical activity recommendations that—not unlike many of the other recommendations discussed in this chapter—vary according to desired health goals. The WHO recommendations suggest a minimum of 30 min of regular, moderate-intensity physical activity on most days of the week in order to reduce the risk for CVD, diabetes, colon cancer, and breast cancer. Muscle-strengthening and balance training are recommended for fall reduction and improved functional status among older adults. Finally, activity above and beyond these minimum levels is suggested for weight control.

After the publication of the 2007 ACSM and AHA updates, the WHO listed updated recommendations on its Web site. Instead of designing de novo guidelines, the WHO simply adapted the evidence-based guidelines released by the ACSM and AHA in 2007. Although it is a subtle difference, the basis of classifying activity intensity by METs varies between the ACSM and AHA and WHO guidelines and the new *PAGA* guidelines. As noted earlier in this chapter, a significant number of activities described as 6.0 METs are classified as vigorous intensity by the newer guidelines but as moderate intensity by the older guidelines. Since *PAGA* provides the most complete synthesis of the existing scientific data on physical activity and theoretically supplants the 2007 ACSM and AHA guidelines, it will be interesting to see if the WHO eventually alters its guidelines to reflect those put out by the PAGAC.

GUIDELINE 2.22

Title: *Global Strategy on Diet, Physical Activity and Health*

Organization: WHO

Year published: 2004

Purpose: To reduce mortality, morbidity, and disability attributable to noncommunicable disease

Location: www.who.int/dietphysicalactivity/strategy/eb11344/strategy_english_web.pdf

Population: Adults

GUIDELINES

- Adults should complete at least 30 min of regular, moderate-intensity physical activity on most days of the week.
- Older adults should add muscle-strengthening exercises and balance training to reduce falls and increase functional status.
- Adults desiring weight control may require more activity.

Reproduced with permission: Global strategy on diet, physical activity and health. Geneva, World Health Organization, 2004. Available: www.who.int/dietphysicalactivity/strategy/eb11344/strategy_english_web.pdf.

GUIDELINE 2.23

Title: *Recommended Amount of Physical Activity*

Organization: WHO

Year published: 2008

Purpose: To promote and maintain health

Location: www.who.int/dietphysicalactivity/factsheet_recommendations/en/index.html

Population: Adults

GUIDELINES

- Adults should complete 30 min of moderate-intensity physical activity 5 d/wk *or* 20 min of vigorous-intensity physical activity 3 d/wk *or* an equivalent combination of moderate-intensity and vigorous-intensity physical activity *and* complete 8 to 10 muscle-strengthening exercises 5 d/wk.

- Alternatively, adults should complete 20 min of vigorous-intensity physical activity 3 d/wk *or* an equivalent combination of moderate-intensity and vigorous-intensity physical activity *and* complete 8 to 10 muscle-strengthening exercises (8 to 12 repetitions) at least 2 d/wk.

Reproduced, with permission from the World Health Organization: www.who.int/dietphysicalactivity/factsheet_recommendations/en/index.html Accessed 10/10/2009.

SUMMARY

There are multiple physical activity guidelines for general health and weight control. Historically these have been issued by the ACSM or by U.S. government entities, though now international and private or community bodies have also presented recommendations. The guidelines covered in this chapter are the most recent recommendations published by these organizations. As time has passed, guidelines have become based on evidence rather than just expert opinion. The overall guidelines targeting all individuals provide the foundation for people to implement physical activity programs and develop their own personal exercise programs, the process of which is discussed in the next chapter.

Guidelines for Personal Exercise Programs

A personal exercise program should incorporate aerobic conditioning, strength training, and flexibility exercises. Any type of exercise can be used as aerobic conditioning; maintaining a heart rate in the aerobic range is recommended. Walking, jogging, swimming, and bicycling are common forms of aerobic exercise. Strength training and flexibility training improve lean body mass, boost bone density, and reduce the risk for injury. While many individuals choose to build muscular strength with weight machines or free weights in a fitness center, there are many other alternatives for resistance training, including the use of elastic bands or body weight for lunges, squats, or other calisthenics.

While there are multiple national physical activity guidelines designed to meet varied goals and purposes, the most important issues to consider when developing a personal exercise program are (1) the current physical activity status of the individual and (2) the health, fitness, and performance goals desired by the individual. Many types of activities, such as walking, can be recommended almost universally, while others may be more appropriate for individuals with certain current fitness levels and lifestyles. Goals for a personal exercise program range from reducing risk of chronic disease to alleviating back or joint pain to improving body weight, mood, or self-esteem to improving various measures of fitness.

This chapter covers the basic components of a personal exercise program, including aerobic exercise, strength training, and flexibility training, along with the parameters that should be considered in formulating a program (the FITT principle.) After these concepts are introduced, specific recommendations are made as referenced by the ACSM and the PCPFS. Guidelines for a personal exercise program vary depending on the experience, fitness, and medical conditions of the individual. In addition, this chapter introduces the physical activity pyramid, which is a unique illustration conveying the components of a physical activity program and relative magnitude for each element.

Basic Definitions and the FITT Principle

Before embarking on a personal exercise program, it is beneficial to become familiar with several concepts related to personal exercise prescription. The major components of an exercise program are aerobic activity, muscular strength and endurance activities, and flexibility exercises. Some people also suggest that body composition is an important component of a physical activity program, although favorable changes in body composition occur naturally with habitual involvement in an effective activity program. Most physical activity programs are based on the FITT principle: *f*requency, *i*ntensity, *t*ime, and *t*ype of exercise. The FITT variables are discussed briefly in the next few sections. *ACSM's Guidelines for Exercise Testing and Prescription* (2010) is an excellent resource for more detailed information on working with the FITT variables.

The following are definitions and commonly accepted abbreviations of many of the terms related to physical activity, exercise, and exercise prescription:

- **$\dot{V}O_2$max (maximum oxygen consumption)**—The maximum oxygen uptake occurring with an increasing workload.

- **HRmax (maximum heart rate)**—The maximum heart rate as measured by maximal stress testing or field testing. An age-predicted maximum is often substituted for the measured maximum if the measured maximum is unknown.

- **HRR (heart rate reserve)**—The difference between HRmax and resting heart rate.

- **$\dot{V}O_2$R (oxygen consumption reserve)**—The difference between $\dot{V}O_2$max and resting oxygen uptake.

- **1RM (1-repetition maximum)**—The maximum amount of weight that can be lifted one time. The 1RM can be measured on any number of different exercises, including the bench press, squat, leg press, and so on.

- **MET**—The ratio of a person's working metabolic rate relative to the person's resting metabolic rate. One MET is the energy it takes to sit quietly, or about 3.5 ml $O_2 \cdot kg^{-1} \cdot min^{-1}$.

- **MET-minute**—A volume of activity, or the number of minutes spent doing a specific physical activity times the metabolic cost of doing that activity. For example, running 1 mi (1.6 km) in 8 min is 12 METs;

doing so for 20 min equals 240 MET-min (12 METs × 20 min).

- **Overload**—The physical stress placed on the body when physical activity is greater than usual in amount or intensity. Overload causes an adaptation that improves efficiency or capacity and increases strength.

- **Periodization**—A method of alternating the volume and intensity of training loads to produce peak performance for a specific competitive event.

- **PAL (physical activity level)**—TEE as a multiple of basal metabolic rate. PAL is a way of representing the average intensity of daily physical activity. Sedentary activity refers to a PAL of 1.4 or less.

- **Progression**—The concept of gradually increasing load after the body adapts to a certain overload with improved fitness.

- **Specificity**—The concept that the benefits of physical activity are specific to the body system being stressed.

Frequency

Frequency is the number of exercise sessions completed in a given amount of time. Most personalized exercise prescriptions recommend a daily or near-daily exercise frequency. Some individuals may choose to participate in more than one session a day, depending on the duration or intensity of their sessions. For example, some people park their cars farther away from work and then walk 15 min to their office at the beginning of the workday and then walk 15 min back to their car at the end of the workday. Many elite athletes (and other athletes aspiring to be elite or simply to improve their physical fitness) exercise more than once each day, varying duration or intensity or even mode of exercise. Triathletes are an excellent example of this, swimming at one point in the day and then bicycling, running, or weight training at other times of the day. Other individuals may exercise every other day or 5 d/wk.

Intensity

Intensity is one of the physical activity variables eliciting the most questions. As discussed previously, physical guidelines for general health have changed in recent years, moving from a recommendation for vigorous activity to one for more moderate activity. Determining what these guidelines translate into for each individual is essential in developing a personal program. There are many ways to quantify the intensity of physical activity, including objective measures such as $\dot{V}O_2$max, percent of HRmax, percent of HRR, or METs achieved and subjective measures such as rating of perceived exertion (RPE) or the talk test.

Physiological Measures

Objective measures can help delineate the desired intensity of physical activity. While oxygen consumption as a percentage of $\dot{V}O_2$max or as a percentage of $\dot{V}O_2$R may be the best objective measurement of intensity, it is (1) difficult to monitor during exercise on a daily basis and (2) the means to measure oxygen consumption are not widely available. Although traditionally the recommendations for exercise intensity have been based on a percentage of $\dot{V}O_2$max, the most recent ACSM position stand advocates using $\dot{V}O_2$R instead of $\dot{V}O_2$max. Calculating $\dot{V}O_2$R requires knowledge of the individual's $\dot{V}O_2$max and $\dot{V}O_2$ at rest. The following is an example of calculating a target oxygen uptake ($\dot{V}O_2$max) as a certain percentage of $\dot{V}O_2$R. The desired percentage of $\dot{V}O_2$R is given by the ACSM recommendations of exercising at an intensity of 50% to 85% $\dot{V}O_2$R. For this example, the measured $\dot{V}O_2$max is 33.5 ml · kg^{-1} · min^{-1} with a resting $\dot{V}O_2$ of 3.5 ml · kg^{-1} · min^{-1}.

$$\text{Target } \dot{V}O_2 = (33.5 - 3.5)(0.50\text{-}0.85) + 3.5,$$
$$\text{Target } \dot{V}O_2 = (15\text{-}25.5) + 3.5,$$
$$\text{Target } \dot{V}O_2 = 18.5 \text{ to } 29.0 \text{ ml· kg}^{-1} \cdot \text{min}^{-1}.$$

It is possible to use either predicted HRmax or measured HRmax to get similar information to use in designing a suitable exercise program. This is because the relatively linear relationship between heart rate and $\dot{V}O_2$ makes heart rate an appropriate surrogate for $\dot{V}O_2$. Individuals who have undergone maximal stress testing can use their actual HRmax; others may use an age-predicted model of HRmax. There are several different tables used to predict HRmax. The most widely used and perhaps easiest model is to calculate HRmax by subtracting age from 220. For example, a person who is 40 y has a predicted HRmax of 220 – 40, or 180 beats/min.

Once the HRmax has been measured or calculated, there are two widely recognized ways of determining the heart rate range within which to exercise. The first is simply to exercise at a given percentage of HRmax. The second method, which takes into account individual differences in fitness through resting heart rate, is the HRR method or the Karvonen method (Karvonen and others 1957). The HRR method is more accurate in depicting the exercise intensity relative to oxygen consumption. Resting heart rate is subtracted from HRmax to come up with HRR, which is essentially the working heart rate range. A percentage of the HRR is then calculated and added back to the resting heart rate to attain a target heart rate for exercise. Table 3.1 provides examples of these calculations. The ACSM recommends values of 70% to 85% of HRmax or 60% to 80% of HRR in the general exercise prescription for improving cardiorespiratory fitness. Other ranges may be more appropriate for individuals with different goals or medical constraints.

TABLE 3.1 Predicting a Target Heart Rate Range for Exercise

Age (y)	Resting HR (beats/min)	HRmax* (beats/min)	% HRmax** (beats/min)	% HRR** (beats/min)
40	60	220 – 40 = 180	126-153	132-156
20	40	220 – 20 = 200	140-170	136-168
40	40	220 – 40 = 180	126-153	124-152
60	80	220 – 60 = 160	112-136	128-144
60	40	220 – 60 = 160	112-136	112-136
20	80	220 – 20 = 200	140-170	152-176

*HRmax may be measured by maximal stress testing or may be estimated by subtracting age from 220.

**Example target heart rate ranges are based on the ACSM recommended target of 70% to 85% of HRmax or 60% to 80% of HRR.

There is significant interindividual variation in HRmax, and the error inherent in estimating HRmax will naturally carry over to any calculated target heart rates. Nevertheless, utilizing the heart rate is a simple method for determining exercise intensity and can help set an initial guideline for intensity that can be modified later as needed. Also, the differences between using HRmax and using HRR to calculate target heart rates are more pronounced at low intensities and decrease as intensity increases. Lastly, cardiac drift also affects calculations based on heart rate. Cardiac drift is the increase in heart rate that occurs over time while a person exercises at a constant workload. For example, when a person exercises at a constant level of intensity for as little as 20 min, the heart rate may increase by as much as 5 to 20 beats/min (Crouter 2005). The easiest way to track and use the heart rate is with a monitor. Choosing and using a heart rate monitor is discussed in chapter 18.

Recommended exercise intensity varies depending on the goal of activity. As stated earlier, the ACSM recommendations of 70% to 85% of HRmax are more in line with improving cardiorespiratory fitness; a lower intensity is appropriate for preventing some chronic medical conditions. As with frequency, there seems to be a dose–response relationship between intensity and health, with additional benefits attained at higher intensities compared with lower intensities. However, at very high intensities there may be an increased risk for illness or injury, particularly orthopedic injury. The ideal intensity for better health, improved cardiorespiratory fitness, and avoidance of exercise-induced complications is yet to be known.

Rating of Perceived Exertion and the Talk Test

There are other ways of qualifying exercise intensity, particularly if physiological variables such as heart rate or maximum oxygen reserve are not known. Additionally, some people may consider heart rate or maximum oxygen reserve too difficult, cumbersome, or interruptive to monitor during exercise and activity. Additionally, exercisers may be on medications that blunt the normal heart rate response to stress and so cannot rely on heart rate monitoring. In all of these cases, RPE is a valuable tool in monitoring and prescribing exercise. There is a reasonably linear relationship among oxygen uptake, heart rate, and RPE, thus making RPE useful as a surrogate when oxygen uptake or heart rate monitoring is not available. RPE is an effective way of monitoring, reflecting, and regulating the physiological response to exercise in order to get the most out of a workout.

One of the most commonly used scales for RPE is the Borg scale. The Borg scale ranges from 6 to 20, with 6 being no exertion at all and 20 being the maximal level of exertion possible. The scale is subjective since perceived exertion is based on the subjective sensations a person experiences during physical activity, including increased heart rate, sweating, increased respiratory rate, and muscle fatigue. In general, an RPE of somewhat difficult (between 12 and 14 on the Borg scale) suggests that physical activity is being performed at a moderate level of intensity. Also, due to the linear relationship between RPE and heart rate, a fairly good estimate of heart rate is to multiply RPE by 10. While this estimate is fairly accurate for many individuals, it is not accurate for individuals who take medications that affect heart rate. Also, accuracy may vary quite a bit depending on the age and physical conditioning of the individual. One way of utilizing this method is to correlate a specific RPE with a specific heart rate during activity; the person can then exercise at the RPE without having to palpate heart rate.

Lastly, several entities have suggested using the talk test to gauge exercise intensity. The talk test reflects an individual's ability to talk while exercising. In general, a moderate level of intensity should allow a person to converse while exercising. An exercise intensity that allows the participant to just respond to conversation fulfills the criteria for a proper talk test. This intensity level has been shown to be within the parameters of established intensity guidelines for many populations, including athletes, patients with stable CVD, and college students (Persinger and others 2004). This is because the talk test approximates the ventilatory threshold, which is the appropriate level of exertion across different modes of exercise, including using the treadmill and cycle ergometer.

Time

The recommended duration of a physical activity session is affected by several variables. Recommended duration generally takes into account the total duration of activity performed in a day or in a week. Also, the recommended time spent in physical activity varies depending on the intensity of the exercise. For example, the 2007 ACSM and AHA guidelines recommend either 30 min of moderate-intensity exercise on 5 d/wk or 20 min of vigorous-intensity exercise on 3 d/wk. The shorter duration in the latter recommendation reflects the higher intensity or higher METs achieved. Total caloric expenditure or MET-minutes are likely to be similar whether moderate or vigorous exercise recommendations are followed. A typical exercise session comprises a warm-up, a central segment of exercise at the desired intensity, and a cool-down. One schematic of a typical exercise session is presented in figure 3.1. Guidelines often recommend a total time for exercise that can be achieved in individual bouts lasting as little as 10 min. There is some evidence that bouts this short may improve cardiorespiratory fitness, although there is no clear-cut scientific evidence that short durations of exercise improve general health. However, shorter bouts have been widely encouraged for several reasons, including the removal of a barrier to exercise due to perceived lack of time and the improvement of fitness to the point where activity duration can be extended to a more beneficial length.

Type

The type, or mode, of activity varies just as much as the other FITT variables do, and even more! Because the recommendations for intensity specify a target heart rate or RPE, any activity that fulfills those criteria is acceptable. In fact,

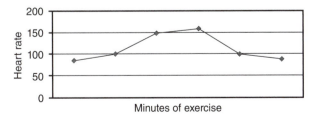

Figure 3.1 An example of exercise intensity and heart rate over a course of time.

some activities such as playing soccer may vary in intensity over the activity duration. In such activities a basal level of intensity is interspersed with higher-intensity surges. Walking is one of the most popular modes of activity, as it is convenient, simple, and inexpensive. It is the most common category of activity for all age groups and both sexes. Other popular modes of activity include gardening or yard work, stretching, weightlifting, biking, swimming, running, and performing aerobics.

Components of Physical Fitness

The health-related components of physical fitness include cardiorespiratory fitness, muscular strength, muscular endurance, and flexibility. Though the general exercise prescription specifies frequency, intensity, duration, and type of activity, it may not improve all the different components of fitness, as some components may improve only when addressed specifically. For example, although brisk walking is sure to improve cardiorespiratory fitness, it will not necessarily provide any gains in flexibility.

When addressing the components of physical fitness, two important principles must be kept in mind: specificity and overload. These are discussed in depth in exercise physiology textbooks; the *ACSM's Guidelines for Exercise Testing and Prescription* also provides a helpful overview. Specificity relates to the fact that training effects derived from an exercise program are specific to the exercise performed and the muscles involved. For example, a swimmer may improve efficiency, time, and $\dot{V}O_2$max during swimming, but these improvements do not necessarily translate to better measures while running. The overload principle states that in order for a tissue or organ to improve its function, it must be exposed to a stimulus greater than that to which it is accustomed (it must be overloaded). Progression and improvement occur through repeated adaptation to stress.

Cardiorespiratory Fitness

The purpose of aerobic exercise is to improve cardiorespiratory fitness, or the ability of the heart and lungs to deliver oxygen to the working muscles and the ability of the muscles to utilize oxygen to generate energy. Swimming, walking, bicycling, jogging, and continuous dancing are all forms of aerobic exercise that improve cardiorespiratory fitness. The terms *cardiorespiratory fitness, aerobic capacity,* and $\dot{V}O_2max$ may all be used interchangeably. In general, a training program can improve cardiorespiratory fitness by 5% to 30%. Differences in improvement depend on the intensity of the program, the initial level of fitness, the initial body weight, and genetics.

Muscular Strength and Endurance

Muscular strength and endurance are improved through resistance training. Improvement decreases the risk for osteoporosis, low back pain, hypertension, and diabetes. It also enhances the ability to perform ADLs and to maintain independence. Improvements in lean body mass and a small increase in caloric expenditure during recovery from resistance training may contribute to a slight increase in resting metabolic rate. Resistance training has a small favorable effect in lowering LDL cholesterol. It also produces mechanical loading on skeletal tissue that can stimulate an increase in bone formation and slow bone loss. Additionally, through the development and maintenance of lean muscle mass, resistance training may reduce insulin resistance. Lastly, muscle-strengthening exercises may attenuate the loss of lean mass that occurs with weight reduction.

Muscular strength and endurance are quantified in several ways, including 1RM, percent of 1RM, 10RM, 15RM, and so on. Muscular strength and muscular endurance can be developed simultaneously. Variables of the resistance training program include the weight (resistance), the number of repetitions, the number of sets, and the duration of the repetition.

Flexibility

Flexibility training, or stretching exercises, may improve range of motion in a joint or a series of joints and is necessary for optimal musculoskeletal function. Increases in flexibility may occur acutely and may enhance the ability to perform ADLs as well as may decrease the risk for injury.

Stretching exercises may be performed as part of a movement series, such as in yoga, tai chi, and Pilates, or may be performed independently as part of a warm-up or cool-down for an aerobic activity. Improvements generally occur after 15 s of stretch, and the optimal number of stretches per muscle group is 2 to 4. Static stretching as opposed to ballistic or bouncing stretching is recommended.

Guidelines for Individualized Exercise Prescription

This section covers the specific recommendations from the ACSM and PCPFS for developing and performing an individualized exercise program. The advice on building an exercise program has changed somewhat over time due to the emergence of evidence about the dose–response relationship of physical activity and health. Earlier exercise prescriptions focused on individuals wanting to develop sport-related or performance-related fitness, while more recent guidelines also emphasize health-related goals for individuals not participating in sport competition. Such goals include weight control and reduction or prevention of chronic disease.

American College of Sports Medicine

The ACSM has developed the standard for exercise prescription. In fact, before the 1996 report from the U.S. Surgeon General, the ACSM's guidelines for exercise prescription were felt to be a national recommendation for all individuals. As already discussed, there is a difference between national physical activity guidelines and an individual exercise prescription. The former focuses on doing the most good for the most people by encouraging a minimum and habitual amount of physical activity to reduce the risk for chronic disease, while the latter may have any number of goals, such as improving cardiorespiratory fitness, reducing blood pressure, aiding in weight loss, improving body composition, increasing bone density, or achieving any combination thereof. Personal exercise prescriptions are often developed by professionals but may also be self-guided by motivated and educated individuals. The FITT principle, detailed earlier, applies to each component of the personal exercise prescription outlined by the ACSM: cardiorespiratory exercise, resistance exercise, and flexibility exercise.

Cardiorespiratory Exercise

The intensity of aerobic exercise is influenced by the desired goal: Higher-intensity exercise of shorter duration may improve cardiorespiratory fitness while lower-intensity training of longer duration may provide health-related benefits. The ACSM currently recommends an exercise intensity corresponding to 30% or 40% up to 85% of $\dot{V}O_2R$ or HRR or corresponding to 57% or 64% up to 94% of HRmax. The lower range applies to extremely deconditioned or sedentary individuals, reflecting the fact that such individuals may achieve improved fitness at lower intensities. On the other hand, individuals who habitually exercise might require an exercise intensity of 65% to 85% of HRR or 80% to 94% of HRmax in order to improve fitness. Individuals who do not want to or have difficulty palpating the heartbeat can use RPE as an acceptable surrogate for determining intensity. The ACSM recommends using a perceived exertion of "somewhat hard" to "hard" for the most fit individuals and "light" to "moderate" for sedentary individuals.

Exercise duration is linked to exercise intensity, as stated previously. A minimum of 20 min is felt to be necessary to improve aerobic capacity. Bouts lasting 10 min or longer may be accumulated throughout the day to reach the total amount of activity. Maximal oxygen uptake can be improved significantly by increasing the overall duration of exercise, whether the exercise is performed continuously or in intermittent short bouts (Murphy and Hardman 1998; DeBusk and others 1990). Performing short bouts of high-intensity exercise is referred to as *interval training* and can also enhance $\dot{V}O_2$max. Whether or not this improved $\dot{V}O_2$max is accompanied by health benefits remains to be seen. Because higher-intensity exercise increases the risk of orthopedic injury and cardiovascular events, interval training is recommended only for specific individuals or athletes with the goal of superior fitness.

The ACSM's recent guidelines for fitness recommend a frequency of 3 to 5 d/wk. Like exercise duration, exercise frequency demonstrates an inverse relationship with exercise intensity. For example, people exercising at the upper end of the recommended intensity range can improve $\dot{V}O_2$max by exercising just 3 d/wk, while people utilizing the lower end of the range may need to exercise 5 d/wk. Although it seems obvious that increasing the exercise frequency brings about greater improvements in fitness, there is a point of diminishing return. In other words, the improvement generally plateaus around a frequency of 3 to 5 d/wk, and additional benefits gained by training more than 6 d/wk are minimal. At this higher frequency, the risk for injury may outweigh any benefits; only competitive athletes or individuals striving for significant performance enhancement should participate at this higher level.

Resistance Exercise

An integral part of a balanced personal exercise program, resistance training offers many health benefits (as already discussed). As opposed to aerobic exercise, which has a more easily quantifiable intensity indicator (measured heart rate), resistance training has an intensity that is less easily quantified. Nevertheless, there are ways of varying and quantifying intensity. Variables in resistance training include the amount of the weight used, the number of repetitions, the speed of movement (momentum), and the maintenance of muscle tension (versus locking out the joint). By changing one of these variables while keeping the others constant, it is possible to increase intensity. In general, RPE may be used for resistance training, with an initial goal of 12 or 13 for beginners and a final goal of 15 or 16 for more advanced individuals. These RPE ranges are considered to be submaximal intensity. For people desiring a high-intensity resistance effort, a target RPE of 19 or 20 is recommended.

There are several considerations in determining the appropriate number of repetitions. The amount of weight lifted should be enough to cause significant muscle fatigue at the end of the set. For example, lifting 4 repetitions at a higher weight or 15 repetitions at a lower weight doesn't matter as much, as long as the muscle is fatigued at the end. According to the ACSM, "there is little evidence to suggest a specific number of repetitions will provide a superior response relative to muscular strength, hypertrophy, or absolute muscular endurance" (Whaley 2006, p. 156). One exception to this guideline, however, occurs when the goal is to increase bone mineral density; superior improvements in bone density are seen with lower (7-10) versus higher (14-18) repetitions (Kerr and others 1996; Vincent and Braith 2002). Thus, women with osteopenia or osteoporosis should train at a higher resistance with fewer repetitions in order to benefit from site-specific improvement in bone mineral density. For the general healthy population, the recommended range is 8 to 12 repetitions.

The last consideration that is addressed frequently in resistance training is the number of sets. Most of the evidence has shown a similar response in muscular strength, hypertrophy, and endurance regardless of whether single or multiple sets are performed (Hass and others 2000; Starkey and others 1996). Because single-set programs are more efficient and can cause less delayed-onset muscle soreness, single-set programs are recommended for the majority of people. Exercises should be rhythmic, performed at a moderate repetition duration (3 s of concentric movement, 3 s of eccentric movement), involve a full range of motion, and not interfere with normal breathing.

Flexibility Exercise

Stretching should be preceded by a warm-up in order to elevate muscle temperature. Flexibility exercise prescription consists of static stretching that involves the major muscle groups and tendon units. The recommendation is to perform flexibility exercises on a minimum of 2 or 3 d/wk, and optimally 5 to 7 d/wk, for at least 10 min. Stretching should be performed to the end of the range of motion (but should not cause discomfort) at a joint of tightness. Each stretch should be held for 15 to 60 s, with at least 4 repetitions per stretch. Choosing which joints and muscles to stretch is personal and depends somewhat on the type of aerobic exercise or resistance training being performed. Also, some activities such as tai chi, yoga, or Pilates may fulfill standards for flexibility while also providing resistance training or light aerobic conditioning. Another type of flexibility training, called *proprioceptive neuromuscular facili-*

tation (PNF), involves a 6 s muscle contraction followed by a 10 to 30 s assisted stretch. There are many different resources detailing appropriate and easy-to-understand stretches. In general, the safest stretches are those that stretch the muscle or joint but do not stretch the ligaments or nerves. For example, a seated toe touch is safer than a standing toe touch and a modified hurdler's stretch is preferable to a barre stretch or a yoga plow pose.

President's Council on Physical Fitness and Sports

The PCPFS has developed a model for physical activity recommendations that takes into account several of the national guidelines and position statements and allows for a range of baseline physical activity levels. Activity recommendations vary based on an individual's activity level and goals; these range from recommendations for everyone to those for sedentary individuals to those for moderately active individuals interested in health or improved physical fitness to those for vigorously active individuals who are interested in performance improvement for sport or specific physical tasks. Guidelines 3.2 through 3.6 present each set of recommendations.

In order to stay healthy and be able to perform everyday tasks, individuals should make choices that result in greater physical activity. These include walking rather than driving, climbing stairs rather than taking the elevator or escalator, and parking farther away from the store or office and walking a longer distance rather than parking right by the door. Additionally, all individuals should consider getting off the bus or train one stop earlier for a short walk to the office, store, or home. Weight-bearing activities, which expend more energy and boost bone health, should be part of everyday routine. Because low back problems are so common, a regular routine of daily stretching is recommended, as it may prevent pain by reducing muscle strain or tension.

Sedentary individuals stand to gain significant benefits by becoming more active. Guidelines for sedentary individuals build on the daily activities recommended for everyone and include additional health-improving activities such as walking, yard work, cycling, slow dancing, and low-impact exercise performed to music. Because

Title: "General Principles of Exercise Prescription"

Organization: ACSM

Year published: 2010

Purpose: To improve cardiorespiratory fitness, body composition, and muscular fitness

Location: *ACSM's Guidelines for Exercise Testing and Prescription, Eighth Edition*

Population: Everyone (healthy adults)

GUIDELINES

Component	Frequency	Intensity	Duration	Activity
Cardiorespiratory conditioning	3-5 d/wk	Lower fit individuals: 30% to 45% HRR or 30% to 45% $\dot{V}O_2R$ or 57% to 67% HRmax; Fitter individuals: 70% to 85% HRR or 70% to 85% $\dot{V}O_2R$ or 84% to 94% HRmax	20-60 min	Work the large muscles; perform dynamic activity
Resistance training	2-3 d/wk	60%-80% 1RM	2-4 total sets of 8-12 repetitions	Work each major muscle group, including multijoint or compound exercises
Flexibility training	Minimum of 2-3 d/wk	Stretch to tightness at the end of the range of motion but not to pain	15-60 s; 4 or more repetitions per muscle group; 6 s contraction followed by 10-30 s assisted stretch for PNF	Use static, dynamic, or ballistic; PNF; and dynamic range-of-motion techniques

Level of habitual physical activity	Physical fitness classification	Frequency		Intensity			Time		
		kcal/wk	d/wk	HRR, $\dot{V}O_2R$	HRmax	RPE	Total daily min	Total daily steps	Weekly min
Sedentary, no habitual exercise, extremely deconditioned	Poor	500-1,000	3-5	30%-45%	57%-67%	Light to moderate	20-30	3,000-5,000	60-150
Minimal physical activity, no exercise, moderately to highly deconditioned	Poor to fair	1,000-1,500	3-5	40%-55%	64%-74%	Light to moderate	30-60	3,000-4,000	150-200
Sporadic physical activity, no or suboptimal exercise, moderately to mildly deconditioned	Fair to average	1,500-2,000	3-5	55%-70%	74%-84%	Moderate to hard	30-90	3,000-4,000	200-300

>continued

GUIDELINE 3.1 >*continued*

		Frequency		Intensity			Time		
Habitual physical activity, regular moderate- to vigorous-intensity exercise	Average to good	>2,000	3-5	65%-80%	80%-91%	Moderate to hard	30-90	3,000-4,000	200-300
High amounts of habitual activity, regular vigorous-intensity exercise	Good to excellent	>2,000	3-5	70%-85%	84%-94%	Somewhat hard to hard	30-90	3,000-4,000	200-300

Reprinted, with permission from "General principles of exercise prescription", in ACSM's Guidelines for Exercise Testing and Prescription (8th Ed.), Thompson WR, senior editor, Baltimore: Lippincott, Williams and Wilkins, p 166-167, 2010.

GUIDELINE 3.2

Title: *Personalizing Physical Activity Prescription*

Organization: PCPFS

Year published: 1997

Purpose: To develop and maintain cardiorespiratory fitness, body composition, muscular strength and endurance, and flexibility

Location: www.fitness.gov/personalizing.pdf

Population: Everyone (healthy adults)

GUIDELINES

- Activities should be the type that can be done as part of home, work, and leisure-time routines.
- Whenever possible, individuals should make decisions to be active within the realm of daily activities.
- Weight-bearing activities should be emphasized, as they use more energy and enhance bone health.
- A daily routine of stretching should be done to prevent low back problems.

many sedentary individuals have not been active to date and cannot walk continuously for 30 min, the guidelines allow activity to be broken up into segments. Suggestions include taking two 10 min exercise breaks during the workday and another 10 min exercise break in the morning or at night, with a 30 min walk on weekend days. Because the goal is to move from a sedentary lifestyle to a routine of regular activity, the emphasis is on being active and accumulating activity as opposed to pushing the intensity level.

People with fitness goals, including the goal of enhanced performance, have a more vigorous exercise prescription. These individuals may utilize heart rate monitoring to ensure that their exercise is intense enough to improve aerobic fitness and potentially also performance. Guidelines for strength training include training 2 or 3 times a week, performing 1 or 2 sets of each exercise and 10 to 15 repetitions for each set. For people wanting to improve fitness, there is an emphasis on improving body leanness through increased caloric expenditure. For people wanting to improve performance, exercise consisting of interval training as well as sport-specific performance skills is recommended.

GUIDELINE 3.3

Title: *Personalizing Physical Activity Prescription*

Organization: PCPFS

Year published: 1997

Purpose: To develop and maintain cardiorespiratory fitness, body composition, muscular strength and endurance, and flexibility

Location: www.fitness.gov/personalizing.pdf

Population: Sedentary individuals (people who do no regular physical activity or who cannot walk for 30 min without experiencing discomfort or pain)

GUIDELINES

- Inactive individuals should continue to find ways to include activity in their daily routine and should accumulate at least 30 min of moderate-intensity activity daily.
- Activity can be broken into 2 to 4 segments.
- Weight-bearing activities should be included; the emphasis should be on being active or accumulating activity and not on intensity.
- Activities include walking, yard work, cycling, slow dancing, and low-impact exercise performed to music.

GUIDELINE 3.4

Title: *Personalizing Physical Activity Prescription*

Organization: PCPFS

Year published: 1997

Purpose: To develop and maintain cardiorespiratory fitness, body composition, muscular strength and endurance, and flexibility

Location: www.fitness.gov/personalizing.pdf

Population: Moderately active individuals (people who can walk 30 min continuously without experiencing pain or discomfort but cannot jog 3 mi or 4.8 km) with health goals

GUIDELINES

- The specific activities performed should be tailored to the desired health goal.
- For general health promotion, individuals should accumulate 30 min of moderate-intensity activity daily. A longer duration or higher intensity of exercise may provide additional gains.
- For healthy bones, individuals should perform weight-bearing activities and activities designed to improve muscular strength and endurance.
- For a healthy low back, individuals should perform static stretching targeting the midtrunk as well as perform abdominal curl-ups.
- For psychological well-being, individuals should select enjoyable activities and environments.

GUIDELINE 3.5

Title: *Personalizing Physical Activity Prescription*

Organization: PCPFS

Year published: 1997

Purpose: To develop and maintain cardiorespiratory fitness, body composition, muscular strength and endurance, and flexibility

Location: www.fitness.gov/personalizing.pdf

Population: Moderately active individuals with fitness goals

GUIDELINES

- Individuals should accumulate 30 min of daily activity in addition to performing different types of activity depending on fitness goals.
- For aerobic fitness, individuals should do 20 to 40 min of vigorous activity (at 70% to 85% of HRmax) 3 or 4 d/wk.
- To achieve relative leanness, individuals with too much fat should increase energy expenditure by increasing activity duration to more than 30 min each session or increasing intensity to include vigorous-intensity activities. Individuals should also include resistance exercise.
- To enhance muscular strength and endurance, individuals should engage in resistance training 2 or 3 times a week, on alternating days, performing 10 to 15 repetitions of the lift for each muscle group and performing 1 or 2 sets of each lift.
- To improve flexibility, individuals should perform daily static stretches, working all the joints, holding each stretch for 10 to 30 s, and repeating each stretch 2 or 3 times.

GUIDELINE 3.6

Title: *Personalizing Physical Activity Prescription*

Organization: PCPFS

Year published: 1997

Purpose: To develop and maintain cardiorespiratory fitness, body composition, muscular strength and endurance, and flexibility

Location: www.fitness.gov/personalizing.pdf

Population: Vigorously active individuals (people who can run 3 mi or 4.8 km continuously) with performance goals

GUIDELINES

- In addition to doing those activities recommended in guideline 3.4, individuals should include sport-specific motor tasks and skills and anaerobic fitness training.
- Interval training should also be included.
- Mental readiness, strategy, and performance-related motor tasks and skills should be emphasized to develop and maintain fitness levels.

American Council on Exercise

The ACE, which is a nonprofit organization committed to enriching quality of life through safe and effective exercise and physical activity, publishes *Fit Facts*, which are handouts containing information pertinent to specific exercise and health topics. The ACE recommends that each exercise program contain the three major components: aerobic exercise, muscular stretch and endurance conditioning, and flexibility exercise. The ACE recommendations for activity duration depend on the goal: For general health maintenance, a 20 min session is appropriate, while a session lasting 45 min or longer is encouraged to promote weight loss. According to the ACE, intensity should be comfortable—that is, 55% to 80% of the estimated HRmax or an intensity amenable to the talk test. Frequency, like duration, depends on the goals of exercise: For people aiming for general health maintenance, 3 or 4 d of weekly activity is recommended, while 4 to 6 d of weekly activity is better for weight loss. The ACE does not recommend a frequency for strength training but does recommend gradually working up to 8 to 12 repetitions. Flexibility exercises consisting of mild stretches held for 10 to 30 s are recommended after a warm-up.

Physical Activity Pyramid

Another practical way of demonstrating physical activity goals and how to formulate an individual exercise prescription is the physical activity pyramid. This pyramid can be used to classify activities by type and associated benefits, making it a useful model for sorting out multiple recommendations. The four levels of activity in the pyramid may be used as categories to simplify the exercise prescription (see figure 3.2). The

---

GUIDELINE 3.7

Title: *Three Things Every Exercise Program Should Have*

Organization: ACE

Year published: 2001

Purpose: For general health maintenance or weight loss

Location: www.acefitness.org/fitfacts/fitfacts_display.aspx?itemid=2627

Population: Adults

GUIDELINES

Component	Frequency	Intensity	Duration	Activity
Aerobic conditioning	General health: 3-4 sessions per week Weight loss: 4 or more weekly sessions (take at least 1 d off each week)	55%-80% HRmax; talk test	General health: 20 min or more Weight loss: 45 min or more	Work large muscles in a continuous, rhythmic fashion
Muscular strength and endurance	Not specified	Not specified	8-12 repetitions per set; add sets to improve strength	Involve every major muscle group (arms, chest, back, stomach, hips, and legs)
Flexibility	Not specified	Mild stretch	10-30 s for each stretch	Stretch after warming up; include all major muscle groups

Adapted, by permission from American Council on Exercise Fit Facts, San Diego 2001. www.acefitness.org/fitfacts Accessed 10/10/2009.

FITT principle may be applied depending on the goals of the individual. As with the USDA food pyramid (see chapter 17), people wanting to improve their health are encouraged to select activities from each of the four levels. The four levels are as follows:

- **Level 1:** The base of the pyramid is the foundation of lifestyle physical activities (LPAs). These are activities such as climbing stairs, walking briskly, or mowing the lawn that serve to promote general health, reduce chronic disease, and help to maintain a healthy weight. These activities reflect the recommendations made by several entities to perform physical activity on most days of the week.

- **Level 2:** This level encompasses active aerobic pursuits, including dance, jogging, bicycling, and some sports such as basketball, tennis, and racquetball. These activities elevate the heart rate to a higher level and represent a moderate to vigorous level of intensity. Working at 50% to 85% of the working heart rate range or HRR can optimize cardiorespiratory fitness. If more time is spent at a vigorous intensity, the frequency or time spent in this activity may be reduced. Because of the dose–response relationship between activity and health, further health and fitness benefits are expected beyond those achieved with LPAs alone.

- **Level 3:** Muscle fitness and flexibility exercises occupy level 3. Frequency of muscle fitness exercises is less than that of activities at the first two levels; 2 or 3 d/wk is appropriate. Utilizing major muscle groups, 8 to 10 exercises should be performed. The number of sets ranges from 1 to 3, depending on the desired goals; 1 set is adequate for people interested in health benefits and 3 may provide superior strength for people aiming for high-level performance (Rhea and others 2002). Flexibility exercises should be performed at a greater frequency—3 to 7 d/wk—but the duration of such exercise generally is short. Each of the body's major muscle groups should be stretched to mild discomfort several times, and each stretch should be held for 15 to 30 s.

- **Level 4:** At the top of the pyramid is inactivity, reflecting the need for some recovery and relaxation time without becoming too sedentary. Because of the increased risk for obesity and CVD

associated with more sedentary time, activities such as watching television, playing video games, working at the computer, and driving should be minimized. In general, screen time (above and beyond what might be required for work) should be limited to no more than 1 h daily. Body composition, or the relative percentages of muscle, fat, and bone, is also represented at the top of the pyramid. By participating in the lower three levels of the pyramid, individuals can balance energy intake and expenditure and maintain a healthy weight.

Other Models

Another way of developing a personal exercise prescription is to sum up the recommended components and attempt to fulfill some combination of them over time. One author has created a body score that takes into account not only the components of a personal exercise prescription but also healthy diet, sleep patterns, and stress manage-

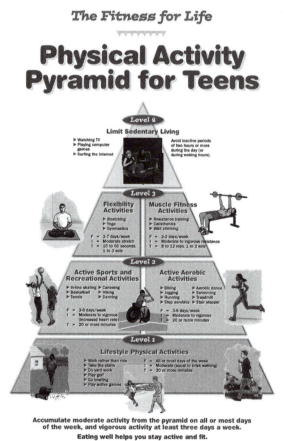

Figure 3.2 The physical activity pyramid.

Reprinted, by permission, from C. Corbin and R. Lindsey, 2007, *Fitness for life*, updated 5th ed. (Champaign, IL: Human Kinetics).

ment (Katzen and Willett 2006). Individuals can chart daily or weekly accomplishments; doing so helps to not only record progress but also identify areas that need attention.

Developing a personal exercise prescription for youths may be challenging due to wide variation in age, maturation, experience, and skill. As discussed in chapters 4 and 5, most physical activity guidelines for children focus more on the total duration of activity—and of inactiv-ity—as opposed to absolute or relative intensity. Nevertheless, there are models for a personal exercise prescription for youths; one is the Activi-tygram developed by The Cooper Institute. The Activitygram helps youths track specific types of activity and sedentary behaviors over time. Using this model, children can become familiar with the components needed for fulfilling activity recommendations. The Activitygram is pictured in figure 3.3 on page 58.

SUMMARY

Individual exercise programs comprise aerobic activity, strength training, and flexibility exer-cises. Each component provides specific health benefits, and the relative emphasis of each component depends on the health- or performance-related goals. Each of these elements can be detailed further by specifying the frequency, intensity, duration, and type of activity. Newer recommendations not only include specific guidelines for these various components but also emphasize a reduction or restriction in sedentary and screen time.

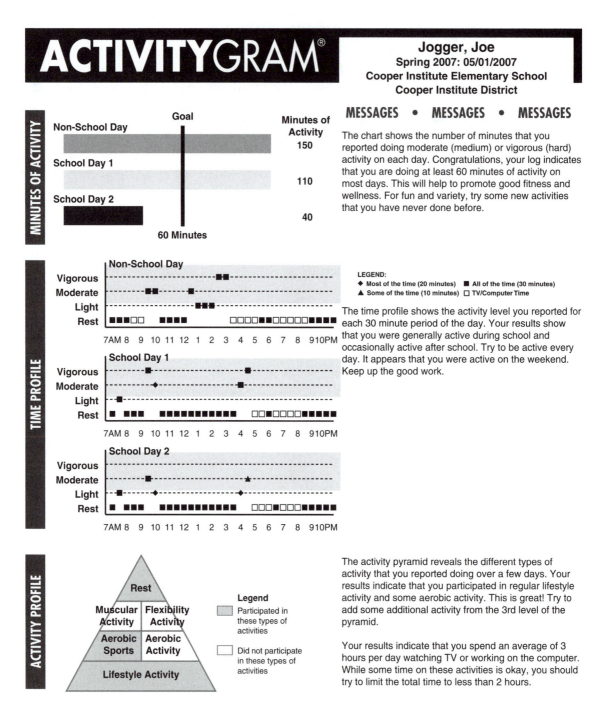

ACTIVITYGRAM provides information about your normal levels of physical activity. The ACTIVITYGRAM report shows what types of activity you do and how often you do them. It includes the information that you previously entered for two or three days during one week.

Figure 3.3 The Cooper Institute's Activitygram.

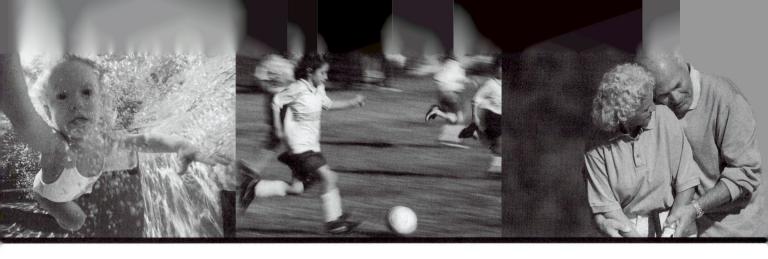

Physical Activity Guidelines by Population

The second part of this book begins by focusing on the guidelines that are specific to individuals of different ages, starting with infancy and advancing to older adulthood. Chapter 4 addresses activity guidelines for the very young, including infants and toddlers. Because of the importance of establishing a lifetime of physical activity, encouraging enjoyment and motor skill development is emphasized for these age groups. Many of the guidelines in chapter 5, which addresses physical activity for youths, pertain to exercise and physical education in schools, since this is where children spend a great deal of time and the potential for influence can be great. Chapter 6 is dedicated to activity guidelines for pregnancy, providing a historical perspective that shows just how much the knowledge base regarding physical activity during pregnancy has changed over the last few decades. Once avoided for fear of harm to the fetus, physical activity is now considered a mainstay in a healthy pregnancy, benefiting both mother and unborn child. The last chapter in part II focuses on the most rapidly growing segment of the American population—older adults. As individuals live longer, there is more of an emphasis on aging gracefully and possibly reversing some of the natural physiological changes that come with aging. The goals for exercise in this age group include preventing disability and falls; maintaining independence, mobility, and functional capacity; and promoting cognitive and behavioral rewards.

Infants and Toddlers

Infants use physical activity to establish movement skills, motor development, and muscle growth. The infant's caregiver not only actively participates in the infant's activities but also fosters emotional and psychosocial development. A great deal of infant–caregiver bonding occurs in the setting of physical games, activity, and movement. This sets the stage for an active daily life and can encourage a healthy lifestyle for both infant and caregiver. For infants, the primary focus of physical activity is to establish large-muscle activities and movement skills in a safe manner. Toddlers need to have both structured and unstructured opportunities for play and activity. The toddler stage should focus on the development of more complex activities.

Because infants, toddlers, and preschoolers have regularly scheduled well-child visits with their pediatricians and other care providers, there should be numerous opportunities for health care providers to promote and discuss physical activity with parents and other caregivers. Guidelines have been issued by the American Academy of Pediatrics (AAP) as well as the USDHHS and the NASPE. Also, other governmental agencies ranging from the state to the international level have made activity recommendations. The emphasis of most of the recommendations is to establish a foundation for regular enjoyable activity through the introduction of motor skills for musculoskeletal development.

Guidelines From the American Academy of Pediatrics

One early statement on exercise for infants was published by the AAP Council on Sports Medicine and Fitness in 1988. The statement noted that the most important aspect for promoting an infant's development is providing a stimulating environment. Given that virtually all of an infant's drive toward activity is based on an intrinsic need to seek arousal and curiosity, the AAP felt that extrinsic activity or structured programs are unlikely to provide any long-term benefit. Actually, the statement suggested that given the susceptibility of infant bones to trauma, parents or other caretakers might inadvertently exceed the infant's physical limitations though a structured exercise program. The conclusion made by the AAP was that (1) structured exercise programs should not be promoted as being therapeutically beneficial for the development of healthy infants and (2) parents should be encouraged to provide a safe, nurturing, and minimally structured play environment for their infants. This policy statement was officially retired by the AAP in 2004 and currently there is no replacement statement regarding activity for infants.

In 1992, the AAP issued a policy statement on physical activity for children. While the guidelines

GUIDELINE 4.1

Title: "Infant Exercise Programs"

Organization: AAP

Year published: 1988

Purpose: To promote healthy infant development

Location: *Pediatrics* 82(5): 800

Population: Infants

GUIDELINES

- Structured exercise programs should not be promoted as being therapeutically beneficial for the development of healthy infants.
- Parents should be encouraged to provide a safe, nurturing, and minimally structured play environment for their infants.

GUIDELINE 4.2

Title: "Fitness, Activity, and Sports Participation in the Preschool Child"

Organization: AAP

Year published: 1992

Purpose: To develop lifelong habits that may help forestall future chronic illness

Location: *Pediatrics* 90(6): 1002-1004

Population: Preschoolers (2-5 y)

GUIDELINES

- All preschool children should participate regularly in physical activity appropriate for their developmental level and physical health status.
- Physical activity should be promoted as a natural and lifelong activity of healthy living. Goals of accelerating motor development to maximize sport ability are inappropriate and futile and should be discouraged.
- Free play designed to provide opportunities for children to develop fundamental motor skills and to reach their potential at their own personal rate is preferable to structured sessions.
- Readiness to participate in organized sport should be determined individually and be based on the child's (not the parent's) eagerness to participate and subsequent enjoyment of the activity. Children are unlikely to be ready before age 6.
- Structured sports should emphasize participation and enjoyment rather than competition and victory. Sessions should be supervised by adults knowledgeable about the specific needs and limitations of preschool children. Setting, format, rules, and equipment should be modified accordingly.
- Pediatricians should assess preschoolers' physical activity level and time spent in passive activities such as television watching by incorporating relevant questions into the medical history taken during health assessment visits. Appropriate physical activity should be promoted by counseling parents, teachers, and coaches.
- Parents and other family members should be encouraged to serve as role models for their children by participating in regular physical activity themselves. In addition, physical activities that parents can do with young children should be encouraged.

Adapted, by permission, from Committee on Sports Medicine and Fitness, 1992, "Fitness, activity, and sports participation in the preschool child," *Pediatrics* 90: 1002-1004.

were very generalized, they did recommend daily activity and emphasize participation and enjoyment rather than competition.

Also in 1992, the AAP recommended that babies sleep on their backs in order to reduce the number of deaths caused by sudden infant death syndrome (SIDS). According to the National Center for Health Statistics, deaths from SIDS have fallen by more than half since then. Because the number of babies who sleep on their backs has grown from 13% to 73% since 1992, it has been stressed that babies should spend some supervised time in the prone position during waking hours so they can develop the abdominal and neck muscles as well as the major extensors. Supervised tummy time can begin as early as the first week of life and can progress from 1 or 2 min initially to 20 to 30 min (as tolerated) by the time the baby is 3 to 4 mo old.

Since 1992, more information has become available regarding the importance of establishing physical activity patterns even at a very young age. The guidelines have become more specific and have been issued by several different entities, including the NASPE and USDHHS.

Guidelines From Other Organizations

Many other organizations—both public and private—have issued guidelines for children in this young age group. As in the AAP guidelines, there is an emphasis on age-appropriate physical activity that is enjoyable and promotes motor development. This sets the stage for a lifetime of activity that is incorporated into the daily lifestyle. Like the AAP guidelines, many other guidelines recommend that parents follow the child's lead rather than dictate activity. Avoiding sedentary time and providing good active role models are also themes from the AAP guidelines that come across in the guidelines provided by many of the following organizations.

National Association for Sport and Physical Education

The NASPE has published *Active Start*, a collection of information and guidelines for parents and other caregivers. Initially published in 1998, it is now in its second edition.

GUIDELINE 4.3

Title: "Guidelines for Infants," *Active Start: A Statement of Physical Activity Guidelines for Children From Birth to Age 5, 2nd Edition*

Organization: NASPE

Year published: 2009

Purpose: To support the NASPE position statement that all children from birth to age 5 should engage in daily physical activity that promotes health-related fitness and movement skills

Location: www.aahperd.org/naspe/standards/nationalGuidelines/ActiveStart.cfm

Population: Infants from birth to 12 mo

GUIDELINES

- Infants should interact with parents and caregivers in daily physical activities that promote exploration of the environment.
- Infants should be placed in safe settings that facilitate physical activity and do not restrict movement for prolonged durations.
- Infants' physical activity should promote the development of movement skills.
- Infants should have an environment that meets or exceeds recommended safety standards for performing large-muscle activities.
- Individuals responsible for the well-being of infants should be aware of the importance of physical activity and facilitate the child's movement skills.

Adapted from *Active Start: A Statement of Physical Activity Guidelines for Children From Birth to Five Years*, 2nd Edition (2009) with permission from the National Association for Sport and Physical Education (NASPE), 1900 Association Drive, Reston, VA 20191.

GUIDELINE 4.4

Title: "Guidelines for Toddlers," *Active Start: A Statement of Physical Activity Guidelines for Children From Birth to Age 5, 2nd Edition*

Organization: NASPE

Year published: 2009

Purpose: To support the NASPE position statement that all children from birth to age 5 should engage in daily physical activity that promotes health-related fitness and movement skills

Location: www.aahperd.org/naspe/standards/nationalGuidelines/ActiveStart.cfm

Population: Children from age 12 to 36 mo

GUIDELINES

- Toddlers should accumulate at least 30 min of structured physical activity every day.
- Toddlers should engage in at least 60 min of—and up to several hours of—daily, unstructured physical activity and should not be sedentary for more than 60 min at a time except when sleeping.
- Toddlers should develop movement skills that are the building blocks for more complex movement tasks.
- Toddlers should have indoor and outdoor play areas that meet or exceed recommended safety standards for performing large-muscle activities.
- Individuals responsible for the well-being of toddlers should be aware of the importance of physical activity and facilitate the child's movement skills.

Adapted from *Active Start: A Statement of Physical Activity Guidelines for Children From Birth to Five Years,* 2nd Edition (2009) with permission from the National Association for Sport and Physical Education (NASPE), 1900 Association Drive, Reston, VA 20191.

GUIDELINE 4.5

Title: "Guidelines for Preschoolers," *Active Start: A Statement of Physical Activity Guidelines for Children From Birth to Age 5, 2nd Edition*

Organization: NASPE

Year published: 2009

Purpose: To support the NASPE position statement that all children from birth to age 5 should engage in daily physical activity that promotes health-related fitness and movement skills

Location: www.aahperd.org/naspe/standards/nationalGuidelines/ActiveStart.cfm

Population: Children from 3 to 5 y

GUIDELINES

- Preschoolers should accumulate at least 60 min of structured physical activity every day.
- Preschoolers should engage in at least 60 min of—and up to several hours of—daily, unstructured physical activity and should not be sedentary for more than 60 min at a time except when sleeping.
- Preschoolers should develop competence in movement skills that are the building blocks for more complex movement tasks.
- Preschoolers should have indoor and outdoor play areas that meet or exceed recommended safety standards for performing large-muscle activities.
- Individuals responsible for the well-being of preschoolers should be aware of the importance of physical activity and facilitate the child's movement skills.

Adapted from *Active Start: A Statement of Physical Activity Guidelines for Children From Birth to Five Years,* 2nd Edition (2009) with permission from the National Association for Sport and Physical Education (NASPE), 1900 Association Drive, Reston, VA 20191.

Government Organizations

The PCPFS, along with the NASPE and Kellogg Company, has published *Kids in Action*, a tool to help parents and caregivers lay the foundation for lifelong health in their children. Because parents serve as role models, they can have an enormous influence on their children's lifetime physical activity patterns. The most recent version of *Kids in Action* was produced in 2003, and its recommendations for daily activity largely mirror the recommendations set forth by the NASPE. The booklet is a good resource, because in addition to providing guidelines, it contains several illustrated examples of activities parents and caregivers can do with children ranging in age from infants to toddlers to preschoolers. Activities include movements that improve health-related fitness components such as cardiorespiratory endurance, muscular strength and endurance, flexibility, and body composition.

Recently, the USDA developed MyPyramid for Preschoolers in order to encourage a healthy, active lifestyle for children aged 2 to 5 y. Like the activity and nutrition pyramids presented in chapters 3, 5, and 17, the pyramid for preschoolers encourages healthy eating and regular physical activity. In addition to recommending 60 min of accumulated daily physical activity, the pyramid emphasizes avoidance of inactivity (less than 2 h of inactivity daily). This new resource is available at http://mypyramid.gov/preschoolers.

One noticeable absence in the group of national guidelines for infants, toddlers, and preschoolers is that of the USDHHS. The comprehensive *Physical Activity Guidelines for Americans* released in October 2008 did not have any specific recommendations for children under the age of 6 y. The PAGAC did not review the evidence for activity and health benefits in this younger population; therefore, keeping in line with its goal of rigorous scientific review and evidence-based guidelines, it did not make recommendations. The report did, however, state that "physical activity in infants and young children is, of course, necessary for healthy growth and development. Children younger than 6 should be physically active in ways appropriate for their age and stage of development." (USDHHS 2008)

Notable State or International Guidelines

The CDHS has issued physical activity guidelines for individuals of all ages, starting with preschoolers. These guidelines are included here for many reasons. California is one of the few U.S. states to issue physical activity guidelines, California has been forward-looking in studying and publicizing the data showing the relationship between physical fitness and academic performance, and

---

GUIDELINE 4.6

Title: *Kids in Action: Fitness for Children Birth to Age Five*

Organization: PCPFS, NASPE, and Kellogg Company

Year published: 2003

Purpose: To help children create the foundation for lifelong healthy behaviors

Location: fitness.gov/funfit/Kidsinactionbook.pdf

Population: Children from birth to age 5 y

GUIDELINES

- Children should never be inactive for more than 60 min.
- Children should engage in several hours of unstructured movement every day.
- Toddlers should have at least 30 min of structured activities.
- Preschoolers should have at least 60 min of structured activities.
- Activity may be broken into smaller units of 10 or 15 min.

GUIDELINE 4.7

Title: *The California Center for Physical Activity's Guidelines for Physical Activity Across the Lifespan: Preschool Children*

Organization: CDHS

Year published: 2002

Purpose: For appropriate development and health

Location: www.caphysicalactivity.org/facts_recomm1.html

Population: Preschoolers

GUIDELINES

- All preschool children should participate every day in physical activity appropriate for their developmental level and physical health status. Activity should occur at home, preschool, day care, or other caregiving settings.
- Free play designed to provide opportunities for children to develop fundamental motor skills and reach their personal potential at their own rate is preferable to structured sessions.
- As much free play as possible should take place in a safe outdoor environment.
- Structured sports should emphasize participation and enjoyment rather than competition and winning.
- Physical activity should be promoted as a natural and lifelong activity of healthy living. Setting, format, rules, and equipment should be modified accordingly.
- Preschool children should not be sedentary for more than 60 min at a time. Sedentary behaviors such as watching television or videos should be kept to a minimum (no more than 1 h/d).

Reprinted by permission, from CA Department of Health 2002. Available: www.nap.edu/books/0309085373/html

GUIDELINE 4.8

Title: *Australia's Physical Activity Recommendations for 5-12 Year Olds*

Organization: Australian Government Department of Health and Ageing

Year published: 2004

Purpose: For healthy growth and development

Location: www.health.gov.au/internet/main/publishing.nsf/Content/9D7D393564FA0C42CA256F97 0014A5D4/$File/kids_phys.pdf

Population: Infants and toddlers

GUIDELINES

- Children should be given plenty of opportunity to move throughout the day.
- Children should not be inactive for prolonged durations, except when they're asleep.
- Allow infants to spend time lying on their front, back, and sides as well as to roll over, creep, and crawl in a space that is safe and hazard free.
- Encourage children's natural instinct to move.
- Take time to play with children to help develop their sensory and motor systems.

Adapted from Australia's Physical Activity Recommendations for 5-12 Year Olds. Australian Government Department of Health and Ageing, Canberra, 2004. © Commonwealth of Australia 2007.

California has the political, athletic, and celebrity prominence of its governor, Arnold Schwarzenegger. California's guidelines are not very different from the NASPE's in terms of overall duration of activity and the recommendation for avoiding sedentary time. On the other hand, California's guidelines emphasize free play in a safe outdoor environment along with participation and enjoyment over competition.

While formal Australian guidelines for children start at age 5 y, they provide information about the appropriate type of activity for younger children. The Australian guidelines suggest providing plenty of opportunities for movement and limiting inactivity.

Guidelines for Prevention of Obesity and Chronic Disease

The AHA has published guidelines applying to children as young as 2 y. Because the AHA has identified physical inactivity as a major risk factor for CAD, stroke, obesity, hypertension, low HDL cholesterol, and diabetes, it has published a position statement on exercise and children. Although older children may need a longer duration of physical activity (see chapter 5), starting at age 2 children should get at least 30 min of moderate-intensity activity every day. Like adults, children can break up activity into smaller chunks; 2 sessions of 15 min or 3 sessions of 10 min may be appropriate, provided that the activity is of vigorous intensity. In order to encourage continued participation, especially for children who are overweight or less coordinated, children should complete a variety of activities appropriate to their age, sex, and stage of both physical and emotional development.

In 2006, the AAP Council on Sports Medicine and Fitness and Council on School Health released a policy statement on physical activity in response to the rising U.S. epidemic of obesity and inactivity. This statement includes recommendations for infants all the way up to adolescents; those for younger children are presented here and the remainder of the recommendations in chapter 5.

In 2004, Carrel and Bernhardt published a review on physical activity, childhood obesity, and the role of exercise in preventing obesity in adolescents. Because overweight or obesity during childhood increases the risk of overweight or obesity during adulthood, it is important to start laying the foundation for exercise and activity when children are young. Though the review's title advises exercise to prevent adolescent obesity, the review included guidelines for children starting at age 3. In addition to moderate physical

---

GUIDELINE 4.9

Title: *Exercise (Physical Activity) and Children*

Organization: AHA

Year published: 2007

Purpose: To produce overall physical, psychological, and social benefits and to decrease the likelihood of becoming an inactive adult

Location: www.americanheart.org/presenter.jhtml?identifier=4596

Population: Children starting at 2 y

GUIDELINES

- All children aged 2 y and older should participate in at least 30 min of enjoyable, moderate-intensity physical activity every day. Activities should be developmentally appropriate and varied.
- If children do not receive a full 30 min activity break each day, then at least 2 breaks of 15 min or 3 breaks of 10 min in which they can engage in vigorous activities appropriate to their age, sex, and stage of physical and emotional development should be provided.

GUIDELINE 4.10

Title: "Active Healthy Living: Prevention of Childhood Obesity Through Increased Physical Activity"

Organization: AAP Council on Sports Medicine and Fitness and Council on School Health

Year published: 2006

Purpose: To prevent childhood obesity

Location: *Pediatrics* 117(5):1834-1842 and www.pediatrics.org/cgi/doi/10.1542/peds.2006-0472

Population: Infants and toddlers

GUIDELINES

- Exercise programs or classes for infants and toddlers are not recommended to promote activity or prevent later obesity.
- Children under 2 y should not watch any television.
- Parents should provide a safe, nurturing, and minimally structured play environment for their infants.
- Infants and toddlers should be allowed to develop enjoyment of outdoor physical activity and unstructured (but supervised) exploration.

GUIDELINE 4.11

Title: "Exercise Prescription for the Prevention of Obesity in Adolescents"

Organization: None

Year published: 2004

Purpose: To prevent and treat obesity by promoting behavioral and environmental changes

Location: *Current Sports Medicine Reports* 3:330-336

Population: Children starting at age 3 y

GUIDELINES

Frequency	Intensity	Duration	Type
Most, preferably all, days of the week	Moderate	At least 30-60 min	Walking, running, swimming, tumbling, and catching

Special considerations include the following:

- Build more physical activity into lifestyle, such as walking or biking to school instead of driving, taking stairs instead of elevators, and helping with active chores inside and outside of the house.
- Activities incorporating gross muscle movements are best to help develop visual and motor skills.
- Reduce sedentary behaviors such as watching television and videos and playing computer games.
- Schools should provide more frequent and effective physical activity programs and more after-school and summer programs that include heart-healthy food choices and physical activity.

activity on most days, lifestyle physical activity is encouraged. As children are not autonomous, both parents and health care providers should serve as advocates for children. They can influence children to spend less time in sedentary pursuits, offer after-school and summer physical activity programs, and push for healthier food offerings at school. Despite their age, young children can and should be included in identifying opportunities to meet activity goals as well as to effect positive and healthy changes in behaviors. Guidelines for preschoolers are presented here; those for older children and adolescents are presented in chapter 5.

Physical Activity as Play and Nutritional Goals for Children

Because infants and young children are not autonomous, guidelines for this population are aimed at parents, caregivers, schools, and the community. In fact, the next chapter outlines multiple sets of guidelines focusing on community or school policy as a way of promoting physical activity for youths. For preschoolers, it is extremely important that age-appropriate opportunities for physical activity and play be available. Playgrounds offer a variety of activities that may be acceptable for a range of ages, making them a sound option for children. Starting with toddlers and extending through older school-aged children, playground equipment allows for independent play with multiple opportunities to challenge the musculoskeletal, neurological, and psychosocial systems. Many of the guidelines for young children include a call to action for communities to continue to provide ample opportunities for outdoor play in the form of playgrounds and outdoor environments. One author summed up the benefits of playgrounds and physical play as allowing children to BECOME FIT (adapted from Clements 2007).

In addition to physical activity, healthy diet and proper nutrition should begin at an early age. The connection between food and fitness is clearly known; parents and caregivers again have an opportunity to play role model for their children by following healthy eating patterns. Sitting down to meals as a family—particularly breakfast—has been shown to improve communication as well as emotional and psychosocial development. Schoolchildren who eat breakfast perform better academically and have a better chance of meeting all their nutritional needs. Getting adequate amounts of macronutrients, vitamins, minerals, and fiber enables children to have the proper energy to live the active lifestyle that is so important for their present and future health status. Recommendations for appropriate activity and nutrition in children appear in both the IOM report on DRIs and the *Dietary Guidelines for Americans*. While the 2005 *Dietary Guidelines for Americans* specifically states that its recommendations apply to children aged 2 y and older, the report on DRIs does not make specific reference to the age at which the youth guidelines apply.

Balance and control the body.
Express emotions and discover physical likes and dislikes.
Carry out imaginary plots that involve chasing and fleeing.
Overcome physical obstacles and develop strength.
Move and stretch groups of muscles and body parts.
Experience the sensation of spinning, twisting, and twirling.
Find novel ways of moving vigorously.
Imagine possibilities and expand upon previous physical tasks.
Transfer the whole body's weight into the air!

SUMMARY

Guidelines for infants and toddlers are important in that they establish the foundation for a lifetime of physical activity and good health. Although exercise per se is not recommended for very young children, there is an emphasis on developmentally appropriate and enjoyable physical activity. Children in this age group use physical activity to build motor skills that will ultimately pave the way for more formalized physical activity as they grow. Properly supervised opportunities for activity, an avoidance of sedentary time, and the ability to observe and pattern healthy physical activity behaviors in parents and caregivers are key components of guidelines for this age group.

School-Aged Children

There are numerous benefits to regular physical activity in school-aged and preadolescent children, including proper musculoskeletal development, reduced adiposity, and better academic performance. Childhood is one of the earliest times to establish a foundation for a lifetime of physical activity; this is the emphasis of the NASPE guidelines for this younger age group. The presumption that activity levels in childhood predict exercise habits in adulthood is not reliable. A literature review evaluating childhood participation in organized school sports and physical education could not find a reliable relationship with future adult physical activity levels (Powell and Dysinger 1987). One possible explanation is that children are taught activities, such as team sports, that are difficult to carry over to adulthood (Sallis and others 1994). Thus it seems worthwhile to focus on stimulating generalized healthy habits and routines as opposed to teaching specific skills. Children who are active with their parents are more active than children who do not exercise with their parents—so modeling a physically active lifestyle may be the best option for parents to encourage activity. In fact, the NASPE feels so strongly about modeling healthy physical activity behaviors for children that it has issued recommendations for physical activity professionals themselves (NASPE 2002)! Although teachers and coaches may not be as powerful role models as parents are, they are still important adult role

models and spend significant time with youths.

This chapter presents several different guidelines for youths. In addition to specifying volumes of activity, these guidelines emphasize limiting screen time. Also, because children spend a great deal of their time in a school environment, some guidelines directly address the need for schools to provide the framework for physical activity and education in order to establish healthy activity patterns going into adulthood. As with the adult guidelines, the recommendations for children are established by state, national, and international organizations.

Basic Facts About Physical Activity and Health in Children

Understanding the activity recommendations for children requires an appreciation of some fundamental facts regarding activity and health in children. Physical activity not only declines with age—starting in childhood—but also declines for every age group, which has led to an increased prevalence of diabetes and obesity in the United States. Young children need exercise and activity as part of their motor skill development, while adolescents focus more on sport skill acquisition and then the transition to adult type of aerobic activity for metabolic health. This background, as well as the varying goals of the different guidelines, helps explain why there is a range of different recommendations across these segments of youths.

Decline in Physical Activity

The progression in physical inactivity that occurs as people age begins in childhood. A study of youth participation in both moderate- and vigorous-intensity physical activity showed that both boys and girls exhibit a steady decline in physical activity starting in grades 1 through 3 and continuing through grades 10 through 12. In this study of objectively measured physical activity, the mean number of minutes spent in vigorous-intensity exercise were higher for boys than they were for girls at all age groups; the mean number of minutes spent in moderate-intensity physical activity were significantly greater for boys at all age groups except grades 1 through 3 (Trost and others 2002). Data from the 2001 Youth Risk Behavior Survey (Grunbaum and others 2002) emphasize the decline in physical activity occurring at each grade level in high school, underscoring the importance of promoting physical activity that is enjoyable and sustainable. Table 5.1 shows the percentage of high school students participating in various types of activities at each grade level.

Despite legislation to mandate physical education and encourage participation in physical activity, recent numbers are striking; in 2007, only 27.3% of girls and 33.2% of boys attended daily high school physical education classes. Physical education is just one way for students to get activity, but still only 25.6% of girls and 43.7% of boys in high school participated in at least 60 min of physical activity in 5 of the preceding 7 d (CDC 2008).

Growth in Obesity and Chronic Disease

The importance of establishing and maintaining a regular physical activity program for youths is not insignificant—for the first time, children born today have a shorter life expectancy than that of their parents. Much of this is attributed to physical inactivity and the resultant risk for obesity, diabetes, and heart disease. The shift to a less-active lifestyle is at the heart of the obesity problem. Childhood obesity has become epidemic over the last two decades; currently there are twice as many overweight children and almost three times as many overweight adolescents as there were in 1980 (USDHHS 2001). More than 9 million youths are overweight or obese, and this is a consequence of both poor eating patterns and not enough physical activity.

Early obesity sets the stage for an increased risk of a myriad of health complications ranging from heart disease to diabetes and cancer to osteoarthritis. Chronic diseases such as hypertension, type 2 diabetes, and CHD are becoming increasingly prevalent in younger populations, including children as young as preteens. This trend is closely correlated with the rapidly increasing prevalence of obesity in this group. In fact, the CDC warns that 1 in 3 U.S. children born in 2000 will become diabetic unless many more people start eating less and exercising more (AAP 2003). The theme of limiting sedentary time surfaces in many physical activity guidelines because it serves two purposes: (1) It encourages an alternative way to spend time that involves physical activity and (2) it limits the time youths spend passively eating excess calories.

Inactivity and poor nutrition also have negative psychological and social effects—for example, they interfere with learning. Children who are less active consistently do not perform as well

TABLE 5.1 Percentage of High School Students Participating in Physical Activity

Grade level	Sufficient vigorous physical activity	Sufficient moderate physical activity	Insufficient physical activity	No vigorous or moderate physical activity
9	71.9%	27.2%	24.3%	7.1%
10	67.0%	24.5%	29.6%	8.5%
11	61.3%	25.8%	34.4%	11.1%
12	55.5%	24.5%	38.9%	12.0%

From www.cdc.gov/mmwr/PDF/SS/SS5104.pdf MMWR 51(SS-4):57, 2002.

on standardized tests. Data from the California Department of Education show that across the board, children who score better on physical tests also score better on academic ones. Regular physical activity is felt to help with the focus needed for learning by allowing children to burn nervous energy during recess or after school so that they are able to concentrate when needed during learning or homework time. For adolescents, physical inactivity not only increases the risk for obesity and chronic disease but also is associated with high-risk behaviors such as truancy, cigarette smoking, sexual intercourse, and delinquency (Nelson and Gordon-Larsen 2006). Nelson's study analyzed self-reported data from about 12,000 adolescents in 1995 and 1996. Interestingly, adolescents who reported participation in skating and gaming activities—traditionally not promoted for fear of other risk factors—had high levels of physical activity and therefore were involved in fewer unfavorable risky behaviors. One other notable observation in this large study was that adolescents involved in a variety of physical activities did not report reduced rates of other necessary activities such as working for pay, doing housework, doing schoolwork, and sleeping. Both cross-sectional and longitudinal studies have found that adolescents who participate in sports are less likely to use cigarettes, marijuana, and other drugs or to participate in risky sexual behaviors (Baumert and others 1998; Pate and others 2000; Pate and others 1996; Sabo and others 1999).

Changes in Physical Activity Needs as Children Develop

During the preschool and early school ages, children need general activity to develop motor skills. As children grow older and their motor skills solidify, the acquisition of new motor skills becomes less important; the focus shifts to the health, fitness, and behavioral components of physical activity. Older children (adolescents) should focus more on health-related activities such as those emphasizing cardiorespiratory and muscular endurance, muscular strength, and weight bearing. Figure 5.1 presents one schematic illustrating this concept. Optimizing bone health is another target of physical activity in this age group. According to the National Osteoporosis

Foundation (NOF), 85% to 90% of adult bone mass is acquired by age 18 in girls and age 20 in boys. For girls aged 11 to 14, the gain of bone mineral mass in the spine is a staggering fivefold increase! Another way of looking at the importance of bone mineralization for girls at this age is to realize that the bone tissue accumulated during ages 11 to 13 approximately equals the amount lost in the 30 y following menopause (Minne and Pfeifer 2005).

It is necessary to encourage an active mind-set in children. While this mind-set can begin at a very young age with modeling by parents and older siblings, it becomes ever-increasing in importance as children become adolescents. When it comes to encouraging adolescents, two factors come to light: autonomy and pleasurability. The first, autonomy, deals with the fact that adolescents want to be independent in their decision making and not necessarily told what to do. By establishing a model of physical activity and providing varied opportunities for such, parents can give teenagers multiple chances to choose healthy activities in their own way. Pleasurability refers to the fact that adolescents do not enjoy participating in activities that are not fulfilling to them. At younger ages, children inherently enjoy activity, learning new skills, moving about freely, climbing, and running. By adolescence, the social dimension of activity becomes more important, as does the concept

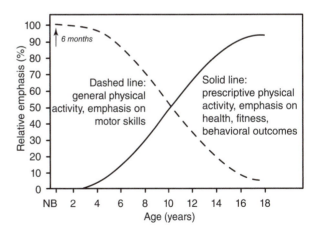

Figure 5.1 The changing emphasis on different kinds of activity from childhood through adolescence.

Reprinted from *The Journal of Pediatrics*, Vol. 146, W.B. Strong, R.M. Malina, C.J.R. Blimkie, S.R. Daniels, R.K. Dishman, B. Gutin, A.C. Hergenroeder, A. Must, P.A. Nixon, J.M. Pivarnik, T. Rowland, S. Trost, and F. Trudeau, "Evidence based physical activity for school-age youth," pgs. 732-737, Copyright 2005, with permission from Elsevier.

of competition. Unfortunately, at this age sport opportunities are actually less available to many teenagers, as the sport systems tend to focus on the elite or highly skilled athletes for team participation (Malina 2007).

Establishment of the Earliest Guidelines for Children

Just a few decades ago, it was not recommended that children participate in regular vigorous physical activity. In fact, it was discouraged because major professors felt that children are incapable of exercise causing high heart rates. This myth persisted in the available literature until the 1960s, when researchers were able to document the cardiorespiratory capabilities of children.

Early recommendations for children's exercise programs were based on those for adults (Rowland 1985). For example, earlier exercise prescription models (EPMs) developed for adults had a central focus on higher-intensity and shorter-duration activity with a goal of improving cardiorespiratory fitness. Using the HRR method to calculate target heart rates, a comparable measure for exercise in children was recommended (Sady 1986). One of the drawbacks with this EPM, however, is that even active children may not be able to meet the standards set for adults. Children tend to enjoy a wide range of activities throughout the day and often have difficulty in persisting with vigorous activity for at least 20 or 30 min, as recommended by the EPM. Researchers found that this EPM deemed a substantial portion of boys and girls as inactive (Armstrong and others 1990; Armstrong and Bray 1991; Sleap and Waburton 1992). Given what is known about the effort-to-benefit ratio and the developmental needs of children, some children may be unmotivated to perform the higher-intensity level for any amount of time at all! Just as the initial ACSM guidelines for adults focused on an exercise prescription for cardio-respiratory fitness (and thus advocated shorter-duration and higher-intensity activity), the initial guidelines for youths had a similar goal. This changed in 1992, when the ACSM and the CDC, in cooperation with the PCPFS, issued a statement acknowledging the importance of lifestyle physical activity for reduction of disease risk. As the approach for adult physical activity shifted to a lower-intensity, longer-duration lifestyle physical activity model that had a goal of reducing chronic disease risk (CDC 1994; Haskell 1994), so did the approach for children.

In 1994, a group of experts from around the world developed the first consensus statement recommending specific activity amounts for adolescents (youths aged 11-21 y). These were subsequently modified by a more recent panel. That same year, Corbin, Pangrazi, and Welk published guidelines for children in the *President's Council on Physical Fitness and Sports Research Digest*. These were called the *children's lifetime physical activity model*, or C-LPAM. The caloric expenditure recommended for adults—3 to 4 kcal · kg^{-1} · d^{-1}—was felt to be a good minimum standard for producing activity-related health benefits in children. With this model, more of the children who were deemed inactive by the EPM standards were considered to be meeting the guidelines and consequently reducing future chronic disease risk. Again, with a nod to the adult guidelines and the knowledge of the dose–response relationship between activity and health, a higher goal of expending 6 to 8 kcal · kg^{-1} · d^{-1} was recommended for optimal health.

The C-LPAM takes into account the following observations:

1. Children learn basic motor skills that serve as the basis for a lifetime of activity. Proper skill formation requires a substantial amount of time and energy.

2. Lifetime physical activities such as walking, riding bicycles, and performing other physical tasks are an essential part of an active lifestyle; learning these at an early age is likely to promote a continued active lifestyle and reduce the risk for obesity later on.

3. Children require activity and exposure to all the components of health-related physical fitness, including aerobic fitness, muscular strength and endurance, flexibility, desirable body composition, and bone strengthening.

4. Most children will choose to be active, given opportunity and proper encouragement.

5. As with adults, the most substantial reduction in risk for chronic disease occurs with a

---

GUIDELINE 5.1

Title: *Toward an Understanding of Appropriate Physical Activity Levels for Youth*

Organization: PCPFS

Year published: 1994

Purpose: To gain health benefits of physical activity

Location: www.fitness.gov/toward.pdf

Population: Children

GUIDELINES

- Activity for children should focus on high volume and moderate intensity and should include sporadic activities such as active play performed in several daily sessions.
- Lifestyle activity such as walking or riding bikes to and from school or performing active physical tasks at home (e.g., doing yard work) should be encouraged.
- The activity program should include opportunities to learn basic motor skills and develop all components of health-related physical fitness through appropriate moderate-intensity activity.
- Children should be afforded opportunities to develop behavioral skills that lead to lifetime activity.
- EPM guidelines can be applied to children who are especially interested in high-level physical performance, but only when it is developmentally appropriate to do so.

Standard	Frequency	Intensity	Duration
Minimum activity standard	Frequent activity sessions (3 or more) each day	Moderate; alternating bouts of activity with rest as needed or moderate activity such as walking or riding a bike to school	Amount necessary to expend at least 3-4 kcal $\cdot$ kg^{-1} $\cdot$ d^{-1}; this equals the calories expended in 30 min or more of active play or moderate sustained activity that may be distributed over 3 or more activity sessions
Optimal functioning standard (a goal for all children)	Frequent activity sessions (3 or more) each day	Moderate to vigorous; alternating bouts of activity with rest as needed or moderate activity such as walking or riding a bike to school	Amount necessary to expend at least 6-8 kcal $\cdot$ kg^{-1} $\cdot$ d^{-1}; this equals the calories expended in 60 min or more of active play or moderate sustained activity that may be distributed over 3 or more activity sessions

caloric expenditure of 3 to 4 kcal $\cdot$ kg^{-1} $\cdot$ d^{-1}. Additional benefits are achieved at levels of 6 to 8 kcal $\cdot$ kg^{-1} $\cdot$ d^{-1} or more; because childhood habits may promote adult habits and peak activity levels tend to occur in childhood and adolescence, aspiring to higher levels during childhood seems appropriate.

The shift from the EPM to the C-LPAM seemed to more accurately reflect the everyday experiences of youths and certainly classified many more youths as active, and in 2001 a British study reached these same conclusions. The study evaluated 25 inner-city English adolescents for 3 d or more and quantified their levels of physical activity by the EPM and C-LPAM approaches. More individuals were found to fulfill the minimum activity criteria for accumulated activity (C-LPAM) versus sustained activity (EPM). The authors suggested that the C-LPAM approach was more viable in terms of establishing a lifelong tendency for participation in activity. Although some children may be able—and willing—to participate habitually in sustained moderate to vigorous activity, the C-LPAM approach is better suited to reaching more children and encouraging at least a minimum of activity with an ultimate goal of building the foundation for a lifetime of activity (Gilson and others 2001).

Landmark Guidelines

In 1998, the NASPE Council on Physical Education for Children (COPEC) in the United States and the Health Education Authority in the United Kingdom each developed physical activity guidelines for children. These were the first national guidelines for children. Both were widely publicized and specified amounts, types, and levels of activity that could provide health benefits. The American guidelines issued by the NASPE were updated in 2004 and are commonly referred to as the *NASPE physical activity guidelines for children* (ages 5-12). These are summarized in guideline 5.4. The guidelines make reference to the physical activity pyramid, which is explained in detail later in this chapter. Emphasizing an accumulation of activity and avoiding inactivity are central to these guidelines.

It wasn't until 2001 that the CDC issued a report recommending physical activity for youths. These recommendations were published in a *Morbidity and Mortality Weekly Report* and included a review of the literature that determined which interventions were effective. Because these guidelines do not specify type or timing of activity they are included in a later section on community and public health recommendations.

In 2003, the AAP issued a policy statement on the prevention of overweight and obesity in children. Although limiting screen time (to no more than 2 h/d) and promoting physical activity were emphasized, no specific guidelines were made regarding frequency or duration of physical activity.

In 2005, an expert panel convened under a contract from the CDC Division of Nutrition, Physical Activity and Obesity and Division of Adolescent and School Health. Using the available scientific evidence, the panel prepared physical activity guidelines for youths aged 6 to 18. While this panel supported the NASPE guidelines for children aged 5 to 12, there were differences between its and the NASPE's guidelines. First, the panel's recommendations extended to youths up to age 18. Second, for this older group of teenagers, the panel recommendations called for 60 min of moderate to vigorous physical activity each day, superseding previous guidelines

---

GUIDELINE 5.2

Title: "Policy Framework for Young People and Health-Enhancing Physical Activity"

Organization: Health Education Authority, United Kingdom

Year published: 1998

Purpose: To enhance health

Location: *Young and Active? Young People and Health-Enhancing Physical Activity—Evidence and Implications*

Population: Youth

GUIDELINES

- All children and young people should accumulate at least 1 h of moderate activity on most days of the week.
- Youths who currently participate in little or no activity should begin to accumulate half an hour of moderate activity on most days of the week and then build up to an hour.
- Activities may be part of transportation, physical education, play, games, sport, recreation, work, or structured exercise

GUIDELINE 5.3

Title: "Prevention of Pediatric Overweight and Obesity"

Organization: AAP

Year published: 2003

Purpose: To prevent childhood obesity

Location: *Pediatrics* 112(2): 424-430

Population: Youth

GUIDELINES

- Promote physical activity, including unstructured play at home, in school, in child care, and throughout the community.
- Limit television and video time to a maximum of 2 h/d.

GUIDELINE 5.4

Title: "Physical Activity Guidelines for Children"

Organization: NASPE

Year published: 2004

Purpose: To achieve good health

Location: *Physical Activity for Children: A Statement of Guidelines for Children Ages 5-12, Second Edition*

Population: Children aged 5 to 12 y

GUIDELINES

- Children in elementary school should accumulate at least 30 to 60 min of age-appropriate and developmentally appropriate physical activity on all, or most, days of the week.
- Children in elementary school are encouraged to accumulate more than 60 min, and up to several hours, of age-appropriate and developmentally appropriate physical activity each day.
- Some of children's activity each day should occur in bouts lasting 10 to 15 min or more and include moderate to vigorous activity. Typically this activity will be intermittent, involving alternating bouts of moderate to vigorous activity with brief times of rest and recovery.
- Extended durations of inactivity are inappropriate for children.
- A variety of physical activities selected from the physical activity pyramid is recommended for elementary school children.

Adapted from *Physical Activity for Children: A Statement of Guidelines for Children Ages 5-12*, Second Edition (2004) with permission from the National Association for Sport and Physical Education (NASPE), 1900 Association Drive, Reston, VA 20191.

of only 30 min of activity each day. While older youths may benefit from more-structured activities, younger children may still need intermittent and less-structured activities.

In 2006, the AAP issued a position statement titled "Active Healthy Living: Prevention of Childhood Obesity Through Increased Physical Activity." The statement outlined not only age-based guidelines for activity but also ways that public health officials and health care providers might encourage physical activity in children. Infant and preschooler guidelines, which focused

GUIDELINE 5.5

Title: "Evidence Based Physical Activity for School-Age Youth"

Organization: CDC Division of Nutrition, Physical Activity and Obesity and Division of Adolescent and School Health and the Constella Group

Year published: 2005

Purpose: To develop a recommendation for the amount of physical activity deemed appropriate to yield beneficial health and behavioral outcomes

Location: *Journal of Pediatrics* 146(6): 732-737

Population: School-age youth

GUIDELINES

- School-aged youths should participate in 60 min or more of moderate to vigorous (5-8 METs) physical activity every day. The activity should be enjoyable and developmentally appropriate.
- The physical activity can be accumulated in school during physical education, recess, intramural sports, and before- and after-school programs.
- Daily quality physical education is recommended from kindergarten through grade 12.
- Sedentary behavior should be reduced to less than 2 h/d.
- Physically inactive youths may gradually increase activity by 10% per week in order to work up to the goal of 60 min daily.

on supervised play and limited television time, were presented in chapter 4. For children in elementary school (children aged 6-9 y), free play is still encouraged, and there is more of an emphasis on the acquisition of fundamental movement skills. Also, organized sports are an appropriate environment for skill acquisition, although the focus should be on enjoyment as opposed to competition, with flexible rules, short instruction time, and ample play time. Children aged 10 to 12 y should also focus on the enjoyment of activity, with additional emphases on skill development, strategy, and factors that promote continued participation. At this stage, the initiation of weight training is appropriate for the more mature. A lower weight with a higher number of repetitions (15-20) is recommended to help children avoid injury and learn proper technique. Specific resistance training guidelines are presented later in this chapter. The activity recommendations for adolescents also focus on fostering long-term participation and take advantage of the fact that adolescents are social and influenced by peers. The AAP rec-ommends a variety of activities for adolescents, depending on personal preference. These include dance, yoga, running, walking, cycling, and both competitive and noncompetitive sport.

Finally, in 2008 the PAGAC report was issued, which was followed shortly thereafter by the 2008 *Physical Activity Guidelines for Americans (PAGA)*. The guidelines were the result of an exhaustive review of the scientific literature on the data for the dose–response relationship of various health and fitness outcomes for children and adolescents. The conclusion was that children need certain types of physical activity, including aerobic exercise, resistance training, and weight-loading activities. These improve cardiorespiratory fitness and cardiovascular and metabolic disease risk factors, enhance muscular strength in the large muscles of the trunk and limbs, and promote bone health, respectively. Each type of activity should be performed on at least 3 d of each week, and children should get at least 60 min of moderate- to vigorous-intensity activity every single day. As with the adult guidelines, the

GUIDELINE 5.6

Title: "Active Healthy Living: Prevention of Childhood Obesity Through Increased Physical Activity"

Organization: AAP Council on Sports Medicine and Fitness and Council on School Health

Year published: 2006

Purpose: To prevent childhood obesity

Location: *Pediatrics* 117(5): 1834-1842

Population: Preschoolers through adolescents

GUIDELINES

Age group	Recommendations
Preschoolers (4-6 y)	• Encourage free play. • Play on flat surfaces with few variables. • Engage in appropriate activities, such as running, swimming, tumbling, throwing, and catching. • Reduce sedentary transportation by car and stroller. • Limit screen time to less than 2 h/d.
Children in elementary school (6-9 y)	• Encourage free play with an emphasis on fundamental skill acquisition. • Make sure organized sports have flexible rules, short instruction time, ample free time, and a focus on enjoyment.
Children in middle school (10-12 y)	• Emphasize skill development and focus on strategy and tactics. • Participate in complex sports such as football, basketball, and ice hockey if desired. • Base placement in contact and collision sports on maturity to minimize risk for injury. • If desired, initiate weight training with small free weights and high repetitions (15-20) to demonstrate proper technique.
Adolescents	• Choose activities that are fun for the individual and include friends. • Choose active transportation and noncompetitive and competitive sports. • Continue weight training, using longer sets with heavier weights and fewer repetitions as maturity develops. • Base participation in competitive contact and collision sports on maturity level.

youth guidelines refer to the health-enhancing physical activity that can be performed above and beyond basic baseline or lifestyle activity. One of the specific goals of the physical activity program is to strengthen bones because of the significant potential gain in bone mass during the years just before and during puberty. One additional strategy indicated in the *PAGA* report is to vary the kinds of activity performed, particularly for youths who already meet or exceed the guidelines. Variety can keep the level of interest high and the risk for overtraining or injury low.

In 2010, the eighth edition of *ACSM's Guidelines for Exercise Testing and Prescription* was issued, with updated recommendations for the pediatric population. Although the book concentrates on exercise testing and fitness improvement for adults, there is a chapter focusing on healthy populations, including children. The ACSM guidelines outlined here are the minimum recommended level of physical activity that will provide health-related fitness. The ACSM recommends a combination of moderate- and vigorous-intensity physical activity that accumulates to 60 min/d. Because children perform physical activity for enjoyment more than for health or exercise benefits, there is an emphasis on choosing a variety of enjoyable activities that are developmentally appropriate. The minimum recommendation is 3 or 4 d/wk; daily physical activity is optimal.

GUIDELINE 5.7

Title: *Physical Activity Guidelines for Americans*

Organization: PAGAC

Year published: 2008

Purpose: To improve cardiorespiratory and muscular fitness, cardiovascular and metabolic disease biomarkers, bone health, and body composition and possibly to reduce symptoms of depression

Location: www.health.gov/paguidelines

Population: Children and youths aged 6 through 17

GUIDELINES

- Children and adolescents should complete 60 min (1 h) or more of physical activity daily.
- Most of the 60 min/d should be moderate- or vigorous-intensity aerobic physical activity and should include vigorous-intensity physical activity at least 3 d/wk.
- As part of their 60 min of daily physical activity, children and adolescents should include muscle-strengthening physical activity on at least 3 d/wk.
- As part of their 60 min of daily physical activity, children and adolescents should include bone-strengthening physical activity on at least 3 d/wk.
- Young people should be encouraged to participate in physical activities that are appropriate for their age, are enjoyable, and offer variety.

Reproduced from 2008 Physical Activity Guidelines for Americans, U.S. Department of Health and Human Services, October 2008. Downloaded from www.health.gov/paguidelines 10/10/2009.

GUIDELINE 5.8

Title: "Exercise Prescription for Healthy Populations"

Organization: ACSM

Year published: 2010

Purpose: To achieve components of health-related fitness, including reducing CVD risk factors and positively influencing academic performance and self-esteem

Location: *ACSM's Guidelines for Exercise Testing and Prescription, Eighth Edition*

Population: Children and adolescents

GUIDELINES

Frequency	Intensity	Duration	Type
At least 3-4 d/wk, preferably daily	Moderate to vigorous	30 min/d *each* of moderate and vigorous intensity	Variety of activities including walking, active play, games, dance, sports, and muscle- and bone-strengthening activities

School and Community Health Guidelines

Many of the guidelines for youths focus on community organizations and schools, as children are influenced not only by their parents but also by their immediate surroundings. Children spend a great deal of time in school, so schools provide a unique opportunity for modifying unhealthy patterns in young people. However, the emphasis on physical education in the public school system has declined noticeably. Although most states mandate physical education, they require only that it is provided; local districts have control over content and format. As a result, the percentage of students taking physical education on a daily basis dropped from 42% to 29% between 1991 and 1999 (CDC 2003). With the adoption of the No Child Left Behind Act of 2001, time allocated to physical education and recess dropped even further (Pate and others 2006). Ironically, intervention studies have shown that spending increased time in structured physical education does not reduce academic achievement and may even contribute to achievement (Sallis and others 1999; Shephard and others 1984). Thus, the attempt by schools and districts to improve academic test scores by reducing physical activity time is unlikely to be effective. The end result is lowered participation rates for physical activity and a proven increased risk for chronic health problems, most notably obesity.

Recently there has been a push in several states to make physical education mandatory at the elementary, intermediate, and high school levels. The goals set in the *Healthy People 2010* initiative reflect this trend. In 1994, the percentage of public and private schools requiring daily physical activity for all students was only 17% for middle and junior high schools and 2% for senior high schools. These compare with goals for 2010 of 25% and 5%, respectively (CDC 2000). Despite requirements for daily physical education, many students simply don't participate, and those who do are not always active a majority of the class time. Objectives in the *Healthy People 2010* reflect this finding as well. The target for the fraction of adolescents who participate in daily school physical education is 50%—a significant increase over the observed 29%. Another target is to increase the proportion of adolescents who spend at least 50% of physical education class time being physically active. Baseline data from 1999 show that only 38% of students in grades 9 through 12 were physically active in physical education class more than 20 min 3 to 5 d/wk. The new goal is for 50% of students to be active for this amount of time (CDC 1998).

Role of Physical Education in Academic Performance

Although some people might argue that substituting physical education or activity for an academic class may hurt academic performance, this simply is not true. Multiple studies have shown that the converse is true; students who are more active through regular physical education score higher on academic tests. A meta-analysis of nearly 200 studies on the effect of exercise on cognitive functioning suggests that physical activity supports learning. The individual studies showed that more school time spent in physical activity—at the expense of academic class time—increased test scores, from mathematics to reading and writing (Shephard and others 1984; Shephard 1997; Symons and others 1997). Additionally, physical activity provided significant benefits with respect to behavior, self-esteem, and mental health (Sallis and others 1999; Keays and Allison 1995).

The relationship between academic achievement and physical fitness has been demonstrated clearly in California schools, where the relationship was proved valid for different academic subjects, both sexes, and each of the three grade levels measured! Nearly 1 million students were given both the Fitnessgram, developed by The Cooper Institute, and the Stanford Achievement Test (SAT-9). The Fitnessgram measures six aspects of fitness: (1) aerobic capacity, (2) body composition, (3) abdominal strength, (4) trunk strength, (5) upper-body strength, and (6) flexibility. A Healthy Fitness Zone (HFZ) is established for each measure, so that a child may achieve the HFZ for none, some, or all of the tests. Students in grades 5, 7, and 9 are tested. For each of these levels, there is a linear relationship between improved physical fitness (as measured by a higher number of HFZs) and academic achievement on both language arts and geometry tests (California Department of Education 2002).

Thus, in addition to providing fitness and health benefits, physical activity improves behavior and academic performance. These data are some of the more compelling information driving the activity guidelines for this age group. Although regular physical activity at this stage certainly decreases the risk for future chronic disease, the absolute number of cases of and predisposition to disease is so low in this young population that it is sometimes difficult to prove physical activity reduces chronic disease at this point in life.

General Guidelines Focusing on the Role of Schools

Starting in 1997, the CDC began publishing guidelines for school and community programs to promote physical activity. These include recommendations on the frequency and duration of physical education as well as recommendations on health education, extracurricular activity programs, and appropriate environments for the development of lifelong physical activity. In 2000, the AAP issued a statement regarding physical activity in schools; this statement reflected both the CDC recommendations and those that had been published in 1998 by COPEC. In 2000 the National Association of State Boards of Education also issued a statement regarding physical activity, healthy eating, and prevention of tobacco use. These guidelines primarily supported and referenced the 1997 CDC guidelines. Interventions include daily physical education for all students, intramural and extracurricular program opportunities, and adequate recess opportunities. In 2005, with the realization that childhood obesity was escalating, the National Academies Press published *Preventing Childhood Obesity: Health in the Balance*. Finally, in 2008, the NASPE issued a position statement recommending that all prekindergarten through grade 12 schools implement a comprehensive school physical activity program (CSPAP). CSPAPs encompass physical activity programming before, during, and after school. Many aspects of all of these guidelines do not

GUIDELINE 5.9

Title: *Guidelines for School and Community Programs to Promote Lifelong Physical Activity Among Young People*

Organization: CDC

Year published: 1997

Purpose: To encourage physical activity among young people so that they will continue to engage in physical activity in adulthood and obtain the benefits of physical activity throughout life

Location: www.cdc.gov/healthyyouth/physicalactivity/guidelines

Population: Children and adolescents

GUIDELINES

- Daily physical education and comprehensive health education should be required for students in kindergarten through grade 12.
- Time during school should be provided for unstructured physical activity.
- Planned, sequential curricula for physical education in kindergarten through grade 12 should promote enjoyable, lifelong physical activity.
- Students in elementary school should develop competence in many basic motor skills.
- Older students should develop competence in a select number of lifetime physical activities that they enjoy and succeed in.
- Students should be physically active not only at school but also at home and in the community.

GUIDELINE 5.10

Title: "Physical Fitness and Activity in Schools"

Organization: AAP Council on Sports Medicine and Fitness and Council on School Health

Year published: 2000

Purpose: To increase physical activity and fitness among students

Location: *Pediatrics* 105(5): 1156-1157

Population: School-age youth

GUIDELINES

- Establish comprehensive (preferably daily) physical education and health education for children in kindergarten through grade 12.
- Provide a safe environment to encourage physical activity, including safe facilities and protective equipment.
- Implement physical education and health education curricula that emphasize enjoyable participation in physical activity and that help students to develop the knowledge, attitudes, motor skills, behavioral skills, and confidence needed to adopt and maintain physically active lifestyles.
- Provide extracurricular physical activity programs that address the needs and interests of all students.
- Include parents and guardians in physical activity instruction and extracurricular physical activity programs.

Adapted, by permission, from Council on Sports Medicine and Fitness and Council on School Health, 2000, "Physical fitness and activity in schools," *Pediatrics* 105(5): 1156-1157.

GUIDELINE 5.11

Title: "Increasing Physical Activity: A Report on Recommendations of the Task Force on Community Prevention Services"

Organization: CDC

Year published: 2001

Purpose: To reduce preventable deaths, morbidity, and disability

Location: *Morbidity and Mortality Weekly Report* 50(RR-18): 1-16

Population: Children in elementary school

GUIDELINES

- Combine enhanced access to places for physical activity with informational outreach activities.
- Build social support interventions to promote physical activity behaviors.
- Modify physical education curricula to increase the amount of moderate or vigorous activity performed in class, the amount of time spent in class, and the amount of active time taking place during class.
- Encourage physical activity through mass media, education, and community events.

delineate specific amounts and types of physical activity but instead focus on the role of schools and public education policies in cultivating an atmosphere encouraging lifelong physical activity. Highlights of each guidelines that are pertinent to this text are presented.

GUIDELINE 5.12

Title: *Preventing Childhood Obesity: Health in the Balance*

Organization: National Academies Press

Year published: 2005

Purpose: To prevent childhood obesity

Location: www.nap.edu/catalog.php?record_id=11015

Population: School-age youth

GUIDELINES

- Children should accumulate a minimum of 60 min of moderate to vigorous physical activity each day.
- Children should participate in at least 30 min of moderate to vigorous physical activity during the school day.

From Kaplan JP, Liverman CT, Kraak VA. Preventing Childhood Obesity: Health in the Balance, National Academies Press, Washington, DC, 2005.

GUIDELINE 5.13

Title: *Comprehensive School Physical Activity Programs*

Organization: NASPE

Year published: 2008

Purpose: To implement a CSPAP to improve public health and the physical, mental, and social benefits of activity for youths

Location: www.aahperd.org/naspe/standards/upload/Comprehensive-School-Physical-Activity-Programs2-2008.pdf

Population: School-age youth

GUIDELINES

- Daily physical education should be provided for at least 150 min/wk for elementary school students and 225 min/wk for middle and high school students.
- School-based physical activity opportunities above and beyond physical education, such as classroom physical activity and breaks, should be provided.
- Elementary school children should have at least one daily recess lasting at least 20 min.
- Schools should offer a wide variety of intramural and interscholastic programs.
- Schools should encourage active transport to school.

Healthy People 2010

As noted in chapter 2, *Healthy People 2010* was the result of hundreds of individuals and organizations coming together to set objectives to improve the health of the United States population. It is recommended that adolescents engage in moderate physical activity for at least 30 min on 5 or more of the previous 7 d, and the *Healthy People 2010* objectives state a target of 35% of adolescents achieving this goal. This compares with a value of 27% of students in grades 9 through 12 who fulfilled this recommendation in 1999. Vigorous activity is also recommended for adolescents; a target of 85% of students in grades 9 through 12 engaging in vigorous physical activity 3 d/wk or more for 20 min or more per occasion was established, in contrast to the measured level of 65% in 1999 (CDC 2000).

Another focus of the *Healthy People 2010* objectives is compelling public and private entities, including different levels of government, health care providers, educators, community leaders, and business executives, to take responsibility in attaining the 2010 goals. This is specifically evident in helping children and adolescents to reach physical activity goals; physical education in schools is a unique opportunity to influence young individuals to embrace activity and improved health not only now but throughout a lifetime. Physical education in school can ensure that adolescents spend a minimum amount of time in physical activity and can provide a substantial portion of the physical activity time recommended for children and adolescents. *Healthy People 2010* goals include increasing physical education requirements at all educational levels, increasing participation in physical education classes, and increasing the number of students who are actually active during physical education classes. Because one study found that 25% of American children watch at least 4 h of television daily, another objective is to decrease the numbers of children watching at least 2 h/d of television.

American Heart Association

In 2003, the AHA published a scientific statement on the prevention of atherosclerotic CVD beginning in childhood. The statement came on the coattails of long-term follow-up studies showing that cardiovascular risk factors (such as obesity, hyperlipidemia, hypertension, and type 2 diabetes) that manifest during childhood seem to persist into adulthood. One of the reasons why the statement was published is that intervention trials showed both safety and success in reducing some of these risk factors in childhood, therefore reducing them in adulthood and hopefully ultimately reducing the development of atherosclerosis. While the statement encompassed more than just physical activity recommendations, it did point to physical activity as an intervention likely to improve several individual risk factors, including hypercholesterolemia, hypertriglyceridemia, obesity, elevated blood pressure, and type 2 diabetes. The statement was designed to complement other existing (and future) guidelines.

Three years later, the AHA issued its position statement on physical activity promotion for youths. Keeping in line with other recent guidelines, the statement recommended 60 min of moderate or vigorous activity. However, this statement focused more on the ways in which schools could further children's physical activity.

GUIDELINE 5.14

Title: *Healthy People 2010*

Organization: USDHHS

Year published: 2000

Purpose: To improve health, fitness, and quality of life through daily physical activity

Location: www.health.gov/healthypeople

Population: Adolescents

GUIDELINES

- Engage in moderate physical activity for at least 30 min on 5 or more days of the week.
- Engage in vigorous physical activity that promotes cardiorespiratory fitness 3 or more days per week for 20 min or more per occasion.
- Participate in daily physical education in school.
- Spend at least 50% of physical education class time in physical activity.

Adapted from U.S. Department of Health and Human Services, Healthy People 2010. www.health.gov/healthypeople Accessed 10/10/2009.

GUIDELINE 5.15

Title: "Primary Prevention of Atherosclerotic Cardiovascular Disease Beginning in Childhood"

Organization: AHA

Year published: 2003

Purpose: To promote cardiovascular health and identify and manage known risk factors for CVD in children and young adults

Location: *Circulation* 107: 1562-1566

Population: Children and young adults

GUIDELINES

- Young people should participate in at least 60 min of moderate to vigorous (and fun) physical activity every day.
- Adolescents can combine resistance training (10-15 repetitions at moderate intensity) with aerobic exercise in an overall activity program.
- Young people should limit sedentary time (e.g., watching television, playing video games, or spending time on the computer) to no more than 2 h/d.

GUIDELINE 5.16

Title: *Promoting Physical Activity in Children and Youth: A Leadership Role for Schools*

Organization: AHA

Year published: 2006

Purpose: To support schools' potential for effectively providing and promoting physical activity

Location: www.americanheart.org or circ.ahajournals.org/cgi/content/full/114/11/1214

Population: School-age youth

GUIDELINES

- All children and youths should participate in at least 30 min of moderate to vigorous physical activity during the school day.
- All children and adolescents should participate in at least 60 minutes of moderate to vigorous physical activity every day.
- At least 50% of time spent in physical education class should be engagement in moderate to vigorous physical activity.
- Physical education programs should deliver 150 min/wk (for kindergarten through grade 8) or 225 min/wk (for grades 9 through 12) of physical activity.
- Childhood development centers and elementary schools should provide children with at least 30 min of recess each day (not including physical education class time.)

Because of the amount of time children spend in schools, the influence schools have, and the numerous benefits of schooltime physical activity, the statement addressed the goal of achieving half of the total recommended activity during the school day itself.

Guidelines Outside of the School Environment

In addition to time spent in schools, time before and after school is a prime opportunity for children to participate in structured or unstructured physical activity. However, instead of walking or bicycling home, playing outside with friends, or participating in structured community center activities, many children spend much of this time in other pursuits such as playing video games, watching television, and talking on the telephone. If each child spent only 30 min of the several after-school hours being moderately or vigorously active, a much greater proportion of children would meet the physical activity guidelines. Because schools and community organizations can provide the infrastructure to encourage and lead children toward meeting physical activity guidelines, much of the efforts to improve childhood activity have focused on improving public policy. The NASPE has issued guidelines for after-school physical activity and intramural sport programs (NASPE 2001), underscoring the importance of extending the opportunities for activity in a safe and professionally supervised environment.

Regular bouts of moderate-intensity exercise, both in structured and unstructured environments, are recommended for children. Such environments include physical education programs. There is an emphasis on avoiding inactivity. A variety of types of exercise is encouraged in order to promote the foundations for wellness, fitness, and disease prevention. New approaches such as restricting sedentary time are complementary to encouraging active exercise. Purposeful activities and games are more effective and enjoyable than is exercising for the sake of getting exercise. Younger children are discouraged from engaging in high-intensity or vigorous exercise, as they generally cannot maintain focus on the activity long enough for it to have a positive effect.

In Australia, physical activity guidelines for youths include a minimum of 60 min of daily physical activity and a maximum of 2 h of daily entertainment with electronic media. A study comparing childhood obesity rates with compliance with national physical activity guidelines (Spinks 2007) found that noncompliance with the electronic media guideline correlated with higher odds for childhood obesity (63% increased odds); noncompliance with the physical activity guidelines correlated with lower odds (28% increased odds). As the rates of obesity in American youths continue to increase, perhaps future U.S. guidelines will quantitatively address the sedentary time associated with electronic media.

Physical Activity Pyramid

As mentioned, participating in a variety of activities is essential for youths striving to meet physical activity guidelines. In its guidelines for young people, the NASPE references the physical activity pyramid. This tool is a method for classifying different types of physical activity, assessing levels of physical activity, and teaching concepts of physical activity. The pyramid provides a visual model that helps people conceptualize components of a physical activity program and the relative weighting of each component. The physical activity pyramid can be a valuable teaching aid to coach individuals—not just children, but people of all ages—on working toward the appropriate amounts of activity for health.

Level 1, the broadest level of the pyramid, includes moderate or lifestyle activities that can be completed during normal daily living. These include walking to school or around the neighborhood, riding a bicycle, helping out with yard work, and play of all types. This type of lifestyle activity is the foundation for a lifetime of physical activity, as it is the type most likely to be performed throughout life.

Level 2 activities become more interesting to children as they grow older; such activities include active aerobics, sports, and recreational activities. Younger children may be introduced to level 2 activities so they can start learning some of the skills involved; however, they should not be expected to be interested in performing continuous aerobic exercises such as running or playing competitive sports. As children mature, develop more skills, and show more interest in more competitive sports and active aerobics, they may gradually spend more time at this level.

Level 3 includes activities geared toward muscular fitness and flexibility. Children may participate in these activities at a developmentally appropriate level, although mild stretching and age-appropriate calisthenics are emphasized instead of heavy overload or long training sessions. In fact, an emphasis on the latter may decrease rather than increase interest in physical activity.

At the top of the physical activity pyramid is inactivity. Long durations of inactivity are not appropriate for children—this is the impetus behind requiring regular recesses and physical education classes in the school system. Parents and other caregivers also have a responsibility to provide children with multiple opportunities to be active and play daily before and after school.

The physical activity pyramid is not the only model of its kind! The USDA has a resource for children aged 6 to 11 called *MyActivity Pyramid*. In addition, the Fitnessgram includes a physical activity pyramid. Youths may find the planning and implementing of a physical activity program overwhelming or simply beyond their means, skills, and desires. On the other hand, participating in daily activity is a necessity for proper growth and establishes the patterns

for appropriate lifelong participation. Finding a way of increasing awareness of the recommendations for physical activity and then encouraging and challenging youths to meet these targets can be formidable, but these models provide a good framework for doing so as well as represent the various guidelines. The last model that deserves mention is the Activitygram, which was presented briefly in chapter 3. The Activitygram is an illustrated method of tracking activity and healthy behaviors that helps young people visually identify appropriate levels and amounts of activity.

Physical Activity as Part of Dietary Guidelines

As with the adult guidelines spelled out in the 2002 IOM report and USDA *Dietary Guidelines for Americans* (2005), physical activity guidelines for children and adolescents were created as part of the overall recommendation for weight management and overall health management. All children and adolescents are encouraged to participate in at least 60 min of physical activity on most days; overweight children may need even more physical activity to reduce weight gain while allowing for growth and development. The IOM and USDA recommendations are based on the limited scientific literature on childhood and weight gain that was available when they were

---

GUIDELINE 5.17

Title: *Dietary Reference Intakes for Energy, Carbohydrate, Fiber, Fat, Fatty Acids, Cholesterol, Protein, and Amino Acids*

Organization: IOM

Year published: 2002

Purpose: To promote health and vigor and to balance calorie intake with expenditure

Location: www.nap.edu/catalog/10490.html

Population: Children and adolescents

GUIDELINES

- Children and adolescents should engage in at least 60 min of moderate- to vigorous-intensity physical activity daily.
- Exercise may be accumulated and should include both lifestyle physical activities and traditional exercise.

GUIDELINE 5.18

Title: *Dietary Guidelines for Americans*

Organization: USDHHS

Year published: 2005

Purpose: To promote healthy growth and development and to help avoid unhealthy weight gain

Location: www.health.gov/dietaryguidelines/dga2005

Population: Children and adolescents 2 y and older

GUIDELINE

Engage in at least 60 min of physical activity on most, preferably all, days of the week.

GUIDELINE 5.19

Title: *The President's Challenge*

Organization: PCPFS

Year published: 2008

Purpose: To encourage lifelong exercise and physical activity to realize health benefits

Location: www.presidentschallenge.org

Population: Children aged 5 to 12 y

GUIDELINES

- Youths should accumulate 60 min or more of daily moderate to vigorous physical activity that is developmentally appropriate and enjoyable and involves a variety of activities.
- Blocks of at least 5 min may accumulate toward the 60 min goal, although bouts of at least 15 min are encouraged.
- An option of accumulating 11,000 daily steps for girls or 13,000 daily steps for boys is roughly equivalent to the physical activity standard.

---

published. Because even though many children at the time were already getting 30 min of regular activity there was still an ever-increasing prevalence of overweight and obesity among this age group, the recommendation was made for 60 min of daily activity.

The President's Challenge

In 2008, the PCPFS developed the Active Lifestyle program and issued criteria to obtain the Presidential Active Lifestyle Award. This group endorsed the same 60 min/d guideline recommended by other entities and aimed to help establish a lifelong pattern of exercise. Although 5 min blocks of activity are acceptable in accumulating time toward achieving the award, the challenge encourages youths to try to complete at least 15 min of activity at a time. Like adults, children are generally both moderately and vigorously active during exercise. However, when compared with adults, children tend to have more rest interspersed with their activity.

Another option for attaining the level of activity needed to receive the award is to accumulate daily steps. A daily step count of 11,000 for girls or 13,000 for boys qualifies for the award. About 12,000 steps corresponds to 60 min of activity; the different standards reflect the fact that data show that girls accumulate fewer steps throughout the day than boys accumulate. Also, because more girls than boys are sedentary, a lower value was chosen for girls in order to persuade more of them to become active in order to achieve the award. Thus setting the lower threshold was in line with the goal of developing an award that previously inactive youths could achieve. Promoting activity through awareness was one of the goals of The President's Challenge.

International and State Guidelines

Though this text focuses primarily on American guidelines, it is useful to recognize and make comparisons with guidelines from corresponding entities in other countries. Recommendations from the United Kingdom are included at the beginning of this chapter, as these were released around the same time those issued by the NASPE were released. Both suggest an hour daily of at least moderate-intensity activity. Like activity guidelines from Canada, the U.K. guidelines include a provision for youths who are inactive.

The U.K. consensus recommends that these individuals work on attaining 30 min of daily activity and ultimately strive for the goal of 1 h. These guidelines also specifically reference the need for activity that strengthens bones, including muscle-strengthening and flexibility exercises.

The Public Health Agency of Canada has issued physical activity guidelines for children aged 6 to 14 y. These were developed shortly after the 1998 *Canada's Physical Activity Guide to Healthy Active Living* for adults. Like the American guidelines, these guidelines are evidence based and developed through scientific review by a committee of experts. While the Canadian recommendations were directed at caregivers, parents, and teachers for the 6 to 9 y age group, they were delivered directly to the youths aged 10 to 14 y. One unique aspect of these guidelines is that they take into account the finding that many children currently fall considerably short of the ideal amount of physical activity. With that in mind, the guidelines have an ultimate goal of at least 90 min of daily activity but focus on working toward the goal through a combination of adding physical activity and decreasing sedentary time. As

GUIDELINE 5.20

Title: *Canada's Physical Activity Guide to Healthy Active Living: Family Guide to Physical Activity for Youth 10-14 Years of Age*

Organization: Public Health Agency of Canada, Canadian Society for Exercise Physiology, The College of Family Physicians of Canada, and Canadian Pediatric Society

Year published: 2002 (developed in 2001)

Purpose: To build awareness and understanding of the importance of physical activity to healthy growth and development

Location: www.phac-aspc.gc.ca/pau-uap/paguide/child_youth/pdf/yth_family_guide_e.pdf

Population: Youths aged 10 to 14 y

GUIDELINES

- Increase time spent on physical activity, starting with 30 min *more* each day.
- Reduce nonactive time spent on watching television and videos, playing computer games, and surfing the Internet, starting with 30 min *less* each day.
- Build up physical activity throughout the day in bouts of at least 5 to 10 min.
- Combine endurance, flexibility, and strength activities for best results.
- Gradually progress to at least 90 min of daily physical activity.

Adapted from Family Guide to physical activity for youth 10-14 years of age, © Public Health Agency of Canada, 2002.

GUIDELINE 5.21

Title: *Canada's Physical Activity Guide to Healthy Active Living: Family Guide to Physical Activity for Children (6-9 Years of Age)*

Organization: Public Health Agency of Canada, Canadian Society for Exercise Physiology

Year published: 2002 (developed in 2001)

Purpose: To encourage healthy growth and development and to reduce weight gain and obesity

Location: www.phac-aspc.gc.ca/pau-uap/paguide/child_youth/pdf/kids_family_guide_e.pdf

Population: Children aged 6 to 9 y

GUIDELINES

- Increase the time currently spent on physical activity, starting with 30 min *more* each day.
- Reduce nonactive time spent on watching television and videos, playing computer games, and surfing the Internet, starting with 30 min *less* each day.
- Build up physical activity throughout the day in bouts of at least 5 to 10 min.
- When increasing physical activity, include moderate activities such as brisk walking, skating, biking, swimming, and playing outdoors and vigorous activities such as running and soccer.
- Gradually progress to at least 90 min of daily physical activity.
- Gradually reduce screen time by 90 min/d.

Adapted from Family Guide to physical activity for youth 6-9 years of age, © Public Health Agency of Canada, 2002.

GUIDELINE 5.22

Title: *Australia's Physical Activity Recommendations for 5-12 Year Olds*

Organization: Australian Government Department of Health and Ageing

Year published: 2004

Purpose: For healthy growth and development

Location: www.health.gov.au/internet/main/publishing.nsf/Content/9D7D393564FA0C42CA256F97 0014A5D4/$File/kids_phys.pdf

Population: Children aged 5 to 12 y

GUIDELINES

- Children need at least 60 min (and up to several hours) of moderate to vigorous physical activity every day.
- Children should not spend more than 2 h/d using electronic media for entertainment (e.g., computer games, television, Internet), particularly during daylight hours.
- Inactive individuals should build up to 30 min of daily moderate activity and then steadily increase active time until reaching the goal of 60 min/d.

with other guidelines for children, a specific exercise prescription is not made; instead, there is an overall concentration on achieving a total volume of activity that includes endurance, strengthening, and flexibility exercises. The approach of both increasing activity and decreasing inactivity is the most ambitious of all youth guidelines worldwide.

Australian guidelines are broken down into recommendations for children aged 5 through 12 y and recommendations for young people aged 12 to 18 y. The guidelines recommend at least 60 min of activity for each group; exercise should include both moderate-intensity and vigorous-intensity activities. The younger age group is

GUIDELINE 5.23

Title: *Australia's Physical Activity Recommendations for 12-18 Year Olds*

Organization: Australian Government Department of Health and Ageing

Year published: 2004

Purpose: For healthy growth and development

Location: www.getmoving.tas.gov.au/RelatedFiles/Physicalguidlinesfor12-18yrolds.pdf

Population: Children aged 12 to 18 y

GUIDELINES

- Adolescents need at least 60 min of moderate to vigorous physical activity every day.
- Adolescents should not spend more than 2 h/d surfing the Internet, watching television, or playing noneducational video games.
- For additional health benefits, adolescents should try to include 20 min or more of vigorous activity 3 or 4 d/wk.
- Inactive individuals should build up to 30 min of daily moderate activity and then steadily increase active time until reaching the goal of 60 min/d.

GUIDELINE 5.24

Title: "Physical Activity Guidelines for Adolescents: Consensus Statement"

Organization: International Consensus Conference

Year published: 1994

Purpose: To establish age-appropriate physical activity guidelines and to consider how these guidelines may be implements in primary health care settings

Location: *Pediatric Exercise Science* 6(4): 302-314

Population: Adolescents

GUIDELINES

- All adolescents should be physically active every day or nearly every day for at least 30 min, as part of play, games, sport, work, transportation, recreation, physical education, or planned exercise, in the context of family, school, and community activities.
- Adolescents should engage in 3 or more activity sessions per week; sessions should last 20 min or more and require moderate to vigorous levels of exertion.

Reprinted, by permission, from J.F. Sallis and K. Patrick, 1994, "Physical activity guidelines for adolescents: Consensus statements," *Pediatric Exercise Science* 6, 302-314.

encouraged to limit sedentary time, particularly when they could be outside and active. Sedentary time should not exceed 2 h daily.

In May of 2002, experts and scientists in the fields of physical activity, energy expenditure, and weight regulation met to set a consensus on preventing obesity worldwide. The resulting consensus statement acknowledged that while the current (at that time) guidelines recommending a minimal amount of activity were appropriate for preventing some chronic diseases, additional activity—whether higher intensity or longer duration—would be needed to prevent overweight or obesity or weight regain. The consensus statement also recommended that children

achieve activity levels beyond those suggested for adults. The specific amount by which children's daily activity should exceed that recommended for adults was not delineated. A further suggestion, which echoed many other recommendations for this age group, was to limit sedentary time. In 2003, the International Association for the Study of Obesity released its guidelines. These guidelines recommended daily activity, including a PAL of 1.7, and, like other international guidelines, emphasized limiting sedentary behavior.

The WHO cited physical activity guidelines as part of its *Global Strategy on Diet, Physical Activity and Health*. Instead of developing different or new guidelines for children, the WHO simply

---

GUIDELINE 5.25

Title: "How Much Physical Activity Is Enough to Prevent Unhealthy Weight Gain? Outcome of the IASO 1st Stock Conference and Consensus Statement."

Organization: International Association for the Study of Obesity

Year published: 2003

Purpose: To prevent weight gain and the transition to overweight or obesity

Location: *Obesity Reviews* 4(2): 101-114

Population: Children

GUIDELINES

- Complete at least 45 to 60 min of moderate-intensity activity per day or attain a PAL of 1.7.
- Reduce sedentary behavior by incorporating more incidental and leisure-time activity into the daily routine.

GUIDELINE 5.26

Title: *Physical Activity and Young People*

Organization: WHO

Year published: 2008 (updated)

Purpose: To ensure healthy development

Location: www.who.int/dietphysicalactivity/factsheet_young_people/en/index.html

Population: Young people

GUIDELINE

Accumulate at least 60 min of moderate- to vigorous-intensity physical activity each day.

Reproduced, with permission from the World Health Organization: www.who.int/dietphysicalactivity/factsheet_young_people/en/index.html Accessed 10/10/2009.

referenced existing guidelines. Recommendations for youths aged 5 to 18 y were based on Strong and others' 2005 paper: 60 min of daily moderate- to vigorous-intensity physical activity that is developmentally appropriate and involves a variety of activities. The WHO strategy reiterates that this activity level is the minimum required to promote and maintain health as well as includes the principles of the dose–response relationship among exercise, health, and accumulation.

The CDHS has also issued guidelines for physical activity in youths. The reasons why these guidelines are included in this text are discussed briefly in chapter 4. The California guidelines address ages 5 to 12 and 13 to 17 as two separate groups.

Guidelines Focused on Specific Health Goals

In addition to the general goals for children and youths, there are guidelines with specific goals in mind, such as preventing obesity, improving bone health, or improving sport performance. In fact, sport participation has become very popular, so much so that the AAP felt the need to issue guidelines regarding overuse injuries and burnout. The primary goal of engaging in sports at this age is moderating activity and participation so that it is fun for the child while still promoting the benefits such as improved body composition and bone health.

Obesity Prevention

Chapter 4 presented Carrel and Bernhardt's review on exercise for the prevention of adolescent obesity, which suggests starting with children as young as 3 y. At the younger ages, the onus is more on the parents and health care providers to encourage healthy behaviors. However, by adolescence, most individuals are more autonomous and choose behaviors independently—hopefully at this age the established inclusion of physical activity in everyday activities will encourage

---

GUIDELINE 5.27

Title: *The California Center for Physical Activity's Guidelines for Physical Activity Across the Lifespan: Children (5-12 Years)*

Organization: CDHS

Year published: 2002

Purpose: For appropriate development and health

Location: www.caphysicalactivity.org/facts_recomm1a.html

Population: Children aged 5 to12 y

GUIDELINES

- Children in elementary school should accumulate at least 30 to 60 min of age-appropriate and developmentally appropriate physical activity on all, or most, days of the week.
- Children in elementary school are encouraged to accumulate more than 60 min, and up to several hours per day, of age-appropriate and developmentally appropriate activities.
- Some activity each day should be in bouts lasting 10 to 15 min or more and should include moderate to vigorous activity. Intermittent activity involves alternating bouts of moderate to vigorous activity with time for rest and recovery.
- Extended durations of inactivity are discouraged for children. Sedentary behaviors such as watching television or videos, playing video games, and leisure surfing of the Internet should be kept to a minimum (no more than 1 h/d total).
- A variety of physical activities are recommended for children. As many of these activities as possible should take place in a safe outdoor environment.

GUIDELINE 5.28

Title: *The California Center for Physical Activity's Guidelines for Physical Activity Across the Lifespan: Youth (13-17 Years)*

Organization: CDHS

Year published: 2002

Purpose: For appropriate development and health

Location: www.caphysicalactivity.org/facts_recomm1b.html

Population: Youths (13-17 y)

GUIDELINES

- All adolescents should be physically active every day or nearly every day, as part of play, games, sport, work, transportation, recreation, physical education, or planned exercise, in the context of family, school, and community activities.
- Adolescents should engage in at least 60 min of moderate to vigorous physical activity on most days of the week; 30 min of daily activity should be viewed as a minimum, whereas 1 h represents a more favorable level.
- Physical activity can be performed continuously or intermittently throughout the day.

---

lifelong exercise patterns. When starting school, children are improving their balance and visual tracking and thus may be ready to participate in team sports. By the age of 10 to 12 y, motor skills are almost fully mature and both variety and intensity of activities can increase, approximating those available for adults. As with younger children, nonorganized sport activities and behaviors are an important component to accumulating overall physical activity time. These include limiting sedentary behaviors and incorporating physical activity in the daily routine (walking to school or on errands, taking the stairs, walking the dog, and so on).

Bone Health

Both the ACSM and the International Osteoporosis Foundation (IOF) have developed guidelines to help children optimize skeletal bone growth and peak bone density. Exercise is as important as diet in achieving peak bone mass. The process of laying down bone occurs throughout childhood and adolescence, particularly during the first 2 y of life and during the pubertal growth spurt. For girls, the amount of bone tissue accumulated from age 11 to 13 y approximately equals the amount of bone lost during the 30 y after menopause (Bon-

jour 2008)! Higher-intensity loading forces, such as those created by jumping, gymnastics, and resistance training, have the potential to increase bone mineral accrual in youths, leading to a higher peak bone mass. For young people who regularly participate in competitive sport or intense activity, these gains in bone mass are maintained even after the activity has been discontinued.

The primary goal of the following ACSM exercise prescription is to augment bone mineral accrual in children and adolescents. As a result, the volume of recommended activity is significantly less than that of other guidelines—this makes sense in that a greater load during activity and not necessarily a greater total volume of activity is what is needed for greater bone density. Although it is not specifically mentioned, these guidelines should be taken as part of an overall physical activity program. That is, the 10 to 20 min sessions performed 3 times each week should be part of the total weekly program (which also includes 60 min of daily activity)—the part of the program that focuses on building bone mass.

The IOF guidelines are part of an overall program for building bone mass in young people and maintaining bone mass in adults. Like the ACSM guidelines, these recommendations are meant to

GUIDELINE 5.29

Title: "Exercise Prescription for the Prevention of Obesity in Adolescents"

Organization: None

Year published: 2004

Purpose: To prevent and treat obesity by promoting behavioral and environmental changes

Location: *Current Sports Medicine Reports* 3: 330-336

Population: Children and adolescents

GUIDELINES

Frequency	Intensity	Duration
Most, preferably all, days of the week	Moderate	At least 30-60 min

Special considerations include the following:

- Incorporate more lifestyle physical activities such as active chores, walking or biking to school, and taking the stairs.
- Reduce sedentary behaviors such as watching television and videos and playing computer games.
- Provide healthy snacks and adequate opportunities for frequent and effective physical activity in schools.
- For adolescents, encourage participation in team sports such as soccer, softball, and others.
- Consider adding strength training for older children and adolescents; follow the strength training guidelines (see guideline 5.32).

Based on Carrel and Bernhardt 2004.

GUIDELINE 5.30

Title: "Physical Activity and Bone Health"

Organization: ACSM

Year published: 2004

Purpose: To augment bone mineral accrual in children and adolescents

Location: *Medicine and Science in Sports and Exercise* 36(11): 1985-1996

Population: Children and adolescents

GUIDELINE

At least 3 d/wk, higher-impact activities such as jumping, plyometrics, and gymnastics should be performed for 10 to 20 min/d. Moderate-intensity resistance training (<60% 1RM) should be included for bone-loading forces. Other recommended activities include running and jumping.

optimize bone mineral accrual. As opposed to the ACSM guidelines, there is no specific exercise prescription. Instead, qualitative guidelines are made with respect to the type and intensity of exercise.

Resistance Training

Historically, there has been controversy regarding resistance training in youths. Specifically, there was a concern that excessive stress on muscle and

GUIDELINE 5.31

Title: *Invest in Your Bones. Move It or Lose It*

Organization: IOF

Year published: 2005

Purpose: To build up bone mass in young people and maintain bone mass in adults

Location: www.iofbonehealth.org/publications/move-it-or-lose-it.html

Population: Children and adults

GUIDELINES

- Weight-bearing and high-impact exercise is required.
- Sports that involve lifting weights, running, sprinting, jumping, and skipping are good.
- Start slowly and progress gradually.
- Short-duration, high-intensity exercise builds bone most efficiently.
- Two short exercise sessions separated by 8 h is better than one long one.
- It is better to reduce the length of each session rather than the number of sessions performed per week.
- Maintain a balanced, healthy diet and lifestyle, as exercise alone cannot prevent osteoporosis.

From Minne HW, Pfeifer M. Invest in your bones. Move it or lose it. International Osteoporosis Foundation, 2005. www.iofbonehealth.org Accessed on 10/10/2009.

bone might be counterproductive and damage the epiphyseal growth plates. These uncertainties apply more to individuals who exercise intensely at the extremes; for most American children, these theoretical concerns pale in comparison to the far greater risk of being sedentary. In observations of the growth patterns of younger athletes who engage in intense training and competition, intense activity has not been found to restrict statural or somatic growth (Malina 1998, 2000). The major exception to this finding is in gymnasts and dancers, who probably have a combination of genetic predisposition, training stress, and undernutrition that slows growth rates and delays skeletal and sexual development (Rowland 2005).

On the other hand, a sedentary lifestyle and obesity are both escalating problems in American youths. Over the past several years there has been more research into the effects of resistance training for youths, and now there is a universal acceptance of its benefits and place in a physical activity program. Resistance training especially helps people who are overweight to gain confidence and experience success with physical activity, potentially motivating them to add other forms of exercise to their program. In addition to creating favorable changes in body composition, regular resistance training can improve bone density, cardiorespiratory fitness (although not as profoundly as an aerobic training program), lipid profile, and psychosocial functioning.

The NSCA has issued a position statement on youth resistance training, although the statement contains no specific guidelines. The NSCA believes that a properly designed and supervised resistance training program is safe for children. The benefits listed in the position statement include the following:

- Increased strength
- Enhanced motor fitness skills and sport performance
- Injury prevention for youth sport and recreational activities
- Improved psychosocial well-being
- Enhanced overall health

The AAP issued a policy statement on strength training; the most recent update was made in 2008. It recommends that youths undergo a medical evaluation before participation in order to identify possible risk factors for injury. Also, because of the escalating use of anabolic steroids at the high school and even younger levels, a preparticipation evaluation allows for discussion of training goals, expectations, and the risks of steroid use. A low-resistance program is recommended initially, until proper technique is established. Weight can then be added gradually, with a target of performing 8 to 15 repetitions. In order to see strength gains, exercise duration should be at least 20 to 30 min and exercise frequency should be at least 2 or 3 times a week. More than 4 times a week is unlikely to confer additional benefits and could theoretically be harmful by increasing injury risk.

Sport Participation

With the ever-increasing emphasis on over-planned, specialized, and programmed lifestyles for children, many youths are participating in team or individual sports at a higher level at a younger age. Parents are eager to get a good start and have their children get ahead or in some way distinguish themselves; thus they push their

---

GUIDELINE 5.32

Title: "Strength Training by Children and Adolescents"

Organization: AAP Council on Sports Medicine and Fitness

Year published: 2008

Purpose: To enhance strength, improve sport performance, prevent or rehabilitate injuries, and enhance long-term health

Location: *Pediatrics 121: 835-840*

Population: Children and adolescents

GUIDELINES

- Strength training programs for preadolescents and adolescents can be safe and effective if proper techniques and safety precautions are followed.
- Preadolescents and adolescents should avoid competitive weightlifting, powerlifting, body building, and maximal lifts until they reach physical and skeletal maturity.

When pediatricians are asked to recommend or evaluate strength training programs for children and adolescents, they should consider the following issues:

- Before beginning a formal strength training program, preadolescents and adolescents should receive a medical evaluation by a pediatrician. If indicated, referral may be made to a physician in sports medicine who is familiar with various strength training methods and the risks and benefits of such training in preadolescents and adolescents.
- Aerobic conditioning should be coupled with resistance training if general health benefits are the goal.
- Strength training programs should include a warm-up and cool-down.
- Specific strength training exercises should be learned with no load (resistance). Once the exercise skill has been mastered, incremental loads can be added.
- Progressive resistance exercise requires the successful completion of 8 to 15 repetitions in good form before weight or resistance is increased.
- A general strengthening program should address all major muscle groups and exercise throughout the complete range of motion.
- Any sign of injury or illness due to strength training should be evaluated before continuing the exercise in question.

GUIDELINE 5.33

Title: "Exercise Prescription for Healthy Populations: Resistance Training"

Organization: ACSM

Year published: 2010

Purpose: To safely increase muscular strength

Location: *ACSM's Guidelines for Exercise Testing and Prescription, Eighth Edition*

Population: Children and adolescents

GUIDELINES

- Perform 8 to 15 repetitions per exercise, to the point of moderate fatigue with good mechanical form, before increasing the resistance.
- In general, adult guidelines for resistance training are appropriate.

GUIDELINE 5.34

Title: *Fit Facts: Strength Training for Kids: A Guide for Parents and Teachers*

Organization: ACE

Year published: 2001

Purpose: To safely include strength training for children and adolescents

Location: www.acefitness.org/fitfacts

Population: Children and adolescents

GUIDELINES

- Starting at age 7 or 8, children can begin to train with weights.
- Start slowly, leaning toward underestimating the strength of children.
- Focus on developing good form and learning the basics.
- Introduce children to a variety of exercises and types of resistance, including medicine balls, resistance tubing, free weights, and machines.
- Address all major muscle groups in a balanced, full-body workout.

Adapted, by permission from American Council on Exercise Fit Facts, San Diego 2001. www.acefitness.org/fitfacts Accessed 10/10/2009.

children to participate in school, club, or elite programs (or even combinations of all three). In these situations, the objective of increasing physical activity is replaced by the objective of ensuring that children do not overdo it. Manifestations of too much activity include fatigue, overuse injuries, decreased motivation, and lost enjoyment of the activity. In 2007 the AAP issued a position statement to address this concern. Recommendations include allowing for adequate time off (both on a weekly and on a seasonal basis) as well as being aware of signs of overuse or burnout. Suggestions for educating parents, athletes, and coaches are made so that children may achieve good health and sport performance while minimizing injuries. The AAP statement emphasizes safety, growth, and fun.

For adolescents who end up participating in team or individual sports at the high school level and beyond, there are guidelines for preparticipation clearance. Most high school, college, and club athletic teams require a preparticipation physical

GUIDELINE 5.35

Title: "Overuse Injuries, Overtraining, and Burnout in Child and Adolescent Athletes"

Organization: AAP

Year published: 2007

Purpose: To identify and counsel at-risk children and their families on intensive training and sport specialization

Location: *Pediatrics* 119: 1242-1245

Population: Children involved in structured sports

GUIDELINES

- Take time off from organized or structured sport participation 1 or 2 d/wk to allow the child to rest or to participate in other activities.
- Gradually increase training volume by no more than 10% each week. Training volume is calculated by sport and can be based on number of pitches, miles run, or hours of exercise.
- Encourage the young athlete to participate on only one team during a season.
- Encourage the athlete to take at least 2 or 3 mo away from a specific sport during the year.
- Emphasize fun, skill acquisition, safety, and good sporting behavior.
- Allow for educational opportunities on appropriate nutrition and fluid intake, sport safety, and the avoidance of overtraining to achieve optimal performance and good health.

Reprinted, by permission, from J.S. Brenner and the Council on Sports Medicine and Fitness, 2007, "Overuse injuries, overtraining, and burnout in child and adolescent athletes," *Pediatrics* 119:1242-1245.

---

exam every year. Some require an initial exam with subsequent exams repeated annually only if indicated. The purpose is not only to maintain the health and safety of the athlete in training and in competition but also to address preventative medicine topics and optimal nutrition goals. Athletes who tend to place themselves or other athletes at risk for injury require follow-up testing and evaluation. Some of these athletes may then be cleared for participation, while others may be allowed only limited participation, with the suggestion for pursuing other and safer activities. The ACSM, along with many state-based interscholastic organizations, supports the concept of the preparticipation physical exam in order to help athletes safely enjoy the benefits of sport participation. Following are lists of the primary and secondary objectives of the preparticipation exam.

Primary Objectives

- Detect conditions that may predispose to injury
- Detect conditions that may be life threatening or disabling (such as single organs or cardiac anomalies)
- Meet legal and insurance requirements

Secondary Objectives

- Determine general health and fitness level
- Counsel on health-related issues, including drug use, mood disorders, sexually transmitted diseases, and pregnancy prevention

The following are common components of the preparticipation physical exam:

Medical History

- Review of past injuries
  - Dental
  - Orthopedic
  - Neurological
- Surgical history
- Menstrual history

- Illness history
  - Heat illness
  - Weight issues or disorders
- Medications and allergies
- Protective devices
- Immunization history
- Cardiopulmonary signs and symptoms
- Family history of cardiac disease

Physical Examination

- Height and weight
- Visual acuity
- Hearing ability
- Blood pressure
- Resting pulse
- Cardiopulmonary exam
- Neurological exam
- Musculoskeletal exam

Note that the extent of the head, ear, nose, throat, abdomen, genitalia, and skin exam depends on the sport.

SUMMARY

One concept that has been universally established is that physical activity begins to decline in adolescence. Childhood is the most physically active time of life, and activity patterns steadily drop off with advancing age. This decline seems to be more prevalent for males than for females, although the proportion of males who are physically active is still usually greater than that of females. The single greatest decline in activity—in overall exercise duration as well as in strength training and regular vigorous activity—occurs from ages 15 to 18 y. Future recommendations for physical activity during this time should take this commonality into account and attempt to reverse the decline by encouraging some other form of physical activity that is more appealing to today's youths.

Pregnant and Postpartum Women

The first national guidelines for physical activity during pregnancy were issued by the American College of Obstetricians and Gynecologists (ACOG) in 1985. These guidelines reflected the studies conducted up to that point in time, which focused primarily on the potential harm rather than the benefits of physical activity during pregnancy. The guidelines were conservative and included an HRmax of 140 beats/min and time limits for physical activity. Since then, there have been updates not only to these recommendations (in 1992, 1994, and 2002) but also to subsequent recommendations made by the ACSM. Early versions of ACSM guidelines discussed exercise only in the nonpregnant population, but in 2000 the ACSM *Current Comments* specifically addressed exercise in pregnancy. In 2008, the *Physical Activity Guidelines for Americans (PAGA)* presented recommendations for pregnant and postpartum women; this was the first time such national guidelines included the pregnant segment of the population. These newer recommendations are based on research from the past 20 y, which has shown that exercise is not only safe but also beneficial to both the mother and the fetus. More than 1,000 studies during this time have confirmed the safety of exercise during pregnancy. The data suggest the following: Exer-

cise is not detrimental to mother or fetus, exercise may reduce the length of labor, and exercise may improve gestational diabetes (in either incidence or severity). In addition, chronic exercisers experience a relatively minor loss of fitness during pregnancy (Suitor and Kraak 2007).

This chapter details the importance of regular physical activity for both the mother and the fetus as well as looks at contraindications for exercise. Guidelines from both the American and the international obstetrical literature are presented, along with general guidelines that specifically address the pregnant population. Finally, postpartum physical activity is discussed briefly.

Encouraging Pregnant Women to Be Physically Active

Pregnancy represents a unique opportunity for health care providers to influence a woman's present and future health habits. During the 9 mo, a relationship develops between provider and patient that is based on concern and care for the health of both mother and child. Each regular visit provides a chance to impart upon women—and their families—the importance of regular physical activity. The short-term goals of physical activity during the prenatal period include a healthy mother, a healthy pregnancy, and a healthy baby. Long-term goals include the establishment of exercise as a regular routine not only for the mother but also for her child and spouse. For many women who have never been active on a regular basis, pregnancy is the first time they feel motivated to emphasize their health. Thus there is a chance for the mother to lay down a foundation for physical activity at this time. In fact, more than 90% of women who exercise during pregnancy continue with their regular exercise after pregnancy. Additionally, 70% of women who exercise during pregnancy reach or exceed prepregnancy fitness levels (Clapp 2000)!

Physical activity has specific health benefits for pregnant women. Regular physical activity protects against the development of gestational diabetes. In addition, physical activity was associated with a reduced risk of preeclampsia in early pregnancy in two case control studies (Marcoux and others 1989; Sorensen and others 2003) and one cohort study (Saftlas and others 2004). Lastly, there are multiple studies demonstrating the role of physical activity in postpartum weight reduction; although it seems a foregone conclusion, it has been shown that women who continue to perform aerobic exercise postpartum are less likely to become obese and develop risk factors such as hypertension, diabetes, and dyslipidemia (Rooney and others 2005). Interestingly, studies have shown that physical activity alone does not produce postpartum weight loss; it must be combined with dietary restriction.

Despite all of the positive consequences of exercise during pregnancy, the majority of pregnant women do not meet activity guidelines. According to data from the 1994, 1996, 1998, and 2000 Behavioral Risk Factor Surveillance System (BRFSS), pregnant women participate in less leisure-time activity than nonpregnant women participate in. Also, 33% of the pregnant women sampled in 2000 reported no leisure-time activity at all! Although many women are active during pregnancy, it has been found that only 16% achieve the exercise volume recommended by the ACOG (Evenson and others 2004). Another recent study suggested that even fewer

women (3%-13%) meet the ACOG guidelines (Borodulin and others 2008). By far, the most popular form of physical activity for pregnant women is walking. BRFSS data show that 52% of pregnant women walk for physical activity. Other popular activities include aerobics (8% of pregnant women), swimming (4%), running or jogging (2%), and weightlifting (2%).

Obstetricians, who are in a unique position to motivate and encourage pregnant women to exercise, do not always recommend that they do. A 2006 study found that only about half of obstetricians discuss exercise with their pregnant patients. Interestingly, many of the obstetricians who do discuss exercise often specify a maximum heart rate (which is inconsistent with the ACOG guidelines), do not always recommend resistance exercise, and do not routinely advise sedentary women to initiate exercise during pregnancy (Entin and Munhall 2006). Also, many advise curtailing exercise (in intensity or duration) in the third trimester even in the absence of complications—again, a deviation from ACOG guidelines. As a result, significantly fewer women exercise appropriately in the third trimester. Depending on how the amount and intensity of activity are quantified, only between 3% and 34% of women meet exercise guidelines during this trimester (Borodulin and others 2008)!

Benefits and Risks of Physical Activity for Pregnant Women

The goals of physical activity during pregnancy include maintaining a sense of maternal well-being, avoiding fetal harm, and establishing a pattern of regular activity that will ultimately prevent the onset of chronic disease associated with a sedentary lifestyle. Most women can participate in a wide range of recreational activities and maintain an active lifestyle during pregnancy, although those with medical or obstetrical complications may require modifications. Modifications may also be necessary for women who regularly participated in strenuous recreational or competitive physical activities before their pregnancies. In addition, women who are active before pregnancy may choose to modify their exercise programs some-what; for example, weight gain or joint laxity may prompt women to walk instead of jog or to exercise indoors instead of outdoors because of improved environmental control. Because of all the physi-ological changes that occur during pregnancy (see sidebar), activity recommendations for pregnant women are subject to modifications. Finally, concerns for fetal health dictate other changes in order to optimize fetal outcomes.

Nevertheless, the benefits of regular physical activity throughout pregnancy generally out-weigh the risks. Women who exercise routinely in the prenatal period have a lower incidence of babies with low birth weight. Starting with the first trimester, regular activity may help ease nausea and lessen fatigue. Body image and mood—both during pregnancy and in the postpartum period—are positively influenced

by regular physical activity. Although there are not yet disorder-specific guidelines, it is felt that regular physical activity during pregnancy benefits preeclampsia, hypertension, and gestational diabetes, as mentioned earlier. It is not yet clear whether a regular physical activity program affects gestational length, quality or duration of labor, or incidence of chronic disease in the offspring.

Although the benefits of physical activity for pregnant women are significant, there are still precautions to take and health risks to keep in mind. Because of the increased metabolic rate during pregnancy, thermoregulation during exer-cise and the potential effect on fetal temperature are a concern. It is important for pregnant exer-cisers to stay well hydrated in order to improve cooling and maintain blood volume. While the benefits of chronic exercise belong to the mother,

Anatomical Changes During Pregnancy

During pregnancy, the following physiological changes occur:
- Blood volume increases by up to 45%.
- Cardiac output increases by 30% to 50 % (higher in multiple gestations).
- Resting heart rate increases by 20%.
- Mean arterial pressure decreases by 5 to 10 mmHg.
- Hyperventilation causes primary respiratory alkalosis; minute ventilation increases by 50%.
- Resting oxygen requirements increase.
- Basal metabolic rate and heat production increase.
- Energy intake requirement increases by about 300 kcal/d (varies throughout the course of the pregnancy).
- Joint laxity and lumbar lordosis increase.
- Plasma glucose decreases with 45 min of exercise.

During pregnancy, the following anatomical changes occur:
- The uterus expands from a pelvic organ to an abdominal organ at 12 wk.
- The center of gravity shifts anterior and cephalad.
- The diaphragm is displaced cephalad, causing increased inspiratory capacity and decreased functional residual capacity.
- Lumbar lordosis progresses.
- Joint stress, caused by weight gain, increases.
- Joint laxity increases, with pubic symphysis widening of 10 mm.
- Soft tissue edema occurs.
- Uterine pressure on the inferior vena cava causes relative obstruction of venous return.

the risk of overexercise predominantly affects the fetus (Whaley 2005).

Starting with the second trimester, supine exercises are not recommended, as these can cause the uterus to compress the inferior vena cava and result in decreased cardiac output. Also, prolonged isometric exercise during weightlifting may decrease uterine perfusion. Because of the shift in the expecting mother's center of gravity, exercises that require balance should be avoided later in pregnancy; other anatomical changes of pregnancy are presented in the sidebar. Contact sports and activities that have an increased risk for falls should also be avoided. The new *PAGA* lists several exercises to avoid; table 6.1 includes these as well as others with risk for maternal or fetal harm. Motionless standing results in venous pooling and a significant decrease in cardiac output and thus should be avoided. In fact, one large study of 7,722 pregnancies found that mothers who had jobs requiring predominantly standing had babies with lower birth weights (Naeye and Peters 1982).

Some women with specific conditions should seek the approval of their physicians before engaging in physical activity, while other women with some conditions should not participate in regular physical activity at all. Examples of relative contraindications to exercise in pregnancy include poorly controlled chronic diseases such as hypertension, hyperthyroidism, seizure disorders, type 1 diabetes, extreme underweight or overweight, chronic bronchitis, and anemia. Absolute contraindications pose more of a risk for the mother or fetus; these include conditions that might limit blood flow or oxygen delivery to the fetus—such as hemodynamically significant heart disease or restrictive lung disease—and other conditions such as pre-

mature labor (or multiple gestation increasing the risk for premature labor), ruptured membranes, third-trimester placenta previa, preeclampsia, incompetent cervix, and persistent second- and third-trimester vaginal bleeding. A complete list of absolute and relative contraindications is available through the ACOG at www.acog.org.

While exercise and physical activity are generally endorsed for fetal and maternal well-being, exercise should be discontinued in specific situations. These include preterm labor, leakage of amniotic fluid, vaginal bleeding, dizziness, and decreased fetal movement. A complete list is available through the ACOG at www.acog.org. These conditions should prompt pregnant women to terminate exercise and seek medical advice.

Primary Concerns of Exercise and Physical Activity During Pregnancy

Despite all of the benefits of physical activity in pregnancy, concerns remain. Although there is not a large body of data regarding these theoretical concerns, some information has come to light over the last several years. This evidence has been the basis for the changing guidelines and generally more permissive attitudes for physical activity in pregnancy. However, potential risk is associated with exercise during pregnancy, and this section discusses the main concerns.

Thermoregulation

Thermoregulation is an oft-cited concern regarding vigorous or even moderate exercise in

TABLE 6.1 Exercises to Avoid During Pregnancy

Absolutely avoid	Cautionary (fall risk)	Cautionary (abdominal trauma risk)
Supine activities after first trimester Scuba diving Vigorous exercise during the first few days of being at high altitude (above 8,202 ft or 2,500 m)	*Downhill skiing* Water skiing Gymnastics *Horseback riding* Skating Hang gliding	Ice or field hockey Boxing Wrestling Weightlifting *Soccer* *Basketball* Vigorous racket sports

Exercises in italics are included in the *PAGA* list of those to avoid.

pregnancy. Although data regarding the effects of exercise-induced hyperthermia are scarce, there is evidence that increased maternal core temperature (induced via hot tub, sauna, or fever) may be teratogenic, particularly during the first 45 to 60 d of gestation. Both basal metabolic rate and heat production increase during pregnancy. Thus hydration status is important in dissipating heat, particularly during prolonged exercise. Increases in body temperature during activity are directly related to exercise intensity. For moderate-intensity aerobic exercise, core temperature rises an average of 2.7 °F (1.5 °C) during the first 30 min of exercise. Heat is then dissipated by a combination of perspiration and conduction of heat from the core to the periphery. During pregnancy, this dissipation occurs more rapidly, although the risk of dehydration increases as well. Studies have shown that 60 min of prolonged exercise at an intensity of 55% of $\dot{V}O_2$max causes the rectal temperature in pregnant women to rise only 1.1 °F (0.6 °C). Exercising in hot, humid environments or for longer durations at higher intensity could cause problems with thermoregulation. A recent study looked at pregnant women participating in low-impact aerobic exercise (at roughly 70% of HRmax) and their core temperatures and peak heart rates (Larsson and Lindqvist 2005). At this level, there was no significant increase in core temperature.

Reduced Fetal Blood Flow and Precipitation of Labor

The uterus and splanchnic organs account for an increasingly significant portion of cardiac output throughout the gestation. Because physical activity—particularly activity at moderate or vigorous intensity—causes selective redistribution of maternal blood flow to exercising muscles and away from the splanchnic circulation, there is a concern that exercise could have an adverse effect on the fetus. However, several studies have found that despite a significant reduction in blood flow to the uterus, there has not been any adverse association between maternal exercise and fetal well-being (Collings and others 1983; Clapp 1985; Carpenter and others 1988; Wolfe and others 1988). In Larsson and Lindqvist's 2005 study of low-impact aerobic exercise, oxygen saturation measured at both 70% of HRmax and

postexercise was reduced from preexercise values but never fell below 95% saturation. A meta-analysis of exercise and pregnancy reported that an exercise program performed for an average of 43 min 3 times a week at a heart rate of up to 144 beats/min was not associated with adverse effects on maternal weight gain, fetal birth weight, length of gestation, length of labor, or Apgar scores in normal pregnancies (Lokey and others 1991).

Also, it has been shown that women who participate in a vigorous exercise program before pregnancy may safely continue their program during pregnancy (Duncombe and others 2006). A meta-analysis published in 2005 evaluated the effect of maternal physical activity on birth weight stratified by exercise intensity. Although there was a slightly lighter mean birth weight for infants born to women who engaged in vigorous activity, the birth weight was not classified as low.

Finally, although anecdotal information suggests that strenuous physical activity may cause preterm labor, at present there are no data suggesting that this is the case. There is, however, an association between physical activity and a small increase in uterine contractions (Grisso and others 1992).

Concerns for Competitive Athletes

For women who have been or are competitive athletes, there are additional questions related to pregnancy: What are the effects of pregnancy on competitive ability? What are the effects of training and competition on pregnancy? Most athletic women find that pregnancy adversely affects performance. Both the anatomical and physiological changes, such as altered center of gravity, weight gain, relaxation of pelvic and other ligaments, and anemia (see sidebar), are likely to decrease competitive ability. The latter question has been studied in some detail because the age of childbearing overlaps significantly with that of peak competitive performance.

Many athletic women continue to train throughout their pregnancy in order to return to competition as soon as possible postpartum. Theoretical concerns of training include thermoregulatory complications, preterm labor, and babies of low birth weight. As noted earlier, hydration plays an essential role in dissipating heat from the fetus and uterus. Also, there is no evidence demonstrating an increase in preterm labor in pregnant women who exercise at high intensities or for long durations. Lastly, there is no clear consensus on the relationship between maternal exercise patterns and fetal birth weight (Hatch and others 1993). In the past, it has been suggested that exercise during pregnancy causes both baby and mother to have lower weights and weight gain than that of sedentary individuals. The lower birth weight, however, has been attributed to decreased neonatal fat accumulation and not preterm birth (Clapp and Capeless 1990). One prospective Australian study of 148 pregnant women looked at intensity, duration, and frequency of vigorous exercise and birth outcomes. The authors concluded that there were no significant differences in birth weight or gestational age among the different exercise groups (Duncombe and others 2006). A 1998 study looked at pregnant elite athletes who sustained heart rates in the 170 to 180 beats/min range during interval training. Even though women in the study exercised at high intensity an average of 8.6 h/wk, even in the week preceding delivery, there was no decrease in the birth weight of their babies (Krandel and Kase 1998). Some of the increase in cardiac output and blood volume as well as the heart remodeling that occurs during pregnancy actually persist into the postpartum period. Essentially this means that pregnancy causes a training effect; this effect has led to lifetime best performances after the index pregnancy (Clapp 2005). One recent example is Paula Radcliffe, who won the 2007 New York City Marathon just 9 mo after giving birth to her daughter.

Guidelines for General Exercise Prescription

In 2008, *PAGA* was published and became the first government-issued guidelines to contain information on pregnant and postpartum women. Like the guidelines issued for nonpregnant individuals, children, older adults, and other segments of the population, those for pregnant women were based on a rigorous review of the scientific evidence on the risks, benefits, and details of physical activity as it relates to health.

Before these government-issued comprehensive guidelines were issued, those put out by

the ACOG were considered to be the standard. In 2002, the ACOG released its most updated guidelines for exercise during pregnancy and the postpartum period. Pregnant women should aim to be physically active in a manner similar to that of nonpregnant individuals, with the caveats discussed previously. Both aerobic exercise and exercise that improves musculoskeletal fitness enhance the health of the mother. These also appear to be safe for the fetus. Activity recommendations for pregnant women follow the FITT principle (which was discussed in chapter 3):

- **Frequency:** For health and well-being, exercise should be performed on most if not all days of the week. A frequency of 3 to 5 d/wk is adequate to maintain fitness. Like sedentary individuals who are not pregnant, pregnant women who have not been regular exercisers should gradually increase activity up to an accumulation of 30 min/d.

- **Intensity:** With respect to activity intensity, the same guidelines used for nonpregnant individuals are again appropriate for pregnant women. Exercising at 3 to 4 METs, which is considered to be moderate intensity, is advisable to establish a basic level of fitness and to improve well-being in pregnancy. Improving fitness requires a somewhat higher intensity. According to the ACSM, fitness improvement corresponds to 60% to 90% of HRmax or 50% to 85% of HRR or $\dot{V}O_2$max. Which end of the range is needed depends on the prepregnancy level of fitness; women who have not been engaging in regular activity should aim for the low end of the ranges while those wishing to maintain fitness may aim for the higher end. Pregnancy is not a time for markedly improving physical fitness. Rather, overall fitness level is expected to decline somewhat as pregnancy progresses.

- **Time:** The general recommended duration is 30 min daily. The ACOG recommends beginning each session with a 5 to 10 min warmup that includes slow walking and stretching. After the exercise session, a gradual cool-down is advocated in order to return the heart rate to a normal level. The cool-down includes light activity and stretching again.

In 2003, The Society of Obstetricians and Gynaecologists of Canada (SOGC) issued guidelines with similar recommendations and even went a step further by suggesting that failure to exercise during pregnancy may be associated with some risks. These first-ever guidelines by the society were developed in conjunction with the Canadian Society for Exercise Physiology (CSEP) and are apparently the first example of obstetricians and exercise physiologists collaborating on the exercise advice for the general public. These and the ACOG guidelines represent the shift from the past restrictive attitude to what is now a permissive and even highly recommended one. In fact, the Canadian guidelines point out that the risks of not participating in an exercise program during pregnancy include loss of muscular and cardiovascular fitness, excessive maternal weight gain, higher risk of gestational diabetes or pregnancy-induced hypertension, development of varicose veins and deep vein thrombosis, a higher incidence of physical complaints such as dyspnea or low back pain, and poor psychological adjustment to the physical changes of pregnancy (Davies and others 2003b).

Important National and International Guidelines

As noted previously, the most recent national guidelines for pregnant women in the United States are those contained in *PAGA*. These recommendations include (1) being under the care of a health care provider with whom physical activity and proper adjustment thereof can be discussed; (2) if inactive, gradually increasing the amount of activity over time; and (3) not necessarily curtailing vigorous-intensity or high-volume activity just because of pregnancy or recent birth. As with other recommendations for pregnant women, there are precautions; table 6.1 lists exercises that *PAGA* recommends avoiding.

Although the 2002 ACOG guidelines are in essence the national specialty guidelines for pregnant and postpartum women, they do not contain much specific information with respect to the FITT principle. The Canadian guidelines delve deeper into the acceptability of exercise and the downside of not participating, but they also do not offer a specific exercise prescription.

The ACE guidelines offer a few details on an exercise program for pregnant women, but the suggestion of a frequency of 3 times a week is inconsistent with the ACOG and SOGC and CSEP recommendations. According to the ACE, pregnant women should talk with their doctors to obtain approval before starting exercise. As stated, the recommended frequency of exercise is only 3 times per week, or as often as every other day.

GUIDELINE 6.1

Title: *Physical Activity Guidelines for Americans*

Organization: USDHHS

Year published: 2008

Purpose: Not stated

Location: www.health.gov/paguidelines

Population: Pregnant and postpartum women

GUIDELINES

- Healthy women who are not already highly active or doing vigorous-intensity activity should get at least 150 min (2 h and 30 min) of moderate-intensity aerobic activity per week during pregnancy and the postpartum period. Preferably, this activity should be spread throughout the week.
- Pregnant women who habitually engage in vigorous-intensity aerobic activity or are highly active can continue physical activity during pregnancy and the postpartum period, provided that they remain healthy and discuss with their health care provider how and when activity should be adjusted over time.

An exercise program should start with warm-up exercises and stretching focusing on the hips, low back, and cervical region. Water-based activities such as swimming or water aerobics are recommended due to the buoyancy and relatively lower risk for injury. The ACE recommendations do not specifically address the intensity of exercise, although walking is recommended as one of the best aerobic activities. The statement also suggests, "don't push it; if you're feeling exhausted, don't try to exercise." (ACE 2001). Because these guidelines predate the 2002 ACOG and 2003 SOGC and CSEP updates, which were measurably different from their previous guidelines, it is possible that future updated ACE recommendations will reflect the newer thinking that daily exercise in pregnancy is not only safe but also healthy.

British guidelines were initially published in January of 2006 and take into account a review of the available evidence. Interestingly, the Royal College of Obstetricians and Gynaecologists (RCOG) statement cites both American and Canadian guidelines. However, it also sets an upper limit of acceptable heart rate, which is in direct divergence with both ACOG and CSEP guidelines, which have not specified an upper limit. The RCOG guidelines suggest that an upper limit of 60% to 70% HRmax is appropriate for women who were sedentary before pregnancy while 60% to 90% of HRmax is acceptable for active women wishing to maintain fitness during pregnancy (see table 6.2).

ACSM guidelines for the pregnant population are included in the widely used *ACSM's Guidelines for Exercise Testing and Prescription*, which was recently updated with the release of the eighth edition. The ACSM guidelines include both objective (40%-60% of $\dot{V}O_2R$) and subjective

(12-14 RPE or the talk test) options for intensity, both of which are considered a moderate activity level. The ACSM guidelines also call for a total accumulation of 150 min/wk and recommend women start with 15 min daily and increase to 30 min daily, using dynamic, rhythmic physical activities working the large-muscle groups.

One paper on appropriate pregnancy exercise prescription (Paisley 2003), written after both the Canadian and the American guidelines were published, has been cited multiple times in the literature, including as part of the RCOG guidelines (see guideline 6.8). The practical approach outlined in this paper uses an exercise prescription adapted from a text on obstetrics; the resulting exercise prescription has been cited as the basis for the ACSM's Health and Fitness Journal paper and has also been quoted in several texts. The prescription offers FITT guidelines for sedentary individuals, recreational exercisers, and elite athletes.

Details on the origin and derivation of the *Dietary Guidelines for Americans* are presented in chapter 2. In these guidelines, a combination of doubly labeled water experiments, epidemiological studies, and expert panel recommendations was used to establish the physical activity recommendations for pregnant women. These are in line with the ACOG guidelines, which had been issued 3 y earlier. For pregnant women without medical or obstetric complications, at least 30 min of moderate-intensity physical activity on most days is recommended. A specific note regarding lactation indicates that acute and regular exercise has no adverse effects on the mother's ability to breast-feed. Lastly, this section includes guidelines from the American Pregnancy Association and KidsHealth, as these are referenced alongside the pregnancy recommendations on the ACSM Web site.

TABLE 6.2 Heart Rate Target Zones for Aerobic Exercise in Pregnancy

Maternal age (y)	Heart rate target zone (beats/min)
<20	140-155
20-29	135-150
30-39	130-145
>40	125-140

Reproduced from RCOG Statement No. 4. Exercise in Pregnancy, 2006, with the permission of the Royal College of Obstetricians and Gynaecologists.

GUIDELINE 6.2

Title: "Exercise During Pregnancy and the Postpartum Period"

Organization: ACOG

Year published: 2002

Purpose: To provide exercise guidelines for pregnancy and postpartum

Location: *Obstetrics and Gynecology* 99: 171-173

Population: Women who are pregnant

GUIDELINES

- In the absence of either medical or obstetric complications, pregnant women should perform 30 min or more of moderate exercise on most, if not all, days of the week.
- Previously inactive women and women with medical or obstetric complications should be medically evaluated before beginning an exercise program.
- Physically active women with a history of or a risk for preterm labor or fetal growth restriction should be advised to reduce their activity in the second and third trimesters.

Note that there is no recommendation to monitor or restrict HRmax during exercise.

GUIDELINE 6.3

Title: "Exercise in Pregnancy and the Postpartum Period"

Organization: SOGC and CSEP

Year published: 2003

Purpose: To improve health-related outcomes during pregnancy

Location: *Journal of Obstetrics and Gynaecology Canada* 129: 1-7, 2003

Population: Women who are pregnant

GUIDELINES

- All women without contraindications should be encouraged to participate in aerobic and strength conditioning exercise as part of a healthy lifestyle during their pregnancy.
- Reasonable goals of aerobic conditioning in pregnancy should be to maintain a good fitness level throughout pregnancy without trying to reach peak fitness or train for an athletic competition.
- Women should choose activities that minimize the risk of loss of balance and fetal trauma.
- Adverse pregnancy or neonatal outcomes are not increased by exercise.
- Initiation of pelvic floor exercises in the immediate postpartum period may reduce the risk of future urinary incontinence.
- Moderate exercise during lactation does not affect the quantity or composition of breast milk or affect infant growth.

Adapted, with permission, free of charge courtesy of the Society of Obstetricians and Gynecologists of Canada. Reprinted, by permission, from G. Davies et al., 2003, "Joint SOGC/CSEP clinical practice guideline exercise in pregnancy and the postpartum period," *Journal of Obstetrics and Gynaecology Canada* 129: 1-7.

GUIDELINE 6.4

Title: *Exercise and Pregnancy*

Organization: ACE

Year published: 2001

Purpose: General physical and emotional benefits

Location: www.acefitness.org/fitfacts/fitfacts_display.aspx?itemid=2597

Population: Women during pregnancy

GUIDELINES

- A frequency of 3 sessions per week or exercising every other day is recommended.
- Low-impact aerobic activity such as walking, swimming, or other water exercise is recommended; the exercise session should include a gradual warm-up as well as stretching before and after the aerobic conditioning component.

Adapted, by permission from American Council on Exercise Fit Facts, San Diego 2001. www.acefitness.org/fitfacts. Accessed 10/10/2009.

GUIDELINE 6.5

Title: *Exercise in Pregnancy*

Organization: RCOG

Year published: 2006

Purpose: To derive health benefits associated with exercise during pregnancy

Location: www.rcog.org.uk/womens-health/clinical-guidance/exercise-pregnancy

Population: Pregnant and postpartum women

GUIDELINES

Frequency	Intensity	Duration
Start with 3 times a week, advance to 4-7 times a week	60%-70% HRmax for sedentary women and 60%-90% HRmax for fit women; 12-14 on RPE scale of 20; can also use the talk test	Start with 15 min, advance to 30 min

Other considerations include the following:

- All women should be encouraged to participate in aerobic and strength conditioning exercise as part of a healthy lifestyle during pregnancy.
- Women should choose activities that minimize the risk of loss of balance and fetal trauma.
- Women may initiate pelvic floor exercises in the immediate postpartum period to reduce the risk of future urinary incontinence. This exercise may begin immediately after an uncomplicated pregnancy and delivery.
- Women should maintain adequate hydration and avoid exercising in very hot, humid environments.
- Women should consume adequate calories and limit exercise sessions to less than 45 min.
- Exercise in the supine position should be avoided after 16 wk of gestation.
- Women should avoid overexertion in altitudes greater than 8,202 ft (2,500 m) until after 4 to 5 d of exposure.

- Women should exercise caution in engaging in levels of fitness activities higher than those accustomed to before pregnancy.
- Women should avoid horseback riding, downhill skiing, ice hockey, gymnastics, cycling, and scuba diving.
- Water exercise is recommended, but water temperature should not exceed 90 °F (32 °C).

Reproduced from RCOG Statement No. 4 Exercise in Pregnancy, 2006, with the permission of the Royal College of Obstetricians and Gynaecologists.

GUIDELINE 6.6

Title: "Exercise Prescription for Healthy Populations: Pregnancy"

Organization: ACSM

Year published: 2010

Purpose: To reduce the risk of developing pregnancy-induced hypertension and gestational diabetes and to provide health and fitness benefits to the mother and child

Location: *ACSM's Guidelines for Exercise Testing and Prescription, Eighth Edition*

Population: Pregnant women

GUIDELINES

Component	Frequency	Intensity	Duration	Type
Aerobic conditioning	At least 3 but preferably all days of the week	Moderate (40%-60% $\dot{V}O_2R$), talk test, or RPE of 12-14 on scale of 20	At least 15 min/d, increasing to 30 min/d; total accumulation of 150 min/wk	Dynamic, rhythmic physical activities using the large-muscle groups
Resistance exercise	Not specified	A resistance that permits multiple repetitions to a point of moderate fatigue	12-15 repetitions	Activities working all the major muscle groups; avoid isometric actions and Valsalva maneuver

Other considerations include the following:

- Women with obesity, gestational diabetes, or hypertension should consult their doctor before starting an exercise program.
- Avoid exercise in the supine position after the first trimester, as such exercise may facilitate orthostatic hypotension.
- Drink ample water to prevent dehydration.
- Avoid heat stress.
- Avoid sport activities involving a risk of abdominal trauma, falls, and excessive joint stress, including soccer, basketball, ice hockey, and racket sports.
- Increase caloric intake to meet demands of both pregnancy and exercise.
- Postpartum exercise may begin 4 to 6 wk after delivery.

Title: *ASCM Current Comment: Exercise During Pregnancy*

Organization: ACSM

Year published: 2000

Purpose: To improve maternal fitness, restrict weight gain without compromising fetal growth, and hasten postpartum recovery

Location: www.acsm.org/AM/Template.cfm?Section=current_comments1&Template=/CM/Content-Display.cfm&ContentID=8638

Population: Women during pregnancy

GUIDELINES

Frequency	Intensity	Duration	Type
Not specified	Not to exceed prepregnancy levels; moderate to hard is safe	Check with physician	Weight-bearing or non-weight-bearing activity; avoid supine or prone positions after first trimester

Special considerations include the following:

- Heavy lifting and activities that require straining should be avoided.
- Bicycle riding should be avoided.
- Extremes of air pressure, such as those caused by scuba diving and high-altitude exercise by unacclimatized women, should be avoided.
- Ensure adequate fluid intake before, during, and after exercise.
- Wear loose-fitting clothing and avoid high heat and humidity to protect against heat stress, especially during the first trimester.
- Modify exercise programs that pose a significant risk of abdominal injury or fatigue.

GUIDELINE 6.8

Title: "Exercise During Pregnancy: A Practical Approach"

Organization: None

Year published: 2003

Purpose: To derive health benefits during and after pregnancy

Location: *Current Sports Medicine Reports* 2: 325-330

Population: Women during pregnancy

GUIDELINES

Level of fitness prepregnancy	Frequency	Intensity	Duration	Type
Sedentary individual	At least 3 times per week	65%-75% HRmax or RPE of moderately hard	30 min	Low-impact activities such as walking, bicycling, swimming, aerobics, and water aerobics

Level of fitness prepregnancy	Frequency	Intensity	Duration	Type
Recreational athlete	3-5 times per week	65%-80% HRmax or RPE of moderately hard to hard	30-60 min	Activities listed for sedentary individuals, plus running, jogging, tennis, and cross-country skiing
Elite athlete	At least 4-6 times per week	75%-80% HRmax or RPE of hard	60-90 min	Activities listed for recreational athletes, plus some competitive activities, depending on gestational age

Special considerations include the following:

- Avoid heat stress and maintain proper hydration.
- Nutritional and caloric requirements for exercise will exceed those that are already increased in pregnancy.
- Avoid high-altitude activities, scuba diving, and activities with a risk for abdominal trauma or falling.

GUIDELINE 6.9

Title: *Dietary Guidelines for Americans*

Organization: USDHHS (jointly with the USDA)

Year published: 2005

Purpose: To promote health, psychological well-being, and a healthy body weight

Location: www.health.gov/dietaryguidelines/dga2005/report/

Population: Pregnant and breast-feeding women

GUIDELINES

- Incorporate 30 min or more of moderate-intensity physical activity on most, if not all, days of the week. Avoid activities with a high risk of falling or abdominal trauma.
- Be aware that neither acute nor regular exercise adversely affects the mother's ability to breast-feed.

Benefits and Risks of Physical Activity During the Postpartum Period

Although the ACSM, ACOG, and SOGC and CSEP guidelines all address not only the time during pregnancy but also the postpartum period, specific exercise guidelines for the postpartum period are virtually nonexistent. In 2002, guidelines for exercise in the postpartum period were published, and a suggestion was made to extend the definition of the postpartum period to 1 y (Mottola 2002). The benefits of a postpartum physical activity program are manifold. It improves weight loss, increases energy and mood, decreases potential urinary stress incontinence, and has other effects that include the following:

- More relaxed relationship between mother and child
- Less depression and anxiety
- Improved perception of new relationship between mother and child
- Prevention of diastasis recti abdominus
- Improved bladder control
- Less pregnancy-induced weight retention
- Less lactation-induced loss of bone mass

GUIDELINE 6.10

Title: *Exercise Guidelines During Pregnancy*

Organization: American Pregnancy Association

Year published: 2008

Purpose: Not given

Location: www.americanpregnancy.org/pregnancyhealth/exerciseguidelines.html

Population: Women during pregnancy

GUIDELINES

- Start slowly and do not overexercise to the point of exhaustion or breathlessness.
- Wear comfortable exercise footwear that provides strong ankle and arch support.
- Take frequent breaks and drink plenty of fluids during exercise.
- Avoid exercising in extremely hot weather.
- Avoid rocky terrain or unstable ground when running or cycling.
- Avoid contact sports.
- During weight training, focus on improving tone, especially in the upper body and abdominal area.
- Avoid lifting weights above the head and using weights that strain the low back muscles.
- During the second and third trimesters, avoid exercise that involves being flat on the back.
- Include relaxation and stretching before and after exercise.
- Eat a healthy diet that includes plenty of fruits, vegetables, and complex carbohydrate.

Adapted, with permission, from the American Pregnancy Association. www.americanpregnancy.org/pregnancyhealth/exerciseguidelines.html. Accessed 10/10/2009.

GUIDELINE 6.11

Title: *Exercising During Pregnancy*

Organization: KidsHealth

Year published: 2007

Purpose: To feel better, look better, prepare the body for birth, and regain prepregnancy body more quickly

Location: http://kidshealth.org/parent/pregnancy_newborn/pregnancy/exercising_pregnancy.html

Population: Women during and after pregnancy

GUIDELINES

- Start gradually, adding 5 min/d each week until reaching 30 min.
- Dress comfortably in loose-fitting clothes and wear a supportive bra.
- Drink plenty of water to avoid overheating and dehydration.
- Opt for a walk in an air-conditioned mall on hot, humid days.
- Listen to your body and skip exercise when sick.

From a public health standpoint, the connection between increased exercise and enhanced weight loss in the first year after delivery is noteworthy in light of the steady increases in obesity among all Americans. Women who are diagnosed with gestational diabetes are at an increased risk for future development of type 2 diabetes; postpartum physical activity and the establishment of a consistent exercise program attenuate this risk. Thus it might be assumed that the discovery of gestational diabetes during pregnancy may be a stimulus for participating in physical activity. Unfortunately, a recent study found that only 34% of women diagnosed with gestational diabetes participated in adequate physical activity postpartum; an additional 27% were classified as sedentary (Smith and others 2005).

Women who work on pelvic floor and abdominal strength experience less urinary stress incontinence and diastasis recti than do women who do not. In fact, women who engage in pelvic floor and abdominal exercises during pregnancy are less apt to develop postpartum structural issues (Sampselle and others 1998; Morkved and Bo 1996; Boissoneault and Blashack 1998). Interestingly, no specific type of physical activity is more effective than another in preventing the separation of the abdominal musculature (Scott 2006).

Hormonal changes associated with pregnancy—including the hormone relaxin—cause some ligamentous and core instability that may persist into the postpartum period and may extend for as long as the new mother is nursing. Because of this, postpartum women may need to avoid specific exercises, including lunges, thigh adduction and abduction, and squats. These exercises should be avoided for at least the first 6 wk after delivery—until the new mother is cleared by the obstetrician—and maybe longer depending on the core and pelvic strength as well as nursing status of the mother.

Safety

The safety of postpartum physical activity has been demonstrated. Areas of concern have included not only the potential risk of injury to mother but also the compatibility of exercise with lactation and infant growth. Depending on the type of delivery, the safe time to return to an exercise regimen varies. Women with an uncomplicated pregnancy may resume exercise immediately, while a complicated vaginal delivery or a surgical delivery necessitates waiting for surgical wounds or pelvic floor muscles to strengthen. Women with specific complications during pregnancy, such as the following, should obtain medical clearance for postpartum exercise:

- Preeclampsia or toxemia of pregnancy
- Extensive vaginal or rectal repair after delivery
- Severe pregnancy-induced or essential hypertension
- Pregnancy-related liver disease
- Pregnancy-induced renal failure
- Uncontrolled seizure disorder
- Hemodynamically significant heart disease
- Preexisting medical conditions that restrict exercise capacity or ability
- Breast-feeding

In addition to Paula Radcliffe, there was an elite marathoner who continued to train for the Olympic trials throughout pregnancy and completed the race several months after delivery. This athlete restarted marathon training 8 d after delivery and won a marathon at 6 mo postpartum!

Some women are afraid to resume prepregnancy physical activity, fearing that it will adversely affect breast-feeding. In fact, mild-to moderate-intensity exercise during the postpartum period does not affect milk volume or cause accumulation of lactic acid—nor does it cause any change in infant growth. Breast-feeding women who exercise demonstrate improved $\dot{V}O_2$max, body composition, lipid profile, and insulin response (Lovelady and others 1990; Dewey and others 1994). The composition of breast milk is not significantly altered with respect to vitamin B_6, long-chain polyunsaturated fatty acids, and lactic acid during moderate-intensity exercise (Lovelady and others 2001; Wright and others 2002; Bopp and others 2005), although lactic acid levels may be higher after a maximum exercise test. There is some evidence that IgA levels in breast milk are decreased for 1 h after maximal aerobic exercise (Gregory and others 1997); this effect can be attenuated by avoiding nursing in the first 60 min after a strenuous bout of exercise. Another study did not find any difference in IgA levels between resting women and women who exercised at moderate intensity (Lovelady and others 2003). Guidelines for exercise and breast-feeding include the following:

- Exercise after the baby is fed or the breasts are empty.
- Choose mild to moderate exercise.
- Avoid strenuous and exhaustive exercise.
- Ensure adequate hydration before, during, and after activity.
- Ensure adequate caloric intake that supports both exercise and lactation.
- Take calcium and vitamin B_6 supplements if daily intake is not adequate.
- Wear a good support bra during exercise to support the breasts; avoid sports bras because of breast compression.

Postpartum Physical Activity Guidelines

The 2002 ACOG statement suggests that prepregnancy exercise routines may be resumed slowly in the postpartum period. Specific exercise guidelines are absent, however. More specific recommendations are presented in Mottola's 2002 paper. An emphasis is made on returning to exercise gradually in terms of frequency, intensity, and duration. A woman may be cleared by a physician for return to a moderate aerobic exercise program (see the previous section for specific situations requiring medical clearance). Vaginal bleeding should be absent or minimal. A program can start slowly, at 3 times a week, and gradually increase to 5 times a week as tolerated based on fatigue.

Intensity may be 70% to 85% of estimated HRmax, depending on the baseline fitness of the woman—women are who less fit or breast-feeding should start toward the lower end while women who are more fit may begin toward the upper end. The duration of aerobic activity also depends on fitness and can be advanced as tolerated. A low-intensity warm-up and cool-down, lasting about 5 min each, should surround the higher-intensity exercise. Strength training can be started slowly and advanced as tolerated; it should focus on functional exercises for the new mother. Improving abdominal strength is a goal, although torque exercises should be avoided to avert injury. Babies can be included in the muscular-strengthening exercises, either by being placed on the mother's body or nearby. In fact, mothers can incorporate their infants in aerobic activity as well by pushing them in strollers or setting them in a seat or crib nearby while using stationary exercise equipment. Kettles and others (Kettles 2006) have published an excellent book with lots of ideas and pictures for exercises during the postpartum period.

The ACE has published a *Fit Facts* handout on postpartum health and exercise. Although specific guidelines for a personal exercise prescription are not described, other suggestions include short, slow walks and Kegel exercises the first few weeks. ACE recommends returning to prepregnancy aerobic activities at roughly 6 wk postpartum. In addition, women after 6 wk are encouraged to add low back exercises, range-of-motion movements for the chest and back, pelvic tilts, and abdominal compression exercises.

A primary goal for many in the postpartum period is to lose weight and return to prepregnancy weight or even achieve a weight lower than prepregnancy weight. The most efficient way to lose weight is to combine caloric restriction with exercise. Additionally, because breast-feeding demands an added 500 kcal/d, it may help to create a negative energy balance. As long as the combined energy balance creates a weight loss of 2 lb/wk (1 kg/wk) or less, there should be no negative effect on breast milk, in either quantity or nutrition. Women who exercise after delivery are more successful in losing the weight gained in pregnancy. Once the desired weight is achieved, women who want to continue breast-feeding should take in an extra 500 kcal/d in order to have enough energy to continue to exercise.

GUIDELINE 6.12

Title: "Exercise in the Postpartum Period: Practical Applications"

Organization: None

Year published: 2002

Purpose: To improve cardiorespiratory fitness, facilitate weight loss, improve energy, decrease anxiety and depression, and increase positive mood

Location: *Current Sports Medicine Reports* 1: 362-368

Population: Women during the first year after delivery

GUIDELINES

Component	Frequency	Intensity	Duration	Examples
Aerobic exercise	At least 3 times per week, up to 5 times per week	Moderate; talk test or 70%-85% HRmax	15 min, advance by 5 min/wk	Activities involving the baby, such as walking or pushing a stroller
Muscular conditioning	Not specified	Not specified	Not specified	Exercises incorporating the baby as a resistance tool; pelvic floor exercises
Warm-up and cool-down	Not specified	Not specified	5-10 min of each	Not specified

Special considerations include the following:

- Stretching, pelvic floor exercises, and relaxation and breathing exercises are safe to perform immediately after delivery.
- Abdominal exercises should be started slowly, and torque exercise should be avoided.
- Beginning a moderate-intensity aerobic exercise program should be delayed until after the first postpartum checkup.

SUMMARY

Because of the compelling benefits of physical activity and the decreasing activity levels observed among youths and adults alike, finding ways to reverse the trend toward inactivity are important. Encouraging pregnant women to start and maintain a regular exercise program is one way of helping not just the woman and her unborn child but also the baby's father, siblings, or other nearby individuals. It is critical to suggest to women to perform physical activity not only during pregnancy and the postpartum period but also during the remainder of their lives. Focusing public health and community efforts on the promotion of physical activity for pregnant women is likely to have far-reaching effects; those who can establish healthy activity habits will lead healthier lives, have more active children, and hopefully serve as role models for their contemporaries.

Older Adults

The U.S. population continues to grow, and the percentage of older Americans is increasing as well. Currently, approximately 12% of the American population is aged 65 y or older. By 2030 this age group is projected to increase to 20% due to the baby boomers, who make up a significant portion of this group. In actual numbers, the proportion of the population who are aged 65 or older is expected to nearly double by the year 2030, increasing from 37 million in 2005 to more than 70 million in 2030 (IOM 2008). In fact, the number of people who are more than 85 y old is estimated to grow from 5 million today to more than 20 million by 2030.

The aging of America has significant economic implications, as the cost of providing health care for an older American is 3 to 5 times greater than it is for people younger than 65 y (CDC 2007). Due solely to the demographic shifts, health care spending is projected to increase by 25% by 2030 (CDC 2007). Anything that reduces the disease burden among older adults is likely to reduce health care costs, and the 1996 *Physical Activity and Health: A Report of the Surgeon General* concluded that regular physical activity greatly reduces the risk for CVD and is effective even if started later in life. This finding is particularly important, as the risks for CVD increase independently with age; even incremental improvements in blood pressure, cholesterol, or glycemic control significantly reduce mortality.

This chapter discusses the benefits, risks, and obstacles of exercise in the older population. The benefits extend into the seventh, eighth, and ninth decades of life and include metabolic, cardiovascular, musculoskeletal, psychological, and neurological gains. Risks may include injury or cardiac event, although these are lessened by moderation in intensity and volume, with adjustments as needed for any underlying conditions. This chapter covers several national and international guidelines; in light of the ever-increasing older population in the United States, it is important to not only be aware of these recommendations but also support and promote them.

Benefits of Physical Activity for Older Adults

As noted in earlier chapters, participation in regular physical activity declines with age, starting even before adolescence. Data from *The State of Aging and Health in America*, published in 2007, showed steady age-related increases in the number of inactive individuals. While only 19% of Americans aged 18 to 29 y get no leisure-time physical activity, more than one-third of all Americans aged 75 y and older do not get any such activity, despite the myriad of health benefits associated with even a minimal amount of exercise.

The American Association of Retired Persons (AARP) issued the results of a 2006 survey on physical activity (Keenan 2006); the findings demonstrated that inactivity is a major problem across all age groups, although the reasons given for exercising (or not) varied. More than 1,000 individuals 18 y and older were included in the survey. The most commonly reported reasons for not engaging in regular activity included lack of energy, lack of time, and poor weather; other common reasons were fear of injury, self-consciousness, and lack of access to appropriate exercise venues. Older individuals cited lack of energy as a primary reason for not exercising, while younger individuals cited lack of time. The reasons listed for exercising included benefits to self-esteem, overall health, stress reduction, weight control, fighting aging, and socializing. For individuals of all ages, walking is the most popular form of exercise (see figure 7.1). In fact, the percentage of people who listed walking as their favorite type of exercise was 36%, more than the next three favorites combined! As people age, walking becomes even more popular. Walking is the favorite activity for 22% of people aged 18 to 34 y, 37% of people aged 35 to 44 y, 45% of people aged 45 to 54 y, 55% of people aged 55 to 64 y, and 43% of people aged 65 y and older.

As with younger adults, older adults who participate in a regular exercise program improve cardiovascular function, enhance endurance and muscle strength, improve functional capacity, and reduce risk factors for heart disease, diabetes, and osteoporosis. Exercise has also been shown to decrease all-cause mortality and improve mood. Because older individuals are at higher risk for CVD and many also possess individual risk factors for CAD, they should obtain medical clearance before initiating an exercise program. This is addressed in more detail in chapter 16.

Aerobic Benefits

$\dot{V}O_2$max generally declines with age, starting at 25 to 30 y. From that point, age-related reductions in cardiac output, stroke volume, left ventricular contractility, muscle oxidative capacity, and vascular capacity contribute to a 5% to 15% drop in $\dot{V}O_2$max per decade (Adams 2008). Older adults who take part in regular aerobic endurance training can expect to achieve the same 10% to 30% increase in $\dot{V}O_2$max that young adults achieve. As with younger adults, the improvement depends on the intensity of training—higher intensity leads to a higher magnitude of improvement.

HRmax also decreases at the rate of approximately 1 beat/min each year. Due to the wide variability of HRmax, an individual's HRmax is more dependent on what it has been in the past than it is on HRmax among age-matched peers. Other physiological alterations that accompany age include an attenuated rise in ejection fracture and a more rapid increase in the systolic hypertensive response to exercise (Fletcher and others 2001).

Strength Benefits

Loss of muscle mass and strength also occurs with age. This loss is attributed to reduction in growth factors, changes in muscle protein turnover, and neuromuscular realignment. Both the size and number of muscle fibers decrease, contributing to a gradual loss in muscle cross-sectional area over time. By age 50, 10% of maximal muscle area is gone. Further muscle mass loss accelerates after age 50; muscle strength diminishes 15% per decade in the 60s and 70s and an additional 30% per decade in the 80s and beyond. Flexibility, balance, and bone density decrease as well. The latter issue is addressed in chapter 11. A resis-

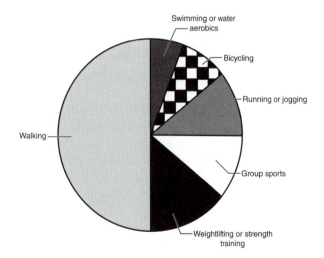

Figure 7.1 Percentage of individuals favoring a certain type of exercise.

Adapted from AARP Physical Activity Survey, 2006.

tance training program can result in significant strength gains. The training stimulus enhances both muscle size and muscle fiber recruitment, resulting in improved insulin sensitivity, bone density, energy metabolism, and functional status. Individuals of all ages who have increased muscle strength participate more in spontaneous activity, and thus improved muscular strength sets the stage for activity-related benefits independent of muscle-related gains.

Older adults who participate in resistance training experience significant gains in strength. There does not appear to be a point at which strength gains plateau, at least not in the first year of training (Morganti and others 1995; Fiatarone and others 1990; Frontera and others 1988). Both resistance training and aerobic exercise improve measures of functional status in older adults; one study found that resistance exercise is more cost efficient than aerobic exercise in improving physical function, although the magnitude of difference is not large (Sevick and others 2000).

Stability and Flexibility Benefits

With age comes physiological change affecting mobility, postural stability, muscle soreness, stiffness, and susceptibility to injury. Physiological changes including the following (Klein 2003; Mazzeo 1998):

- Increased crystallinity of collagen fibers with increased fiber diameter
- Increased calcification, fraying, or cracking in cartilage and ligaments
- Erosion of cartilage in heavily used joints, particularly joints of the knees and hands
- Decreased elasticity in joint capsules, tendons, and ligaments, with the development of cross-linkages between adjacent fibrils of collagen
- Increased dehydration and loss of joint lubricants in connective tissue
- Changes in the chemical structure of the tissues
- Alterations in sensory and motor systems, reducing ability to respond appropriately
- Alterations in processing of information from basal ganglia and cerebellum

- Diminished feedback to postural control centers via vestibular, visual, and somatosensory systems

In particular, fall risk is often increased for older adults. Individual risk for falling depends on many factors, including frailty, visual acuity, balance, arthritis, and underlying neurological status. Evidence suggests that regular physical activity not only is safe for individuals who are susceptible to falls but also may reduce fall rate by about 30% (PAGAC 2008). For individuals not at risk for falling, regular activity does not seem to affect fall rate.

The increasing amount of physical inactivity in the older age group contributes to loss of flexibility over time. Regular flexibility and stretching sessions can stimulate production and retention of connective tissue lubricants, which in turn can help to reduce losses in flexibility. The flexibility component of an activity program should be balanced with the strength training component in order to prevent connective tissues from becoming too loose and weak or being overstretched by muscular contractions. It is recommended that flexibility work be completed after resistance training so that muscles are warmed before stretching.

Neurological Benefits

One of the primary concerns of older adults is preserving memory, cognition, and executive function. Neurocognitive decline occurs with age due to atrophy of cortical gyri, decreased brain vascularization, and reduction in central nervous system neurotransmitters. Regular physical activity reverses some of these processes and positively affects a number of cognitive processes. Additionally, the prevalence of dementia is reduced in people who are active regularly.

In preventing cognitive decline, there is a dose–response relationship between exercise and benefit; that is, higher exercise intensities, more than simply expending additional calories, protect against loss of executive function (Bixby and others 2007). In the Nurses' Health Study, women aged 70 to 81 y were evaluated, over time, on several tests of cognition. Compared with the women who were more physically active (those who performed easy walking or the equivalent for at least 1.5 h/wk), the least active women (those walking less than 40 min/wk) had a 20% higher risk of cognitive impairment. Also, women who were more active were found to experience less cognitive decline (Weuve and others 2004). The protective effect of physical activity with respect to cognition appears to extend well into the 10th decade, implying that there is no end point for pursuing physical activity for older adults.

Major National Guidelines

National guidelines include the recent 2009 update to the 1998 ACSM guidelines for older adults, the 2008 *Physical Activity Guidelines for Americans (PAGA)*, and other ACSM guidelines found in different sources, including the recommendations in the 2010 edition of the *ACSM's Guidelines for Exercise Testing and Prescription* and the 2007 ACSM and AHA guidelines. Because of the growing older population and the increasing evidence that physical activity has so many benefits for health, a common theme of these guidelines is that, in the absence of contraindications, subjectively moderate-intensity exercise should be continued into the later decades of life.

Physical Activity Guidelines for Americans

In the 2008 *PAGA*, the PAGAC issued specific recommendations for the older population. The guidelines are similar to those for younger adults in terms of frequency and duration. In light of the general decrease in physical fitness and exercise capacity that occurs with age, the recommended absolute intensity for older adults is lower than that for younger adults. However, in relative terms, the intensity level should be similar between the two age groups. For example, a moderate-intensity level measured by either the talk test or RPE may correspond to a lower MET level in an older adult than it corresponds to in a younger adult. Both individuals achieve similar health-related and fitness benefits. *PAGA* calls a relative intensity of 5 or 6 out of 10 *moderate* and 7 or 8 *vigorous*.

PAGA contains additional guidelines specifically for older adults, with a focus on maintaining health and reducing the risk of falling and losses in functionality. For example, balance exercises

are recommended, particularly if an individual is already susceptible to falls. Because many older adults have chronic conditions that alter their level of fitness or ability to perform physical activity for a significant amount of time, the guidelines allow for some flexibility in terms of total time or effort level. The guidelines emphasize an overall volume of activity that may be accumulated throughout the week; for those who cannot achieve this amount of activity, a lesser amount is still indicated as likely to be helpful. Safely improving functional ability and being as physically active as ability and condition allow are emphasized.

The 683-page *Physical Activity Guidelines Advisory Committee Report* preceded *PAGA* by 3 mo, and the details contained in this report were used to create *PAGA*. Within this report were several chapters addressing various populations and the effects of physical activity on their health outcomes. Although *PAGA* did not give a specific exercise prescription for older adults who have a significant risk for falling, the PAGAC report noted that both balance and muscle-strengthening training are beneficial and should be performed for 30 min 3 times a week. This is in addition to at least 2 weekly walking sessions. Although this prescription for fall prevention did not make it into *PAGA*, it deserves discussion here due to the significance of fall risk among older adults.

American College of Sports Medicine Position Stand

In 1998, the ACSM released a position stand with guidelines specifically addressing older adults. This was a milestone report in that, by issuing a separate report dedicated to older individuals, the ACSM sent an implicit message highlighting the importance of exercise for this population. The older adult population was experiencing considerable growth at the time, and older adults now constitute the fastest-growing segment of the U.S. population. The ACSM position stand on exercise and physical activity for older adults was updated in 2007 as a joint statement with the AHA and again in 2009 as a stand-alone statement. The most recent ACSM recommendations agree with and support those referenced in *PAGA* and expand on the issues and data related

---

GUIDELINE 7.1

Title: *Physical Activity Guidelines for Americans*

Organization: USDHHS

Year published: 2008

Purpose: To lower the risk for all-cause mortality, CHD, stroke, hypertension, and type 2 diabetes

Location: www.health.gov/paguidelines

Population: Older adults

GUIDELINES

- Adults should get at least 150 min of moderate-intensity aerobic activity or 75 min of vigorous-intensity aerobic activity, or an equivalent combination of moderate-intensity and vigorous-intensity activity, per week.
- Aerobic activity should be performed in episodes lasting at least 10 min and should be spread throughout the week.
- For additional and more extensive health benefits, adults should increase to 300 min/wk (5 h/wk) of moderate-intensity aerobic activity or 150 min/wk of vigorous-intensity aerobic activity or an equivalent combination of moderate-intensity and vigorous-intensity activity. Additional health benefits are gained by engaging in physical activity beyond even these amounts.
- Adults should also perform muscle-strengthening activities that are moderate or high in intensity and that involve all the major muscle groups, as these activities provide additional health benefits. Muscle-strengthening exercises should be performed 2 d/wk or more; 1 to 3 sets per exercise and 8 to 12 repetitions per set are effective.
- When older adults cannot do 150 min/wk of moderate-intensity aerobic activity because of chronic conditions, they should be as physically active as their abilities and conditions allow.
- Older adults at risk of falling should engage in exercises that maintain or improve balance.
- Older adults should determine their level of effort for physical activity relative to their level of fitness.
- Activity should be performed on at least 3 d/wk in order to reduce the risk of injury and avoid excessive fatigue.

---

to the benefits of exercise and physical activity in older adults. There are, however, a few differences between recommendations endorsed by the ACSM and recommendations endorsed by the USDHHS (*PAGA*), and these reflect the differing goals between the two organizations. While *PAGA* strives to do the most good for the most people and hopes to reduce age-related functional decline, loss of bone density, increased CAD risk (and increased CAD), sarcopenia, depression, and frailty, the ACSM guidelines also aim to not only maintain but also improve cardiovascular function. A higher level of activity—whether in frequency, intensity, or duration—is expected to lead to a higher degree of benefit. The dose–response relationship between activity and health remains valid in the older adult age group!

The 1998 ACSM position stand also included information on exercise for the frail and the very old. Goals for older adults vary from CVD, cancer, and diabetes prevention in younger individuals to reversing or minimizing the biological changes of aging, reversing disuse syndromes, maximizing psychological health, increasing mobility and function, and assisting in the rehabilitation from acute and chronic illnesses in older individuals.

GUIDELINE 7.2

Title: *Physical Activity Guidelines Advisory Committee Report*

Organization: PAGAC

Year published: 2008

Purpose: To reduce fall rate

Location: www.health.gov/paguidelines

Population: Older adults at risk for falls

GUIDELINES

Type	Duration	Frequency
Moderate-intensity walking activities	30 min each session	2-3 times per week
Muscle-strengthening exercises	30 min each session	3 times per week
Balance training	As part of muscle-strengthening program	3 times per week

Additional comments in the 2009 ACSM position stand that are not found in the 2008 *PAGA* guidelines include the following: (1) Exercise need not be of high intensity to reduce risks for chronic disease, although some *established* diseases such as depression, muscle weakness, or type 2 diabetes may benefit from a higher level of intensity. (2) In this age group, acute effects of exercise are short lived and even the long-term adaptations of chronic exercise may be reversed with cessation of exercise; thus regularity and consistency are key. (3) There is significant variability attributable to age, sex, and hormonal status in terms of response to exercise, so outcomes may be quite varied. In summary, the latest ACSM position stand for older adults reiterates the powerful role of exercise in preventing and even reversing physiological decline naturally associated with aging. The ACSM recommendations and their updates are presented here and later in the chapter.

In 2007, the ACSM and AHA issued an update to the 1995 physical activity recommendations for all adults that had been issued by the ACSM along with the CDC. Concurrently issued was the update to the 1998 ACSM position stand on physical activity for older adults. This 2007 update was triggered by the significantly increased number of studies detailing the benefits of various types of exercise for older adults. The update, which was published in both *Circulation* and *Medicine & Science in Sports & Exercise*, reflected the ACSM and AHA guidelines for all adults but included some differences related to addressing the specific needs and medical issues of older adults as well as adults with chronic conditions. The updated guidelines are more detailed, reflecting the new data on the frequency, intensity, duration, and type of exercise necessary to improve various health parameters. Major differences between general recommendations for all adults and recommendations for older adults include the way in which intensity is quantified, the number of repetitions recommended for muscle-strengthening activities, and the suggestion that older adults include both flexibility and balance exercises in their activity programs.

Like the general guidelines for all adults, the guidelines for older adults recommend moderate- and vigorous-intensity aerobic exercise: the minimum recommendation is 5 30 min sessions of moderate-intensity physical activity, 3 20 min sessions of vigorous activity, or some combination thereof. While the general adult recommendations define aerobic intensity in absolute terms, those for older adults use a relative scale. The older adult should exercise at an RPE of 5 or 6 (out of 10) for moderate intensity and an RPE of 7 or 8 for vigorous intensity. These RPE values reflect the ACSM recommendation to exercise at 50% to 85% of $\dot{V}O_2R$, which is relative to any individual's fitness. When resistance training, older adults are recommended to perform 10 to 15 repetitions (as opposed to 8-12) for each set. Although the health

---

GUIDELINE 7.3

Title: "Position Stand: Exercise and Physical Activity for Older Adults"

Organization: ACSM

Year published: 1998 (updated in 2009)

Purpose: To improve health, functional capacity, quality of life, and independence

Location: *Medicine and Science in Sports and Exercise* 30(6): 992-1008 and *Medicine and Science in Sports and Exercise* 41(7): 1510-1530 (update)

Population: Individuals aged 65 y and older

GUIDELINES

- Walking, running, swimming, and cycling should be added to habitual lifestyle.
- Light- to moderate-intensity lifestyle physical activities optimize health, while moderate- or high-intensity exercise may be required to elicit adaptations in the cardiovascular system and in CVD risk factors.
- Range of motion can be increased through a general exercise program that includes exercises such as walking, aerobic dance, and stretching.
- General ACSM guidelines for exercise testing and medical supervision are imperative.

2009 Updates

- A combination of aerobic and resistance training is more effective than either modality alone is, so both should be included in an exercise program, along with flexibility exercises.
- Programs do not need to be high intensity in order to achieve metabolic and cardiovascular benefits.
- Individuals at risk for falling or mobility impairment should also perform exercises that improve balance.

---

benefits of flexibility training have not been well documented, such training is nevertheless recommended for older adults to help them maintain range of motion needed for daily activities and be able to participate in physical activity. Lastly, balance exercises (not balance activity such as dancing) are recommended for older adults, particularly for those at risk of falling. A specific balance exercise prescription is not available because the optimal pattern has not been established.

Additional American College of Sports Medicine Guidelines for Older Adults

In 2003, the ACSM published an edition of the *ACSM Fit Society Page* that focused on exercise and the older adult. In it, the ACSM emphasized obtaining approval from a physician before initiating an exercise program, due to the general increase in both cardiovascular risk factors and CVD in the older population. The ACSM also highlighted the concept of slow progression, or starting with a low frequency, intensity, and duration and gradually working toward higher levels as adaptation occurs. Lastly, the ACSM suggested that older individuals consider working with an exercise specialist, at least initially.

ACSM's Guidelines for Exercise Testing and Prescription was initially published in 1976 and is now in its eighth edition. The guidelines have gradually grown to include different recommendations for different segments of the population. The latest edition included guidelines for elderly adults, although the exact age at which elderly begins was not defined. These guidelines differ somewhat from other guidelines, in that the ACSM has set as one of its primary goals the improvement of fitness and not merely the reduction of chronic disease risk. This explains

GUIDELINE 7.4

Title: "Position Stand: Physical Activity and Public Health in Older Adults"

Organization: ACSM and AHA

Year published: 2007

Purpose: To improve health, functional capacity, quality of life, and independence

Location: *Medicine and Science in Sports and Exercise* 39(8): 1435-1445 and *Circulation* 116: 1094-1105

Population: Individuals aged 65 y and older as well as individuals aged 50 to 64 y who have clinically significant chronic conditions or functional limitations that affect movement ability, fitness, or physical activity

GUIDELINES

Type	Frequency	Intensity	Duration
Aerobic exercise	At least 5 d/wk for moderate intensity or 3 d/wk for vigorous intensity	On a 10-point RPE scale, 5-6 for moderate intensity and 7-8 for vigorous intensity	Accumulate at least 30 min/d of moderate-intensity exercise or 20 min/d of vigorous-intensity exercise
Resistance exercise	At least 2 d/wk (nonconsecutive)	Moderate to high effort for 10-15 repetitions; same intensity levels as those used for aerobic conditioning	8-10 exercises
Flexibility exercise	At least 2 d/wk (preferably on all days that aerobic or muscle-strengthening activity is performed)	Not specified	At least 10 min each session
Balance exercise	3 times per week	Not specified	Not specified

Special considerations include the following:

- Older adults should exceed the minimum recommended amounts of physical activity if they have no conditions that preclude higher amounts of physical activity and if they wish to improve fitness, improve existing disease management, or further reduce risk for premature chronic health conditions related to physical activity.
- Older adults at risk for falling should be particularly sure to include balance exercises.

some of the variation in the physical activity guidelines issued by the ACSM when part of an exercise prescription versus national guidelines issued with and endorsed by the AHA.

Other National Guidelines

The *National Blueprint: Increasing Physical Activity Among Adults Age 50 and Older* was created by the Robert Wood Johnson Foundation in 2001. This was meant to serve as a guide not just for individuals but for health organizations and associations to help plan for older individuals to increase their physical activity. Several organizations including AARP, ACSM, American Geriatrics Society, CDC, and the National Institute on Aging also sponsored the summary. Much of the guide details community and public health initiatives which can encourage and enable the participation of

Title: *Exercise and the Older Adult*

Organization: ACSM

Year published: 2003

Purpose: To reduce the risk of heart disease, decrease the chance of illness or death from all causes, decrease anxiety and depression, and slow the process of aging

Location: www.acsm.org/AM/Template.cfm?Section=ACSM_Fit_Society_Page&CONTENTID= 1263&TEMPLATE=/CM/ContentDisplay.cfm

Population: Older individuals

GUIDELINE

Start and finish each session with stretching, choosing exercises that minimize the stress on the most painful joints.

Component	Frequency	Intensity	Duration	Examples
Endurance exercise	Every day not in strength training	At a level that increases heart rate and breathing but allows talking	30 min, may gradually work up to this level	Walking, swimming, jogging, rowing, and bicycling
Strength training	2-4 d/wk, with at least 48 h between sessions	Low intensity to moderate intensity (65%-75% of 1RM)	1-3 sets of 10-15 repetitions per exercise for a total of 20-45 min each session	Weightlifting, lifting items, or going from sitting to standing; primarily multijoint exercises
Flexibility exercise	Every day as part of warm-up and after strengthening exercises	Start with lower intensity and progress	Hold stretches for 10-30 s for a total of 15-20 min each session	Stretches for legs, back, neck, shoulders, and ankles; tai chi, yoga, Pilates
Balance exercise	Not specified	Not specified	Not specified	Standing on one leg or leg-strengthening exercises

Title: "Exercise Prescription for Healthy Populations: Elderly People"

Organization: ACSM

Year published: 2010

Purpose: To attenuate some of the observed changes in aging and to develop sufficient muscular fitness to enhance an individual's ability to live a physically independent lifestyle

Location: *ACSM's Guidelines for Exercise Testing and Prescription, Eighth Edition*

Population: Elderly individuals

>continued

GUIDELINE 7.6 *>continued*

GUIDELINES

Component	Frequency	Intensity	Duration	Type
Cardiorespiratory fitness	A minimum of 3 d/wk for vigorous intensity or 5 d/wk for moderate intensity (or some combination of both)	RPE of 5-6 out of 10 for moderate intensity; RPE of 7-8 out of 10 for vigorous intensity	Accumulate at least 30 min/d or up to 60 min/d; activity may be accumulated in 10 min bouts throughout the day	Activities causing low orthopedic stress
Resistance training	At least 2 d/wk	Between moderate intensity (RPE of 5-6 out of 10) and vigorous intensity (RPE of 7-8 out of 10)	8-10 exercises, 10-15 repetitions each	Exercises working the major muscle groups; stair-climbing
Flexibility exercise	At least 2 d/wk	Moderate intensity (RPE of 5-6 out of 10)	Not specified	Sustained static stretches for each major muscle group

Special considerations include the following:

- If activity is moderate intensity, it should total 150 to 300 min/wk; if activity is vigorous intensity, it should total 75 to 100 min/wk; an equivalent combination is fine.
- More than the minimum amount of activity may be needed by older adults with chronic conditions for which a higher level of activity is known to be beneficial.
- Initial strength training sessions should be supervised and monitored.

Adapted, by permission, from American College of Sports Medicine, Exercise prescription for healthy populations, in ACSM's Guidelines for Exercise Testing and Prescription (8th Ed.), Baltimore: Lippincott, Williams and Wilkins, 2010, 192-194.

older adults in regular physical activity. Additionally, barriers to physical activity were addressed and strategies to tackle these were described. These included research into medical, social and behavioral aspects; encouraging worksites to serve as community resources; promoting activity through the medical community; coordinating public policy at local, state, and national levels; and taking into account the nature of individuals' normal daily tasks. The specific guidelines for older individuals reflect those of the major organizations which participated in its creation, with a summary below.

Within the 2005 *Dietary Guidelines for Americans* there is a specific mention of physical activity for older adults as it relates to the maintenance of a healthy weight. As with the guidelines for all adults, the recommended activity amount may vary, depending on the goal for weight maintenance or sustained weight loss. One subtle difference for this older population is the sug-

gestion that physical activity not only helps in attaining an appropriate and healthy weight but also reduces the decline of functionality that is so often associated with aging—particularly among older adults who are inactive. Recommendations for older adults are very similar to those for all adults.

The NIH, through the National Institute on Aging (NIA), has also released guidelines for physical activity and exercise in older adults. These were updated and reprinted most recently in 2009. The NIA has created a book for older adults, *Exercise & Physical Activity: Your Everyday Guide From the National Institute on Aging*, that is accessible online at www.nia.nih.gov. The NIA recommends endurance exercise at a minimum of 30 min on most or all days of the week. The NIA also emphasizes, however, building up gradually. Older individuals may start out with as little as 5 min of activity at a low level of effort. Intensity for this age group is based on the Borg

GUIDELINE 7.7

Title: "The Elderly"

Organization: ACSM

Year published: 2009

Purpose: To preserve and enhance skeletal muscle strength and endurance, flexibility, cardiorespiratory fitness, and body composition

Location: *Clinical Exercise Physiology, Second Edition*

Population: Elderly individuals

GUIDELINES

Component	Frequency	Intensity	Duration	Type
Cardiorespiratory fitness	Daily at low to moderate intensity or 3-5 times per week for moderate to high intensity	ADLs at comfortable pace; low to moderate intensity is 40%-70% of $\dot{V}O_2$max or HRR; moderate to high intensity is greater than 70% $\dot{V}O_2$max or HRR	At least 30 min and up to 60 min of continuous activity; intervals may be as short as 8-10 min	Walking, cycling, pool activities, seated aerobics, ADLs, jogging, swimming, rowing, aerobic dance
Resistance training	2-3 times per week	60%-80% of 1RM	8-20 repetitions; 20-30 min each session	Multistation machines, elastic bands, hand weights
Flexibility exercise	Daily following an aerobic or resistance training session	Mild stretch without pain	5-30 min total, with 2 bouts lasting 30 s each on every muscle group	Exercises working all major muscle groups (neck, shoulders, arms, lower back, quadriceps, hamstrings, calves, ankles)

Special considerations include the following:

- Assistance should always be available.
- Individuals may need to start with short bouts and gradually build up to longer bouts. Progress duration and frequency before intensity.

GUIDELINE 7.8

Title: *National Blueprint: Increasing Physical Activity Among Adults Age 50 and Older*

Organization: Robert Wood Johnson Foundation

Year published: 2001

Purpose: To enjoy health and quality of life

Location: www.agingblueprint.org/PDFs/Final_Blueprint_Doc.pdf

Population: Adults 50 y and older

>continued

GUIDELINE 7.8 *>continued*

GUIDELINES

- Build up to at least 30 min of endurance exercise on all or most days.
- Perform strength and endurance exercises for all major muscles at least twice a week.
- Incorporate balance exercises into strength training exercises.
- Stretch regularly before and after exercise.

GUIDELINE 7.9

Title: *Dietary Guidelines for Americans*

Organization: USDHHS (jointly with the USDA)

Year published: 2005

Purpose: To promote health, psychological well-being, and a healthy body weight

Location: www.health.gov/dietaryguidelines/dga2005/report/

Population: Older adults

GUIDELINES

- Participate in regular physical activity to reduce functional decline associated with aging and to achieve the other physical activity benefits identified for all adults.
- Engage in regular physical activity and reduce sedentary activities to promote health, psychological well-being, and a health body weight.
- To reduce the risk of chronic disease in adulthood, engage in at least 30 min of moderate-intensity physical activity, above usual activity, at work or at home on most days of the week.
- Obtain greater health benefits by engaging in physical activity of more vigorous intensity or longer duration.
- To help manage body weight and prevent gradual, unhealthy body weight gain in adulthood, engage in approximately 60 min of moderate- to vigorous-intensity activity on most days of the week while not exceeding caloric intake requirements.
- To sustain weight loss in adulthood, participate in at least 60 to 90 min of daily moderate-intensity physical activity while not exceeding caloric intake requirements. Some people may need to consult with a health care provider before participating in this level of activity.
- Achieve physical fitness by including aerobic exercise for cardiorespiratory conditioning, stretching exercises for flexibility, and resistance exercises or calisthenics for muscle strength and endurance.

scale. The NIA recommends achieving an RPE 13 on the Borg scale; alternatively, older adults may use the talk test. A warm-up and cool-down are recommended, as is a postexercise stretching session. Strength training is recommended twice a week. For older adults, the NIA recommends a higher number of repetitions (8-15 as opposed to 8-12 for younger adults). Also, two sets of repetitions are recommended in order to avoid sarcopenia.

International and State Guidelines

The European office of the WHO has a particular focus on healthy aging. As the world population of elderly individuals continues to grow, there is an emphasis on not only preventing disease or death but also improving disability-free life expectancy. There is much to be gained by begin-

GUIDELINE 7.10

Title: *Exercise & Physical Activity: Your Everyday Guide From the National Institute on Aging*

Organization: NIA

Year published: 2009

Purpose: To improve health and independence

Location: www.nia.nih.gov/HealthInformation/Publications/ExerciseGuide

Population: Older individuals

GUIDELINES

Component	Frequency	Intensity	Duration	Type
Endurance exercise	Most or all days of the week	Moderate: RPE of 13 on Borg scale or talk test	30 min, may gradually work up to this level	Walking, swimming, jogging, raking
Strength training	At least 2 times a week; avoid 2 d in a row on the same muscle groups	Moderately hard: RPE of 15-17 on Borg scale	2 sets of 8-15 repetitions per exercise	Exercises using resistance bands, hand weights, or weight machines on all major muscle groups
Flexibility exercise	After regularly scheduled strength and endurance exercises (or at least 3 times a week)	Start with lower intensity and progress to slightly uncomfortable (stretching should never be to the point of pain)	Hold stretches for 10-30 s, doing each 3-5 times at each session for a total of 20 min	Stretches for hamstrings, calves, ankles, triceps, wrists, quadriceps, hips
Balance training	Not specified	Start by holding onto table or chair, progress to one fingertip and then no hands, and then attempt with eyes closed (if steady)	Not specified	Strengthening exercises, such as plantar flexion, hip flexion, hip extension, knee flexion, and side leg raise

Special considerations include the following:

- Be sure to drink liquids to avoid dehydration, especially in hot weather.
- Dress in layers to avoid hypothermia; avoid too much heat exposure.
- Use safety equipment when appropriate (when bicycling, skiing, or skating).
- When progressing, focus on duration before increasing intensity.

ning a regular exercise program, even later in life.

Functional status is a relatively new concept that has received greater recognition as more and more people grow older and the prevalence of chronic disease increases. Functional status is defined as a person's ability to perform the activities necessary to ensure well-being. Three areas—biological, psychological, and social—are integral to overall function. In 1995, a WHO expert group stated that physical exercise is important in preserving functional status and health status but that inactivity, on the

other hand, is a waste of human resources and an important risk factor for poor health and reduced functional ability. Although a reduced level of activity often occurs with age, this does not necessarily have to be the case. Even small amounts of activity may reverse the disabilities caused by illness and concomitant functional declines.

The WHO goals of a physical activity program for older individuals are to preserve mobility, reduce CVD, prevent falls and fractures from osteoporosis, prevent (or treat) type 2 diabetes, and reduce depression and anxiety. In 2002, the WHO published *Keep Fit for Life. Meeting the Nutritional Needs of Older Persons.* This book suggested that adults 60 y and older should participate in both aerobic and strength training.

In the WHO's *Global Strategy on Diet, Physical Activity and Health*, the amount of physical activity recommended for older adults reflects that suggested by the 2007 ACSM and AHA guidelines. That is, older adults should perform 30 min of moderate-intensity physical activity at least 5 d/wk, 20 min of vigorous-intensity physical activity 3 d/wk, or some equivalent combination thereof. In addition, the WHO recommends strengthening and flexibility exercises for older adults. As with the recommendations for all adults, moderate-intensity exercise is defined as that which requires 3 to 6 METs, while vigorous-intensity exercise requires more than 6 METs.

Many guidelines from other countries endorse similar amounts of activity for older adults. Guidelines from Australia, Canada, and the United Kingdom recommend a minimum of 30 min of daily activity. These recommendations are detailed in guidelines 7.13 through 7.15.

Previous chapters presented guidelines from the CDHS for children, youths, and adults. California has issued guidelines for older adults, and they are included here for comparison to national guidelines as well as for completeness. Like those issued by the ACSM and AHA, these guidelines for older adults reflect those made for all adults, with a few additional recommendations. These include consulting with a physician before starting a physical activity program, gradually increasing activity to reduce injury risk, and emphasizing different modalities of exercise, including activities for balance, mobility, agility, and coordination.

Guidelines for Adults Who Are Frail or at Risk for Falling

The term *frailty* is often used for a segment of older adults who may in general participate in

---

GUIDELINE 7.11

Title: *Keep Fit for Life. Meeting the Nutritional Needs of Older Persons*

Organization: WHO

Year published: 2002

Purpose: To achieve healthy and active aging

Location: www.who.int/nutrition/publications/olderpersons/en/index.html

Population: Adults 60 y and older

GUIDELINES

- Build up to at least 30 min of aerobic exercise on most if not all days.
- Engage in strength training 2 or 3 d/wk, with a day of rest between workouts.
  - At the fitness center, perform 1 set of 8 to 12 repetitions on 12 or more machines.
  - At home, perform 2 or 3 sets of 8 to 12 repetitions each using 6 to 8 different exercises.
  - Increase the weight being lifted when repetitions can be completed in good form with ease.

Adapted by permission, from World Health Organization. Keep Fit for Life. Meeting the Nutritional Needs of Older Persons. Geneva World Health Organization, 2002. www.who.int/nutrition/publications/olderpersons/en/index.html

GUIDELINE 7.12

Title: *Recommended Amount of Physical Activity*

Organization: WHO

Year published: 2008

Purpose: To promote and maintain health

Location: www.who.int/dietphysicalactivity/factsheet_recommendations/en/index.html

Population: Older adults

GUIDELINES

- Older adults should complete 30 min of moderate-intensity physical activity 5 d/wk *or* 20 min of vigorous-intensity physical activity 3 d/wk *or* an equivalent combination of moderate-intensity and vigorous-intensity physical activity *and* 8 to 10 muscle-strengthening physical exercises at least 2 d/wk.
- Alternatively, older adults should complete 20 min of vigorous-intensity physical activity 3 d/wk *or* an equivalent combination of moderate-intensity and vigorous-intensity physical activity *and* 8 to 10 muscle-strengthening exercises (8-12 repetitions) at least 2 d/wk.
- Older adults should perform exercises to maintain flexibility.
- Older adults at risk of falling should perform balance exercises.

While these are similar to the WHO recommendations for all adults, it is noted that there should be due consideration for the intensity and type of physical activity appropriate for older people.

Reproduced, with permission, from the World Health Organization. www.who.int/dietphysicalactivity/factsheet_olderadults/en/index.html

GUIDELINE 7.13

Title: *At Least Five a Week: Evidence on the Impact of Physical Activity and Its Relationship to Health*

Organization: Department of Health, United Kingdom

Year published: 2005

Purpose: For general health benefit

Location: www.dh.gov.uk/dr_consum_dh/groups/dh_digitalassets/@dh/@en/documents/digitalasset/dh_4080981.pdf

Population: Older adults

GUIDELINES

- Adults should accumulate at least 30 min/d of at least moderate-intensity physical activity on at least 5 d/wk.
- Older adults should take particular care to keep moving and retain their mobility through daily activity.
- Specific activities that promote improved strength, coordination, and balance are particularly beneficial for older people.

GUIDELINE 7.14

Title: *National Physical Activity Guidelines for Australians*

Organization: Australian Government Department of Health and Ageing

Year published: 1999

Purpose: For good physical health

Location: http://fulltext.ausport.gov.au/fulltext/1999/feddep/physguide.pdf

Population: Older adults

GUIDELINES

- Think of movement as an opportunity, not an inconvenience.
- Be active every day in as many ways as you can.
- Put together at least 30 min of (moderate-intensity) physical activity each day and build up to 30 min over a length of time that suits you.
- If you can, also enjoy some regular, vigorous exercise for extra health and fitness.

GUIDELINE 7.15

Title: *Canada's Physical Activity Guide to Healthy Active Living for Older Adults*

Organization: Public Health Agency of Canada and CSEP

Year published: 2002

Purpose: To improve health, prevent disease, and get the most out of life

Location: www.phac-aspc.gc.ca/pau-uap/paguide/older/phys_guide.html

Population: Older adults

GUIDELINES

- Accumulate 30 to 60 min of moderate physical activity most days.
- Sessions should be at least 10 min long.
- Include a variety of activities from endurance, flexibility, strength, and balance exercises.

very little physical activity. The 1998 ACSM position statement included a section of guidelines for frail adults. In 2003, the second edition of a text on exercise programming for individuals with chronic disease and disability was published; this also included a specific section for frailty. Although older adults certainly are not considered to have chronic disease or disability by virtue of their chronological age, there are certainly older adults who fit into one or more classifications for chronic disease or disability. One such category is that labeled as *frailty*. The definition of frailty includes having one or more of the following features: extreme old age, some type of disability, and the presence of multiple chronic diseases or geriatric syndromes (Bayles 2003). Some of the more common medical disorders contributing to frailty are listed in table 7.1. Because frail adults are at higher risk for acute illnesses, falling, and other injuries in addition to being more dependent in general, they may shy away from physical activity, assuming that they are beyond the age of benefits. On the contrary, small gains in functional ability have the potential for significant effect on maintaining and promoting independence.

GUIDELINE 7.16

Title: *The California Center for Physical Activity's Guidelines for Physical Activity Across the Lifespan: Older Adult (≥60 Years)*

Organization: California Center for Physical Activity

Year published: 2002

Purpose: For general health benefits as well as cardiorespiratory fitness, muscular strength and endurance, skeletal and flexibility benefits, and improved body composition

Location: www.caphysicalactivity.org/facts_recomm1d.html

Population: Older adults (60 y and older)

GUIDELINES

- All adults should meet the recommendations listed in guideline 2.18.
- Exercises improving balance, agility, mobility, coordination, and reaction time should be performed by persons experiencing a diminished capacity in these areas of function.
- Physical activity level should be increased more gradually in older adults to decrease the risk for soreness, discomfort, and injury. Older adults who have been sedentary should start with physical activity sessions of short duration and light intensity.
- Older adults who have existing medical conditions or who are unsure about their safety during physical activity should first consult their physician before embarking on a physical activity program.

From California Center for Physical Activity: Guidelines for Physical Activity Across the Lifespan: Older Adult (≥60 years), Sacramento, CA [Internet] [cited 2009 October 8] www.caphysicalactivity.org/facts_recomm1d.html

For frail individuals, reversing some of the metabolic, physiological, and physical changes of aging are among the goals of a physical activity program. In addition to the usual program components of aerobic exercise, resistance training, and flexibility, the areas of coordination, balance, neuromuscular performance, and functional exercise are addressed. Many individuals participate in walking; older adults who have trouble walking due to joint degeneration or injury may bicycle, swim, and participate in chair activities. Strength training may start with low weights, such as those that can be strapped onto ankles and wrists, with a goal of gradually improving overall strength and decreasing fall risk. Other essential components include flexibility and neuromuscular training to increase general coordination and eye–hand coordination, gait, and balance.

With increasing age, there is an increasing emphasis on balance and flexibility, due to the higher risk for falling. In fact, it is estimated that more than 10 million falls occur each year among Americans older than 35 y (Rogers 2006), and falls are the leading cause of injury-related deaths for older adults. Fall risk depends on many variables, including visual, vestibular, and somatosensory strengths as well as muscular strength. By improving balance and muscle strength, older adults can prevent falls. The National Center on Physical Activity and Disability (NCPAD) developed First Steps to Active Health, a standardized program for health care professionals to use to promote physical activity among older adults. This program includes balance training and flexibility training as part of a well-rounded physical activity program (see Guideline 7.19 for examples of exercises). The program is evidence based and employs the principles of specificity and progression to improve balance and flexibility and ultimately reduce not only falls but also morbidity or mortality from falls. Although specifics regarding cardiorespiratory and resistance training are not given, the program does state that these components should be part of a physical activity program and may be individually prescribed by a health care provider, depending on the individual needs.

TABLE 7.1 Medical Disorders Contributing to Frailty in Older Adults

System	Medical disorder
Cardiovascular	Hypertension, hypotension, CAD, valvular heart disease, heart failure, dysrhythmia, peripheral arterial disease
Pulmonary	Asthma, chronic pulmonary disease, pneumonia
Musculoskeletal	Arthritis, degenerative disc disease, polymyalgia rheumatica, osteoporosis, degenerative joint disease
Metabolic and endocrine	Diabetes, hyperlipidemia
Gastrointestinal	Dental disorders, malnutrition, incontinence, diarrhea
Genitourinary	Urinary tract infection, incontinence, cancer
Hematologic and immunologic	Anemia, leukemia, lymphoma, cancer
Neurological	Dementia, cerebrovascular disease, Parkinson's disease
Eye and ear	Cataracts, glaucoma, hearing loss
Psychiatric	Anxiety disorders, hypochondria, depression, alcoholism

Reprinted from Bayles C. Frailty. In ACSM's Exercise Management for Person with Chronic Diseases and Disabilities, 3rd Ed. Durstine JL, Moore GE, eds. Champaign, IL. Human Kinetics, 2009. page 203.

GUIDELINE 7.17

Title: "Position Stand: Exercise and Physical Activity for Older Adults"

Organization: ACSM

Year published: 1998

Purpose: To improve health, functional capacity, quality of life, and independence

Location: *Medicine and Science in Sports and Exercise* 30(6): 992-1008

Population: Older adults who are frail

GUIDELINES

Muscle groups to be included in strength training programs for frail older adults are the following:

- Hip extensors
- Knee extensors
- Ankle plantar flexors and dorsiflexors
- Biceps
- Triceps
- Shoulder muscles
- Back extensors
- Abdominal muscles

Component	Frequency	Intensity	Duration	Type
Progressive resistance training	2, preferably 3, d/wk	Start low and increase to high	2-3 sets per exercise	Upper and lower extremities and trunk
Balance training	As part of strength training	Not specified	Not specified	One-leg stand, tandem walk, circle turns
Aerobic training	At least 3 d/wk	Moderate: 40%-60% HRR or RPE of 11-13 out of 20	At least 20 min	Walking with or without assistive devices, arm and leg ergometry, water exercises

Special considerations include the following:

- Training and supervision are mandatory for safety and progression.
- Walking intensity may be increased by adding hills, inclines, steps, and stairs; pushing a weighted wheelchair; or adding arm and dance movements. These methods are preferable to increasing velocity or changing to jogging.
- Assistive devices are recommended both for safety and to increase energy expenditure.
- Higher-intensity strength training is more beneficial than and just as safe as lower-intensity training.

GUIDELINE 7.18

Title: "Frailty: Exercise Programming"

Organization: ACSM

Year published: 2009

Purpose: To increase functional capacity and independence, develop overall muscular strength, and decrease the risk of falling

Location: *ACSM's Exercise Management for Persons With Chronic Diseases and Disabilities, Third Edition*

Population: Individuals who are frail

GUIDELINES

Component	Frequency	Intensity	Duration	Type
Aerobic exercise	3-5 d/wk	Monitor RPE (intensity should not be the main focus)	5-60 min for each session	Large-muscle activities (walking, cycling, rowing, swimming, chair exercises)
Resistance training	3 d/wk	Start without weight; add weight slowly	1 set of 10 or fewer repetitions for each major muscle group	Low-level progressive resistance exercises with free weights, machines, isokinetic machines, and balls
Flexibility training	3 d/wk	Maintain stretch below point of discomfort		Stretching or yoga

>continued

Component	Frequency	Intensity	Duration	Type
Neuromuscular training	Not specified	Not specified	Not specified	One-foot stand, stair-climbing, falling techniques, balloon activity, tandem gait and chair stand

Special considerations include the following:

- Some form of exercise testing should be performed before initiating a program (graded exercise test for those who have or are at risk for CAD; 6 or 12 min walk test for others).
- The focus should be on RPE instead of target heart rate.
- Avoid exercises and maneuvers likely to cause injury, such as ballistic exercises, neck circumduction, isometric exercises, and static resistance exercises.

From Bayles C. Frailty. In ACSM's Exercise Management for Persons with Chronic Diseases and Disabilities, 3rd Ed. Durstine JL, Moore GE, eds. Champaign, IL. Human Kinetics, 2009, page 206.

GUIDELINE 7.19

Title: *First Steps to Active Health: Balance and Flexibility Exercises for Older Adults*

Organization: NCPAD

Year published: 2006

Purpose: To improve balance and flexibility as a means of improving health and functional ability as well as to prevent chronic disease and disability in older adults

Location: www.ncpad.org/exercise/fact_sheet.php?sheet=144

Population: Older adults

GUIDELINES

Component	Frequency	Intensity	Duration	Type
Balance training	2-3 d/wk	Not specified	2-3 sets of 8-10 exercises, hold each exercise 15-30 s	Exercises involving major muscle groups
Flexibility exercises	At least 2-3 d/wk	Held at level of mild discomfort	5-10 stretches, holding for 10-30 s, repeated 3-4 times per exercise	Stretches for major muscle or tendon groups

Special considerations include the following:

- These exercises should be part of a well-rounded physical activity program encompassing cardiorespiratory, strengthening, balance, and flexibility activities based on individual needs, abilities, and interests.
- Individuals may start by exercising for 5 min 3 times a day and gradually progressing in frequency, duration, and intensity.
- Individuals should perform a warm-up and cool-down with every exercise session.
- Individuals should perform flexibility exercises before and after an exercise routine.
- Individuals should use the Borg scale or heart rate to monitor intensity as prescribed by a health care provider.

Balance Exercises

- Tandem standing
- Balancing on one leg
- Standing hip raise
- Standing knee bend
- Standing kick
- Standing side kick
- Standing on foam
- Standing on foam on one leg
- Back kick with band
- Side kick with band

Flexibility Exercises

- Overhead stretch
- Chest stretch
- Midback stretch
- Side bend
- Trunk rotation
- Hamstrings stretch
- Calf stretch

Contraindications to Exercise

As noted earlier, physical activity benefits to the cardiovascular, musculoskeletal, endocrine, and neurocognitive systems appear to continue throughout life. Is there a point at which physical activity no longer needs to be performed or at which the risks outweigh the benefits? Is there an upper limit of age, after which exercise should not be undertaken? With few exceptions, the answer appears to be no. In general, extreme age should not preclude participation in an exercise program. Some acute illnesses may warrant modification of the program or further investigation into the illness. Other temporary situations such as injury or surgery may require momentary cessation of specific exercises. Very few conditions preclude long-term or permanent participation in a moderate- or vigorous-intensity exercise program. For these conditions the risks of exercise may outweigh the benefits. However, for the majority of people with diagnoses of osteoporosis, CVD, dementia, renal failure, vascular disease, diabetes, or arthritis, the risk of being sedentary is far more dangerous than being active. Table 7.2 provides examples of contraindications to exercise.

TABLE 7.2   Contraindications to Participation in Regular Physical Activity

Illnesses requiring investigation before exercise regimen is undertaken	Conditions requiring temporary avoidance until situation resolves	Conditions resulting in permanent exclusion from vigorous exercise
Febrile illnesses	Hernia	Inoperable enlarging aortic aneurysm
Unstable chest pain	Cataracts	Malignant ventricular arrhythmia related to exertion
Uncontrolled diabetes	Retinal hemorrhage	Severe aortic stenosis
Hypertension	Joint injury	End-stage congestive heart failure
Asthma		Rapidly terminal illness
Congestive heart failure		
Musculoskeletal pain		
Weight loss		
Falling episode		

SUMMARY

Throughout life, individuals benefit from regular physical activity, and older adults are no exception. Long-term physical activity can prevent and reverse many of the physiological changes associated with aging, including bone and muscle loss, insulin resistance, and loss of functional capacity. Most recommendations for older adults emphasize moderate-intensity activity, with roles for aerobic, resistance, and flexibility training. All three contribute to the physiological benefits of exercise, even among aging adults who are very frail, and can prevent falls as well as preserve muscle mass, cognition, and positive mood.

Physical Activity Guidelines by Disease States

The chapters of part III are directed toward physical activity and exercise guidelines for individuals with specific diseases. Chapter 8 addresses guidelines for disease prevention, while chapters 9 through 14 focus on recommendations for people already diagnosed with disease, including cancer, heart disease, arthritis, osteoporosis, diabetes, asthma, and neuromuscular disorders. These disease states are included not only because they are common but also because physical activity is frequently recommended for their treatment and management.

Cancer Prevention and Optimal Cardiometabolic Health

Lack of physical activity is linked to obesity, and obesity is linked not only to increased cancer incidence but also to the metabolic syndrome and its individual components. Several entities recommend both physical activity guidelines and nutritional guidelines to reduce the risk for cancer. For people who do not smoke cigarettes, activity and diet choices are the most modifiable determinants of cancer risk (McGinnis 1993; Mokdad and others 2004). Roughly one-third of cancer deaths in America can be attributed to poor physical activity and nutritional habits. As early as 1922, it was observed that men involved in physically active occupations experienced lower cancer mortality rates than men engaged in less-strenuous jobs experienced. Many guidelines focus on attaining an optimal weight because there is a clear relationship between obesity and increased rates of cancer and metabolic disease (including the metabolic syndrome, diabetes, and CAD). While general guidelines for weight loss and prevention of weight gain are presented in chapter 2, guidelines issued specifi-

cally to prevent cancer or cardiometabolic risks are presented in this chapter.

Cancer is the second leading cause of death for Americans. The cancers that account for the most deaths in the United States each year are lung, colorectal, breast, pancreas, and prostate. The incidence of some cancers (colorectal, breast, pancreas, prostate) seems to be related to exercise and weight, whereas the incidence of other cancers (lung, brain, skin, leukemia, and lymphoma) unfortunately seems to be unaffected by exercise. Regular physical activity works in a variety of ways to prevent cancer and may also improve outcomes in people who are diagnosed with cancer. The energy expenditure of physical activity helps to maintain a healthy body weight. Another mechanism by which regular physical activity may lower cancer risk is through modulating the immune system. Additionally, regular exercise moderates the levels of several hormones in the body. Lastly, by reducing insulin resistance, physical activity may reduce the risk for several cancers. Details are presented throughout the next section.

How Exercise Reduces the Risk of Cancer

There are many theories about and mechanisms responsible for the relationship between exercise and a reduction in certain types of cancers. Much of the decrease in cancer risk is attributable to improved lean body mass and metabolic effects associated with regular exercise. Also, hormonal changes such as reduced circulating estrogen are thought to be linked to the improved risk of breast cancer in habitual exercisers. Lastly, the beneficial effects of regular physical activity on the immune system are felt to decrease cancer risk.

Energy Expenditure

By utilizing energy, physical activity helps to reduce body weight. Specifically, regular physical activity reduces central adiposity, or abdominal fat. Overweight, obesity, and central adiposity are risk factors for several types of cancer, including the following:

- Brain
- Breast (among postmenopausal women)
- Cervical
- Colorectal
- Endometrial
- Esophageal (adenocarcinoma)
- Gallbladder
- Kidney
- Non-Hodgkin's lymphoma
- Ovarian
- Pancreatic
- Prostate

There are several positive metabolic effects of decreasing central adiposity that serve to reduce overall cancer risk. One of the most important is improved insulin sensitivity. Regular physical activity can improve energy metabolism and reduce concentrations of insulin and related growth factors. Improved insulin sensitivity may slow or arrest the development of type 2 diabetes, and thus, by reducing the risk for type 2 diabetes, physical activity reduces the risk of several cancers that are increased in people with type 2 diabetes (Will and others 1998; Lindblad and others 1999; Calle and others 1998):

- Breast
- Colon
- Endometrial
- Liver
- Pancreatic
- Renal

Hormone Levels

Regular exercise affects the body's hormones, including estrogen, progesterone, and testosterone. Higher levels of these hormones may increase the risk for breast and prostate cancers. Starting in early adolescence, physical activity may reduce the risk for breast and endometrial cancer by both delaying the onset of menses and decreasing menses regularity. Women who exercise regularly have lower levels of estrogen and progesterone as well as more frequent missed periods, or amenorrhea. Earlier age of menarche and higher lifetime total number of menstrual cycles are both associated with an increased risk for breast cancer (Vihko and Apter 1986). Additionally, exercise lowers both estrogen and androgen levels in postmenopausal women. In the Physical Activity for Total Health Study, overweight and sedentary postmenopausal women were randomly assigned to either an exercise group or a control group. The group that exercised had lower levels of total testosterone, free testosterone, estrone, total estradiol, and free estradiol (McTiernan and others 2004). These observed effects were probably mediated at least in part by the reduction of body weight seen in the exercisers. Another possible mechanism relates to localized hormone levels; it is felt that vigorous physical activity reduces the estrogen circulation in breast tissue itself (McTiernan and others 1998).

One study examined the relationship between college athleticism and all reproductive cancers, including breast, uterine, cervical, vaginal, and ovarian. Nonathlete women had 2.5 times the risk of these cancers when compared with women who had been college athletes (Frisch and others 1985). Another study in Europe found that the women who were the most sedentary had a significantly higher risk of endometrial cancer compared with those who were the most active (Levi and others 1993). The risk ranged from 2.4 to 8.6 times higher for sedentary women, depending on the age group.

Among men, regular activity seems to protect against testicular cancer. In a British study, men who spent at least 15 h/wk in exercise had half the risk for testicular cancer that those who did not exercise had (United Kingdom Testicular Cancer Study Group 1994). Conversely, using sedentary lifestyle as the measure, men who spent at least 10 h sitting each day had 1.7 times the risk of testicular cancer when compared with men who spent 2 h or less sitting! The precise mechanism of hormonal modulation is not known, but these examples underscore the importance of physical activity in cancer prevention.

Immune System

There is a close relationship among the immune system, cell senescence, and susceptibility to both the initiation and the promotion of tumors. Regular, moderate levels of physical activity enhance the immune system not only by increasing the absolute number of immune cells but also by enhancing the activity of those cells (Shephard and others 1995). Macrophages, natural killer cells, and lymphokine-activated killer cells and their regulating cytokines, neutrophils, and acute-phase proteins increase in number and activity in response to exercise (Shephard and Shek 1995). Thus, exercise improves the immune system's ability to suppress tumor growth and lyse tumor cells. Because the immune system naturally weakens with age, regular physical activity becomes that much more important for older adults, specifically to boost immune function. One study in particular evaluated this issue and found that older adults who exercised routinely had improved immune cell function when compared with those who did not exercise (Mazzeo 1994).

There is a myriad of other postulated methods by which regular physical activity reduces cancer risk. These vary according to the type of cancer. For example, regular physical activity enhances bowel motility, and the acceleration of food movement through the intestine decreases the length of time the mucosa is exposed to carcinogens (McTiernan and others 1998). Physical activity,

in conjunction with caloric restriction (or sensible nutritional intake), is a crucial part of maintaining a BMI within the healthy range of 18.5 to 25.0 kg/m^2. This in turn significantly reduces the risk for obesity-related cancers.

General Cancer Prevention Guidelines

The American Cancer Society (ACS) encourages people to exercise in order to reduce the risk for developing some cancers. The 2002 ACS guidelines are consistent with guidelines from the AHA for preventing CHD (guideline 2.5) as well as those from the USDHHS 2000 *Dietary Guidelines for Americans* (the most recent at the time) for general health promotion and the 1996 *Physical Activity and Health: A Report of the Surgeon General*. ACS cancer prevention guidelines are updated every 5 y in order to reflect the current state of the scientific evidence linking physical activity and cancer risk. In developing its exercise guidelines, the ACS took into consideration both benefits for cancer risk reduction and benefits for overall health. Additionally, the guidelines include recommendations not only for individuals but also for community organizations. This reflects the role of both public entities and private entities in supporting healthy physical activity that not only reduces cancer risk but also benefits society at large through improved economic efficiency and productivity. Preparticipation exercise clearance is recommended for men 40 y and older, for women 50 y and older, and for people with two or more cardiovascular risk factors.

The new *PAGA* guidelines contain a section dedicated to reducing cancer risk through physical activity. As noted in other chapters, the PAGAC recommendations are built on a thorough review of the scientific literature—in this case, the epidemiological data, observational data, case-control studies, and cohort studies on the association between physical activity and cancer risk. Also, there are randomized controlled trials providing indirect evidence of the benefits of physical activity on markers for cancer risk. Although *PAGA* does not have a specific recommendation for cancer prevention, it does describe the health benefits of physical activity with

GUIDELINE 8.1

Title: "American Cancer Society Guidelines on Nutrition and Physical Activity for Cancer Prevention: Reducing the Risk of Cancer With Healthy Food Choices and Physical Activity"

Organization: ACS

Year published: 2002 (updated in 2006)

Purpose: To decrease cancer incidence and mortality and to improve the qualfy of life of cancer survivors

Location: *CA: a Cancer Journal for Clinicians* 52: 92-119 and *CA: a Cancer Journal for Clinicians* 56: 254-281 (update)

Population: Children, adolescents, and adults

GUIDELINES

Recommendations for Individual Action

- Adopt a physically active lifestyle.
  - Adults should engage in at least moderate activity for 30 min or more on 5 d/wk or more; 45 min or more of moderate to vigorous activity on 5 d/wk or more may further reduce the risk of breast and colon cancer.
  - Children and adolescents should engage in at least 60 min of moderate to vigorous physical activity on at least 5 d/wk.
- Maintain a healthful weight throughout life.
  - Balance caloric intake with physical activity.
  - Lose weight if currently overweight or obese.

Recommendations for Community Action

- Public, private, and community organizations should work to create social and physical environments that support the adoption and maintenance of healthful nutrition and physical activity.
- Public, private, and community organizations should work to provide safe, enjoyable, and accessible environments for physical activity in schools and for transportation and recreation in communities.

GUIDELINE 8.2

Title: *Physical Activity Guidelines for Americans*

Organization: USDHHS

Year published: 2008

Purpose: To reduce the risk of cancer, particularly breast and colon cancer

Location: www.health.gov/paguidelines

Population: Adults

GUIDELINE

Participate in 210 to 420 min/wk of moderate-intensity activity.

GUIDELINE 8.3

Title: *Food, Nutrition, Physical Activity, and the Prevention of Cancer: A Global Perspective*

Organization: World Cancer Research Fund and the American Institute for Cancer Research

Year published: 2007

Purpose: To reduce the incidence of cancer

Location: www.dietandcancerreport.org

Population: Children, adolescents, and adults around the world

GUIDELINES

Personal Recommendations

- Be as lean as possible within the normal range of body weight.
  - Ensure that body weight throughout childhood and adolescence projects toward the lower end of the normal BMI range at age 21.
  - Maintain body weight within the normal range from age 21 on.
  - Avoid weight gain and increases in waist circumference throughout adulthood.
  - During childhood and adolescence, engage in at least 60 min/d of moderate to vigorous physical activity at least 5 d/wk.
- Be physically active as part of everyday life.
  - Be moderately physically active, equivalent to brisk walking, for at least 30 min every day.
  - As fitness improves, aim for 60 min or more of moderate physical activity or for 30 min or more of vigorous physical activity every day.
  - Limit sedentary habits such as watching television.

Public Health Recommendations

- Be as lean as possible within the normal range of body weight.
  - The median adult BMI should be between 21 and 32 kg/m², depending on the normal range for different populations.
  - The proportion of the population that is overweight or obese should be no more than the current level—preferably lower—in 10 y.
- The public should be physically active as part of everyday life.
  - The proportion of the population that is sedentary should be halved every 10 y.
  - Average PAL should be 1.6 or greater.

Adapted, by permission, from World Cancer Research Fund/AICR 2007 guidelines. Available: www.dietandcancerreport.org

respect to cancer risk and the dose–response relationship between physical activity and cancer risk. The data referenced by *PAGA* suggest that 210 to 420 min of moderate-intensity physical activity weekly are required to significantly reduce the risk of breast and colon cancer. Cancer prevention is one of the reasons why *PAGA* suggests going beyond its recommended minimum of 150 min/wk of moderate-intensity physical activity; cancer prevention is one of the significant health benefits achieved at higher activity levels. Physical activity guidelines for Americans as they pertain to cancer survivors are presented in chapter 9.

Together with the World Cancer Research Fund, the American Institute for Cancer Research has published a report with recommendations on nutrition and physical activity designed to reduce the risk of cancer. The first report was published in 1997 and the second was published in 2007. The report was put together by a panel of experts who performed a systematic review of the scientific literature and then made assessments and recommendations. Because of the increasing amount of evidence demonstrating the relationship between physical activity and healthy growth, metabolism, and development beginning at a very young age, the report emphasizes an approach to cancer prevention that includes the whole lifetime. Guidelines include both personal recommendations for individuals and public health goals for health professionals and policy makers. Overall the report makes eight recommendations, two of which are special recommendations (pertaining to breast-feeding and cancer survivors)—only the ones directly pertaining to physical activity are presented here.

The National Cancer Institute, which is part of the U.S. NIH, supports the 2007 AHA and ACSM recommendation of moderate-intensity physical activity for at least 30 min on 5 or more days of the week or of vigorous-intensity physical activity for at least 20 min on 3 or more days of the week (see guideline 2.1 in chapter 2).

Guidelines for Specific Types of Cancer

The guidelines presented thus far in this chapter apply to overall cancer risk reduction; however, site-specific cancers may also be affected by certain physical activity habits. This section discusses these data and the corresponding recommendations.

- **Breast cancer:** Obesity is a risk factor for breast cancer, so staying physically active to expend energy and maintain a caloric balance is essential. The ACS recommends that individuals engage in vigorous physical activity at least 4 h/wk; longer durations and higher intensities are associated with an even greater risk reduction. A recent meta-analysis showed a 25% reduction in cancer risk, and a dose–response relationship was found in 28 of 34 studies (Freidenreich 2008). It has been clearly established that exercise reduces the risk for postmenopausal breast cancer, and now a new study has suggested that premenopausal breast cancer is reduced by 23% (Maruti and others 2008) in women reporting the highest levels of activity (the equivalent of running 3.25 h/wk or walking 13 h/wk).

- **Colorectal cancer:** Colorectal cancer is the most extensively studied cancer with respect to physical activity and risk reduction. Physical inactivity is consistently associated with an increased risk for colon cancer. Several studies have shown that there is a lower risk of colon cancer among people who regularly participate in moderate physical activity compared with individuals who are sedentary. New evidence suggests an even greater risk reduction in people who participate in more vigorous activity—30% to 40% compared with being sedentary (Martinez and others 1997; Slattery and others 1997; Slattery 2004).

- **Endometrial cancer:** Women who are physically active have a 20% to 40% lower risk of endometrial cancer. Risk levels are the lowest among women with the highest levels of physical activity (NCI n.d.).

- **Pancreatic cancer:** Risk factors for pancreatic cancer that are modulated by exercise include type 2 diabetes, impaired glucose tolerance, obesity, and physical inactivity. As such, remaining physically active to maintain a healthy weight and thus reduce the first three of these risk factors is a priority.

Metabolic Syndrome

Physical activity is one of the key lifestyle components for achieving optimal cardiometabolic health. The metabolic syndrome is a cluster of individual cardiovascular risk factors that when combined may impart a significantly higher risk for CAD and type 2 diabetes. Although there are several variations on the definition of metabolic syndrome, the pathophysiology is similar; the most widely recognized criteria are those of the Third Report of the National Cholesterol Education Program (NCEP) Expert Panel on Detection, Evaluation, and Treatment of High Blood Cholesterol in Adults (Adult Treatment Panel III or ATP III). According to these criteria, a person with metabolic syndrome exhibits at least three of the following (NCEP 2002):

- Central obesity as measured by a waist circumference of 40 in. (102 cm) or more in men or 35 in. (88 cm) or more in women

- High fasting blood triglycerides of 150 mg/dl or more
- Low HDL cholesterol of less than 40 mg/dl in men or less than 50 mg/dl in women
- Elevated blood pressure of 130/85 mmHg or higher
- Fasting glucose level of 100 mg/dl or higher

The metabolic syndrome was first recognized by Reaven, who called it *syndrome X* in 1988. The metabolic syndrome, which has also been called *cardiometabolic risk* and *insulin resistance syndrome*, is closely correlated with a lack of regular physical activity. Conversely, initiating a physical activity program can reverse many of the components of the metabolic syndrome, improve cardiometabolic parameters, and greatly reduce the risk for type 2 diabetes and heart disease.

The 2008 *Physical Activity Guidelines for Americans* emphasizes prevention of the metabolic syndrome, in light of the relationship between it and diabetes, cardiometabolic, and CAD risk. In addition to playing a protective role in the metabolic syndrome, physical activity also plays a treatment role. Additionally, there are data from at least one study (Laaksonen 2002) suggesting that there is a dose–response relationship between physical activity and the development of the metabolic syndrome (see figure 8.1). Cross-sectional studies also demonstrate a reduced prevalence of metabolic syndrome for individuals who participate in higher levels of physical activity. It appears that at least 180 min of moderate-intensity physical activity per week is the minimum associated with a reduced risk for the metabolic syndrome (PAGAC 2008). This amount of activity falls between the minimum 150 min and the desired 300 min of weekly activity recommended in the 2008 *Physical Activity Guidelines for Americans.*

Type 2 Diabetes

The basic metabolic abnormalities responsible for high blood glucose levels are insulin resistance and insulin deficiency. The first reflects the body's decreased sensitivity to insulin, while the second reflects the inability of the pancreas to secrete enough insulin to meet glucose demand. In the early stages of type 2 diabetes, or even in the pre-diabetic state, insulin resistance predominates and glucose cannot enter cells, which in turn results in ever-increasing blood glucose levels. As this pathway continues, the pancreas secretes more and more insulin in an attempt to counteract the hyperglycemia and normalize the glucose levels. Eventually, the amount of insulin secreted is no longer adequate to compensate for the insulin resistance, and diabetes results.

Because physical activity improves insulin sensitivity, it is an important component in preventing the onset of type 2 diabetes. According to the International Diabetes Federation, up to 80% of cases of type 2 diabetes are preventable by healthy diet and physical activity. Thus it makes sense that physical activity has a great capacity to improve insulin action in individuals with hyperinsulinemia and insulin resistance as opposed to people with inadequate insulin secretion. Abdominal (or central) fat distribution is another major risk factor for insulin resistance. Physical activity is inversely associated with both obesity and central fat distribution. Thus, regular physical activity plays a vital role in preventing

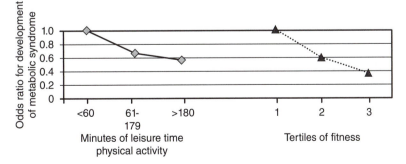

Figure 8.1 Higher levels of physical activity and fitness protect against development of the metabolic syndrome.

Data from Laaksonen 2002.

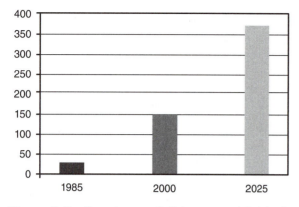

Figure 8.2 Prevalence of diabetes worldwide (in millions), estimated for 2025.

Data from International Diabetes Federation, www.idf.org, accessed 8/19/2008.

weight as well as to engage in at least 150 min/wk of moderate-intensity activity such as walking. Randomized controlled trials for individuals at high risk for developing type 2 diabetes have shown that lifestyle interventions including weight reduction through healthy diet and regular physical activity significantly decrease the rate of onset of diabetes (Knowler and others 2002; Tuomilehto and others 2001; Pan and others 1997; Chiasson and others 2002; Ramachandran and others 2006; Gerstein and others 2006). The magnitude of that reduction (on average, 58% after 3 y) is similar or superior to that seen with pharmacological agents such as metformin, orlistat, rosiglitazone, and acarbose.

The AHA, in conjunction with the NHLBI, developed guidelines for curbing the risk of diabetes that might elevate the risk for CAD. These recommendations include 30 to 60 min of daily aerobic activity plus two weekly sessions of resistance training.

The eighth edition of *ACSM's Guidelines for Exercise Testing and Prescription* has a specific section addressing individuals with metabolic syndrome with the goal of preventing diabetes. For the most part, these guidelines are consistent with the recommendations for all healthy individuals, although there is an emphasis on higher caloric expenditure through a higher volume of activity in order to achieve weight loss.

or at least delaying type 2 diabetes by reducing obesity, intraabdominal fat, and ultimately insulin resistance. On the other hand, because of the rising prevalence of obesity (see figure 8.2), diabetes continues to escalate as a public health problem.

The American Diabetes Association (ADA) makes recommendations not only for individuals with diabetes but also for people with impaired glucose tolerance (IGT) or impaired fasting glucose (IFG) and for people at risk for type 2 diabetes due to other factors. Patients at risk for developing diabetes (with IFG or IGT) are recommended to lose 5% to 10% of their body

---

GUIDELINE 8.4

Title: "Diagnosis and Classification of Diabetes Mellitus" and "Executive Summary: Standards of Medical Care in Diabetes – 2008"

Organization: ADA

Year published: 2008

Purpose: To reduce the risk for developing diabetes

Location: Executive Summary: Standards of Medical Care in Diabetes – 2008, *Diabetes Care*, 31(S1): S5-S11, 2008; Diagnosis and Classification of Diabetes Mellitus *Diabetes Care* S12-S54, 2008.

Population: All adults, including individuals at high risk for type 2 diabetes

GUIDELINE
Aerobic activity should be performed for 150 min/wk.

GUIDELINE 8.5

Title: *Diagnosis and Management of the Metabolic Syndrome*

Organization: AHA and NHLBI

Year published: 2008

Purpose: To reduce the risk for developing diabetes

Location: www.circ.ahajournals.org/cgi/content/full/112/17/e285

Population: All adults, including people at high risk for type 2 diabetes

GUIDELINES

Type	Frequency	Intensity	Duration
Aerobic exercise	At least 5 d/wk; preferably daily	Moderate	At least 30 min and preferably 60 min of continuous or intermittent physical activity
Resistance exercise	2 d/wk	Not specified	Not specified

These recommended amounts of exercise should be supplemented with an increase in lifestyle activities such as taking a greater number of steps each day (as tracked by a pedometer), walking during breaks at work, gardening, performing household work, and so on.

GUIDELINE 8.6

Title: "Exercise Prescription for Other Clinical Populations: Metabolic Syndrome"

Organization: ACSM

Year published: 2010

Purpose: To promote weight loss and improve health and fitness outcomes

Location: *ACSM's Guidelines for Exercise Testing and Prescription, Eighth Edition*

Population: Individuals with metabolic syndrome

GUIDELINES

- Start with a moderate intensity (40%-60% $\dot{V}O_2R$ or HRR) and progress when appropriate to vigorous intensity (50%-75% $\dot{V}O_2R$ or HRR).
- A duration of 50 to 60 min on 5 d/wk or 300 min/wk is likely needed for benefit (activity may be accumulated in multiple daily bouts lasting at least 10 min each).
- Some individuals will need to progress to 60 to 90 min/d to promote or maintain weight loss.

Adapted, by permission, from ACSM, 2010, *ACSM's guidelines for exercise testing and prescription*, 8th ed. (Baltimore, MD: Lippincott, Williams, and Wilkins), 252-253.

SUMMARY

Physical activity is inextricably linked with the risk for cancer and metabolic diseases such as diabetes. There are many ways in which regular physical activity can reduce the risk of cancer, including weight control, hormonal changes, and improved immune function. Physical activity also reduces the risk for the metabolic syndrome and diabetes, primarily through enhanced insulin sensitivity and weight control. National and international organizations have issued guidelines for physical activity and the promotion of such in order to reduce the risk for these largely preventable diseases. For the most part, these guidelines are not materially different from guidelines recommended for all adults.

Cancer

Cancer is the second leading cause of death (behind CVD) in the United States and the leading cause of morbidity and mortality worldwide. Additionally, the rate of death due to CVD has been declining over the past few decades, while the age-adjusted cancer mortality rate continues to rise (Bailar and Smith 1986). Current estimates show that 37.5% of American females and nearly 45% of American males will develop cancer (NCI 2007). However, the number of cancer survivors continues to increase as cancer treatments improve. Nearly two-thirds of patients with cancer stay alive more than 5 y after diagnosis, and currently there are more than 10 million cancer survivors living in America.

As detailed in chapter 8, physical activity reduces the risk of developing many different types of cancer. For people who are already diagnosed with cancer, regular physical activity may still provide benefits, including improvements in energy, functional status, quality of life, and fracture risk. Historically, people being treated for cancer and other chronic illnesses were advised by their physicians to rest and reduce their levels of physical activity. It was felt that too much activity increased weight loss in those who already found it difficult to maintain their weight. However, over the past several years, more information has surfaced showing that regular exercise is an effective way to improve physical functioning and quality of life among people with chronic disease. In fact, too much rest may reduce strength, muscle mass, bone density, and range of motion as well as lead to a general loss of function.

In addition, individuals undergoing adjuvant chemotherapy or radiation therapy may improve their quality of life by participating in home-based exercise programs. In fact, one report suggested that exercise is second only to prayer as a popular complementary therapy participated in by survivors of breast cancer. Despite the heterogeneity of the hundreds of different types of cancer, virtually all individuals with cancer can benefit from regular physical activity. Also, at least for people with breast or colorectal cancer, physical activity improves survival, and—as with many of the benefits of physical activity—this survival demonstrates a dose–response relationship with exercise. Finally, there is evidence that weight loss, independent of physical activity (but perhaps precipitated by activity), reduces the rate of cancer recurrence.

Benefits of Physical Activity

Physical activity provides many benefits for cancer patients both during and after treatment. Due to the improvement in cancer treatments, the population of cancer survivors is growing, so it is important to emphasize the benefits of physical activity on reducing the risks of *other* chronic diseases that can occur in cancer survivors. Additionally, there are improvements in *quality of life* for cancer survivors who participate in physical activity.

Benefits During Treatment

Many studies have looked at the relationship between physical activity and quality of life after cancer diagnosis. Regardless of the exercise prescription, cancer rate, or cancer treatment, studies have demonstrated a consistent array of benefits for a wide range of quality of life outcomes. Psychosocial outcomes may be as important as physiological ones; cancer patients often experience anxiety, depression, tension, anger, hostility, helplessness, and pessimism. Regular exercise during and after cancer treatment has been shown to decrease these negative feelings while improving self-concept, self-esteem, confidence, self-image, sense of personal worth, control, well-being, and self-acceptance (Schairer and Keteyian 2003; Schmitz and others 2005).

Some of the many reasons why the ACS recommends regular physical activity for individuals during cancer treatment include the following:

- Maintained or improved physical abilities
- Improved balance reducing the risk of falls and fractures
- Prevention of muscle wasting caused by inactivity
- Reduced risk of heart disease
- Prevention of osteoporosis
- Improved blood flow to legs and reduced risk of blood clots
- Less dependence on others to perform normal ADLs
- Improved self-esteem
- Reduced anxiety and depression
- Decreased nausea
- Increased ability to maintain social contact
- Reduced symptoms of fatigue
- Better ability to control weight
- Improved quality of life

Because cancer treatment—whether surgery, chemotherapy, or radiation—results in fatigue

in 70% of individuals, one major goal of physical activity is to reduce this fatigue. Multiple studies have evaluated the effects of physical activity interventions on fatigue; because cancer-related fatigue is notable for its severity, any intervention that has a positive effect is noteworthy. Most of the recent studies on physical activity, including walking, cycling, and aerobic exercise, have shown important improvements in fatigue.

Cancer treatment varies significantly depending on the type of cancer, the cancer stage, the patient's age, and the patient's comorbidities. In addition to surgery, radiation, and chemotherapy, treatments may include antimetabolites, immunotherapy, hormones, and steroids. As a result, the potential effects of cancer treatment are very diverse and include physical changes such as loss of bone density (due to hormones and steroids) and muscle mass in addition to lymphedema as well as psychological changes such as depression and loss of cognitive function. Physical activity can reverse many of these changes. Physical activity can improve the following long-term negative effects of cancer treatment (PAGAC 2008):

Decreased
- Pulmonary function
- Cardiac function
- Muscle mass
- Muscle strength or power
- Immune function
- Bone health
- Lymphatic function
- Exercise level
- Cognitive function
- Quality of life

Increased
- Fat mass
- Weight or BMI
- Inflammation
- Trauma and scarring
- Physical symptoms and pain
- Depression

Patients should always discuss physical activity and exercise with their physicians before starting an exercise program during cancer treatment.

Often, persons with cancer may utilize an exercise specialist (exercise physiologist, personal trainer, or rehabilitative specialist) in order to maximize the benefits and minimize the risks of exercise. People who were physically active before cancer diagnosis or treatment may resume the same type of activity they participated in beforehand, although they may have to modify the intensity, frequency, or duration. Cancer treatment affects different individuals differently and may even affect a single person differently at different stages in treatment. For this reason, when it comes to measuring intensity, RPE may be more appropriate than an objective measure such as time, distance, or heart rate. Although people with cancer often require a lower intensity of activity, their goal should be to maintain activity as much as possible. Progression should occur at a slower pace than that recommended for individuals without cancer.

Benefits for Cancer Survivors

A meta-analysis of controlled trials examining physical activity in cancer survivors showed both qualitative and quantitative evidence of a small to moderate benefit in several outcomes. Physical activity improves symptoms and physiological effects during treatment, cardiorespiratory fitness during and after treatment, and vigor after treatment (Schmitz 2005). In the past few years, additional randomized controlled trials have examined the effects of various exercise interventions in cancer survivors. All have shown positive effects in cardiorespiratory fitness achieved through activities such as exercise at home or in fitness centers, walking, yoga, and tai chi (Culos-Reed and others 2006; Pinto and others 2005; Mustian and others 2006; Hutnick and others 2005; Herrero and others 2006; Daley and others 2007).

Muscular strength and endurance also appear to be improved in cancer survivors who participate in resistance training, tai chi, or yoga. Both cancer and its various treatment regimens tend to decrease activity, increase deconditioning, and consequently decrease muscle mass while increasing fat mass. Interventions that reverse these processes are critical not only in improving these elements but also in augmenting survival and decreasing the chance of recurrence. The latter

outcome is particularly worth mentioning for people with breast cancer (the survivors of which make up most of the individuals for which the data on cancer and physical activity are collected) because body composition can play a central role in cancer recurrence.

Lastly, there appear to be improvements in flexibility for cancer survivors engaged in various types of physical activity programs. The drop in activity associated with cancer treatment, along with the potential scarring or tissue trauma from surgery or radiation therapy, can hamper flexibility. In three studies evaluating shoulder range of motion in breast cancer survivors who participated in tai chi, dance and movement, or aerobic exercise and stretching, significant improvements were observed in two of the studies (Mustian and others 2006; Cho and others 2006; Sandel and others 2005).

One very significant marker of success—survival—is enhanced in cancer survivors who participate in regular physical activity. Although the data are most striking for colorectal and breast cancer, there is a suggestion of benefit for other cancers as well. Ways for cancer survivors to increase physical activity including taking the stairs, walking or biking instead of driving, exercising with family and coworkers, planning active vacations, wearing a pedometer daily to monitor and create awareness of daily steps, using a stationary bicycle or treadmill while watching television, and taking short exercise breaks to visit friends or coworkers (instead of sending e-mail). People who previously were sedentary benefit from slowly working up to short walks and gradually increasing physical activity.

American Cancer Society Guidelines

The ACS has published exercise recommendations for patients undergoing cancer treatment. These recommendations stress to obtain approval for starting an exercise program (or continuing one already established before cancer diagnosis and treatment) from the health care provider and suggest to exercise at a moderate intensity level, which is defined as the amount of effort required by a brisk walk. Aerobic, strengthening, and flexibility exercises are all incorporated into the ACS program. Depending on a person's fatigue, side effects, disability, and precancer functional status, one or more of these components may be emphasized. For example, someone with extreme fatigue who is staying at home and doesn't feel up to exercising could focus on doing 10 min of daily stretching. The ACS recommendations stress taking caution in order to avoid overexertion and the consequences of treatment that may be exacerbated by too much activity. Thus the ACS includes the following safety precautions:

- Obtain exercise clearance from the treating physician.
- Avoid exercise if blood count (white or red cell) is low.
- Avoid exercise if experiencing electrolyte disturbances.
- Avoid exercise if experiencing extreme nausea, diarrhea, or unrelieved pain.
- Avoid uneven surfaces that might precipitate falls.
- Avoid excessive weight-bearing activity or lifting heavy weights if cancer has spread to bone or if there is osteoporosis.
- Watch for lower-extremity edema, excessive shortness of breath, bleeding, dizziness, or blurred vision.

Avoiding injury is another important consideration. People with significant handicaps or impairments before their cancer diagnosis need to follow guidelines similar to those for people without cancer who have the same impairments. For example, individuals with peripheral neuropathy or arthritis, for whom balance may be an issue, should avoid activities that increase the risk for falling and injury and engage in functional exercises to improve balance. Working with a personal trainer, physical therapist, or caregiver during exercise sessions might also reduce the risk for harm.

After recovery from cancer, goals include achieving and maintaining an appropriate weight, a physically active lifestyle, and a healthy diet. These promote not only overall health and longevity but also quality of life. While lack of regular physical activity is linked to primary incidence of many cancers, including breast, colon, prostate, pancreas, and endometrial cancers, physical activity after diagnosis improves survival and recurrence rates for many different types of cancer. Combining diet and exercise strategies to achieve an ideal weight imparts significant health benefits and forms the basis of the ACS guidelines on nutrition and physical activity during and after cancer treatment. These guidelines are felt to be the companion guidelines to the ACS guidelines focused on cancer prevention (see chapter 8). Although some cancer survivors (particularly survivors of lung, gastrointestinal, head, and neck cancers) may experience unintentional weight loss and malnourishment, they can still benefit from regular, modest physical activity. Potential advantages for these cancer survivors include stress reduction and increased strength. Higher levels of activity, however, may make appropriate weight gain difficult.

Other Notable Guidelines

PAGAC recommends physical activity for individuals with cancer because of the resultant improvement in health outcomes, although the 2008 report emphasizes that most of the studies showing significant health improvements have been performed on individuals with either colorectal or breast cancer. Most of the studies that created the basis for the PAGAC recommendations used walking as the form of physical activity. There is not a huge body of data with respect to cancer survivors, physical activity, and outcomes, and more studies are underway. Current information suggests that in addition to a survival benefit, there are superior results in quality-of-life measures in people who participate regularly in physical activity. Also, breast cancer survivors who participate in regular physical activity and maintain a healthy weight have fewer recurrences than survivors who do not.

Ehrman's *Clinical Exercise Physiology* presents an exercise prescription for patients with cancer. The prescription emphasizes flexibility in modifying exercise activities, as people with cancer often have fluctuating clinical status and treatment protocols that necessitate temporary suspension of exercise. Intensity level is determined by RPE

GUIDELINE 9.1

Title: *Physical Activity and the Cancer Patient*

Organization: ACS

Year published: 2008

Purpose: To improve physical functioning and quality of life and to counteract the negative effects of inactivity in chronic disease

Location: www.cancer.org/docroot/mit/content/mit_2_3x_physical_activity_and_the_cancer_patient.asp

Population: Individuals undergoing cancer treatment

GUIDELINES

- Work up to 30 min/d of brisk activity.
- Exercise in sessions as short as 10 min to work toward the 30 min goal.
- Include physical activity that uses large-muscle groups.
- Incorporate strength, flexibility, and aerobic fitness.
- Include exercises that maintain lean muscle mass, bone strength, and range of motion.
- Always begin with warm-up exercises for 2 or 3 min and end with stretching or flexibility exercises, holding stretches for 15 to 30 s.
- Always talk to your health care provider before initiating an exercise program.

GUIDELINE 9.2

Title: *Nutrition and Physical Activity During and After Cancer Treatment: An American Cancer Society Guide for Informed Choices*

Organization: ACS

Year published: 2006

Purpose: To improve treatment outcomes, qualify of life, and survival after cancer diagnosis and treatment

Location: http://caonline.amcancersoc.org/cgi/content/full/56/6/323

Population: Children, adolescents, and adults

GUIDELINES

- Achieve and maintain a healthy weight and avoid excessive weight gain throughout life by balancing caloric intake with caloric expenditure through physical activity.
- Engage in moderate to vigorous physical activity, above and beyond usual activities, on 5 or more days of the week.
  - Adults should engage in at least 30 min (and preferably 45-60 min) of intentional physical activity.
  - Children and adolescents should engage in at least 60 min of intentional physical activity.

GUIDELINE 9.3

Title: *Physical Activity Guidelines Advisory Committee Report*

Organization: USDHHS

Year published: 2008

Purpose: To prevent premature death or cancer recurrence and reduce adverse effects of cancer treatment

Location: www.health.gov/paguidelines/committeereport.aspx

Population: Cancer survivors

GUIDELINES

- Cancer survivors should follow the PAGAC activity guidelines for all adults (see guideline 2.2).
- Cancer survivors should consult their health care providers to match their physical activity plan to their abilities and health status.

GUIDELINE 9.4

Title: "Cancer"

Organization: None

Year published: 2009

Purpose: To improve physical functioning and quality of life and to counteract the negative effects of inactivity in chronic disease

Location: *Clinical Exercise Physiology, Second Edition*

Population: Individuals undergoing cancer treatment

GUIDELINES

Component	Frequency	Intensity	Duration	Notes
Aerobic exercise	3-5 d/wk	Progressively increase to 50%-70% HRR or to RPE of 11-13 out of 20	15-30 min (can be divided into 2-3 intervals)	Large-muscle groups
Resistance training	1-2 d/wk	50%-60% of 1RM, 1 set of 12-15 repetitions	20-30 min	Standardized machines are preferred to free weights
Flexibility exercise	Before and after aerobic or resistance training workouts	Stretch to point of challenge but not discomfort	10-30 s per stretch for a total of 5 min before and 5-10 min after each workout	Static stretching

Special considerations include to avoid resistance training for 36 h before laboratory testing.

CATEGORIZING PERFORMANCE LEVELS

Activity level	Category	Exercise duration	Exercise frequency
Active, no limitations	0	15-20 min	Daily
Ambulatory, decreased leisure activity, can perform self-care	1	15-20 min	Daily
Ambulatory >50% of time, moderately fatigued, limited assistance required with ADLs	2	15-20 min	Daily
Ambulatory <50% of time, fatigued upon mild exertion, assistance required with ADLs	3	5-10 min	Two sessions daily
Confined to bed	4	No exercise	No exercise

or a percentage of HRR. Exercise duration and frequency can be determined by classifying activity level based on limitations before initiating the program. Recognizing the limitations posed by either the underlying disease or the treatment plan can help individuals develop an appropriate physical activity program and realize significant short-term and long-term benefits.

The ACSM has issued exercise management guidelines for individuals with cancer. Because of the variability in functional ability of patients with cancer at different stages of diagnosis, treatment, recovery, and remission, individualization of the exercise program is highly recommended. Exercise testing before beginning a program is encouraged and generally can be performed using standard protocols (Schwartz 2003). Because many individuals with cancer may also have hypertension, diabetes, or hyperlipidemia, the exercise testing, if done with 12-lead electrocardiogram (ECG) monitoring, may be helpful in

uncovering comorbid conditions such as CAD. Additionally, many cancer treatments increase the risk for CAD by affecting the heart either directly or indirectly via induction of anemia or diabetes.

The need for personalizing exercise programs is stressed in the ACSM recommendations for cancer survivors. Patients with cancer vary in both premorbid conditioning and active shape after treatment. Those who are receiving chemotherapy may need further assessment and continual review; nausea, diarrhea, dehydration, anemia, and heightened risk for infection, bleeding, or falls are increased during chemotherapy. Thus, according to the current condition of the individual, exercise adjustments may need to be made.

The most recent edition of *ACSM's Guidelines for Exercise Testing and Prescription* also includes a section on exercise prescription for people with cancer. These recommendations are similar to ACS guidelines as well as to the ACSM's general

GUIDELINE 9.5

Title: "Cancer: Exercise Programming"

Organization: ACSM

Year published: 2009

Purpose: To help people undergoing cancer therapy to maintain strength, endurance, and level of function and to return people who have survived cancer to their former level of physical and psychological function

Location: *ACSM's Exercise Management for Persons with Chronic Diseases and Disabilities, Third Edition*

Population: Individuals undergoing cancer therapy or individuals who have survived cancer

>continued

GUIDELINES

Component	Frequency	Intensity	Duration	Type
Aerobic exercise	At least every other day	Moderate intensity; activity should be limited by symptoms	15-40 min each session	Large-muscle activities (walking, cycling, rowing, water aerobics)
Resistance training	2-3 d/wk	50% of 1RM, 2-3 sets of 3-5 repetitions building to 10-12 repetitions	20-30 min	Free weights, machines, exercise bands, isokinetic machines, circuit training
Flexibility exercise	5-7 d/wk	Maintain stretch below point of discomfort	20-60 s for each stretch	Stretching
Functional exercise	Daily	Not specified	Not specified	ADLs, gait and balance exercises

Special considerations include the following:

- Progressive programs may be necessary for people starting in a very debilitated state.
- Frequent adjustments may be necessary due to treatments or other changes in medical condition.

Adapted, by permission, from A. Schwartz, 2009, Cancer. In *ACSM's exercise management for persons with chronic diseases*, 3rd ed., edited by J.L. Durstine et al. (Champaign, IL: Human Kinetics), 216.

GUIDELINE 9.6

Title: "Exercise Prescription for Other Clinical Populations: Cancer"

Organization: ACSM

Year published: 2010

Purpose: To limit physical performance limitations

Location: *ACSM's Guidelines for Exercise Testing and Prescription, Eighth Edition*

Population: Individuals who are cancer survivors

GUIDELINES

Component	Frequency	Intensity	Duration	Type
Aerobic exercise	3-5 d/wk	40%-<60% $\dot{V}O_2R$ or HRR	20-60 min/d (accumulated in shorter bouts if necessary)	Walking, cycling, swimming
Resistance training	2-3 d/wk with at least 48 h of recovery between sessions	40%-60% of 1RM	1-3 sets of 8-12 repetitions per exercise (up to 15 repetitions for individuals who are deconditioned, fatigued, or frail)	Free weights, resistance machines, or weight-bearing functional tasks for all major muscle groups
Flexibility exercise	2-7 d/wk	Slow static stretching to the point of tension	4 repetitions of 10-30 s stretches	Stretching or range-of-motion exercises for all major muscle groups

Special considerations include the following:

- People with cancer should avoid high-impact activities and contact sports to minimize fracture risk.
- Patients who have indwelling catheters, central lines, or feeding tubes or who are receiving radiation therapy should avoid swimming.
- Patients with bone marrow transplants or low white blood cell counts should avoid exercising in public places.
- Frequent modifications to exercise may be necessary during treatment cycles.

Adapted, by permission, from ACSM, 2010, *ACSM's guidelines for exercise testing and prescription*, 8th ed. (Baltimore, MD: Lippincott, Williams, and Wilkins), 231-232.

principles of exercise prescription. In addressing people with cancer, the ACSM stresses the importance of obtaining medical clearance for any exercise that is vigorous intensity.

Guidelines for Specific Cancers

This section describes the physical activity guidelines for site-specific cancers. These recommendations focus on reducing the risk of cancer recurrence, moderating side effects of treatment, and decreasing the later risk for CVD.

- **Breast cancer:** Well-established risk factors for worse prognosis and recurrence of breast cancer are overweight and obesity. Survivors of breast cancer who are physically active have significantly lower rates of cancer recurrence as well as disease-specific and overall mortality when compared with sedentary survivors (Holmes 2005; Enger 2004). Thus, people with breast cancer should avoid weight gain, and while both diet and physical activity play a role in maintaining a normal weight, physical activity should be emphasized regardless of weight concerns.

- **Colorectal cancer:** Although physical activity (or lack thereof) is correlated with colorectal cancer risk, it is unknown whether regular physical activity improves prognosis. Increased body weight is associated with shorter survival (Tartter 2002), so weight control is important. Several studies have shown a positive relationship among physical activity, physical fitness, survival, and quality of life in survivors of colorectal cancer (Meyerhardt 2006a, 2006b; Courneya 2003, 2004).

- **Hematologic cancers:** Most of the studies addressing survivors of hematologic cancer and exercise involve the acute recovery from bone marrow or stem cell transplantation. These studies report some benefits of exercise on functional capacity and aerobic fitness, muscular strength, fatigue, and psychosocial functioning and quality of life. Also, obesity and overweight adversely affect prognosis for patients receiving stem cell transplants (Meloni 2001), so weight control is an important goal of physical activity.

- **Lung cancer:** Because of the aggressive nature not only of lung cancer but also of lung cancer treatment, patients and survivors often have difficulty maintaining their weight and enough energy to participate in regular activity. The potential role of physical activity in improving prognosis for survivors of lung cancer has not been studied.

- **Prostate cancer:** Prostate cancer is unique in that it is slow growing, many patients are older at the time of diagnosis, and thus CAD risk factors often occur concurrently. Recommendations for physical activity focus on preventing recurrence of the cancer and improving overall mortality by reducing CVD risk. It is controversial whether obesity affects outcomes in terms of prostate cancer itself. Two trials have studied the effects of exercise in survivors of prostate cancer, one via resistance training (Segal 2003) and one via a home-based walking program (Windsor 2004). These studies demonstrated improvements in quality of life, fatigue levels, and muscular fitness (Segal 2003) as well as physical functioning (Windsor 2004).

- **Upper gastrointestinal tract and head and neck cancers:** As with lung cancer, little data are available regarding physical activity and outcomes in cancers of the upper gastrointestinal tract, head, and neck. For this reason, it is recommended that survivors of these cancers follow the ACS guidelines for physical activity for cancer prevention (see guideline 8.1).

Side Effects of Cancer Affecting Physical Activity

Although physical activity is recommended for patients with cancer, these patients face caveats and concerns when it comes to exercising safely, just as any individual with a chronic condition might face. These concerns include short-term effects such as decreased blood count or susceptibility to infection and long-term effects such as lymphedema or skin changes. Being mindful of these circumstances allows people with cancer to reap the benefits of physical activity while avoiding potential harm.

Lymphedema

One common side effect of cancer treatment is lymphedema. Although most commonly observed in the upper extremities of survivors of breast cancer, lymphedema is also observed in the lower extremities of people who have had lymph node dissection or radiation therapy in the lower half of the body. Thus lymphedema occurs in up to 50% of breast cancer survivors and in 20% to 30% of patients who have had treatment in the groin or retroperitoneal lymph nodes (PAGAC 2008). Lymphedema is considered a chronic condition. For years, health care providers have cautioned patients who have had axillary lymph node dissection not to lift heavy items with the affected side in order to prevent lymphedema. However, there are no studies showing that patients who exercise increase their risk for lymphedema. Exercise may actually help patients with lymphedema, as it may keep lymphatic fluid circulating. Studies focusing on resistance training by patients with cancer have not observed higher rates or worsening of lymphedema (McKenzie and Kalda 2003; Ahmed and others 2006). It must be noted, however, that in these studies participants all started exercising slowly and gradually built up strength and endurance (Schwartz 2004, p 35-36).

Skin Conditions

In addition to fatigue, vomiting, and diarrhea, radiation therapy may cause skin irritation, redness, or itchiness. Skin irritation is generally localized to the area of focused radiation. If excessive perspiration causes more pain in the irritated area, then exercise intensity may need to be adjusted to avoid causing further irritation or burning. Avoiding swimming is recommended, as chlorination may exacerbate tenderness. Also, patients with indwelling urinary or intravenous catheters should avoid water or other microbial exposures that could result in infections. Skin is a common portal of entry for infections, and so people with compromised immune function should avoid public gyms and other public places until their white blood cell counts return to safe levels.

Blood Cell Counts and Contraindications to Exercise

Although physical activity is generally well tolerated during cancer treatment, it is important to be mindful of possible concerns and to be cautious during exercise. Many studies examining the potentially harmful effects of exercise during cancer treatment have found that exercising with low blood cell counts (white or red) is safe with proper caution. In fact, one study showed that patients who exercise while undergoing bone marrow transplantation actually require fewer blood transfusions (Schwartz 2004). Common sense dictates that people with extremely low blood counts avoid situations placing them at increased risk for infection or bleeding. Avoiding contact sports or exercises with a heightened risk for falling is prudent for individuals with low platelet counts. Finally, patients who exercise during the course of cancer treatment may have one or more of the following contraindications to exercise and should use appropriate caution (Schairer and Keteyian 2003):

- Fever (temperature >100.4 °F or >38 °C)
- Active infection
- Severe nausea
- Severe weight loss (>35% of precancer weight)
- Ataxia

- Bone or joint pain
- Extreme dyspnea on exertion
- Rapid or irregular heartbeat
- Anemia (hemoglobin level of <10.0 g/dl)

- Leukopenia (white blood cell count of <3,000 per microliter or absolute neutrophil count of <500 per microliter)
- Thrombocytopenia (platelet count of <50,000 per microliter)

SUMMARY

Individuals with cancer should not be excluded from a regular physical activity program on the basis of their cancer diagnosis. While there are some precautions to be taken before embarking on a physical activity program, there are many rewards—physical, physiological, and emotional—to participating in exercise and activity both during and after cancer treatment. Guidelines include these precautions and also emphasize the importance of preventing other common chronic conditions such as diabetes, heart disease, and obesity in light of the fact that survivors of cancer commonly live a long time after treatment.

Hypertension and Cardiovascular Disease

Currently more than 50 million Americans—and more than 50% of the American population over the age of 60—have hypertension, or high blood pressure (Hagberg 2005). Worldwide, the number of people with hypertension is estimated to be 1 billion! Furthermore, because blood pressure tends to increase with age, the number of people with hypertension is likely to rise significantly as the U.S. population continues to age. Unfortunately, due to a lack of symptoms a large number of individuals are unaware that they have *the silent killer*, as hypertension is called, and do not realize their hypertension is placing them at a higher risk for a heart attack or stroke. Although hypertension affects nearly every organ, its lack of symptoms means that many individuals do not realize the damage that may be occurring even at modest elevations of blood pressure.

Physical activity is considered one of the mainstays in the nonpharmacological treatment of hypertension. This chapter discusses the current definition of hypertension and introduces the physiological relationship between exercise and blood pressure, as well as provides some background and mechanistic details. Then it presents both nationally and internationally issued activity guidelines for individuals with hypertension. Also, because of the close and causal relationship between hypertension and CAD, this chapter also includes guidelines that pertain to CAD.

Benefits of Exercise for Hypertension

The definition of hypertension has changed over time. Currently, the Seventh Report of the Joint National Committee on Prevention, Detection, Evaluation, and Treatment of High Blood Pressure (JNC 7: NHBPEP 2003) defines normal blood pressure as a pressure less than 120/80 mmHg (see table 10.1). Elevated blood pressure, which at one time was diagnosed only when in excess of 160/100 mmHg, is a primary cardiac risk factor and also increases the risk for stroke, type 2 diabetes, kidney disease, and vascular disease. According to the JNC 7, the risk of CVD begins at only 115/75 mmHg and doubles with each incremental increase of 20/10 mmHg. People who are classified as prehypertensive have twice the risk of developing hypertension. Even individuals with normal blood pressure have a high lifetime risk of developing hypertension—an estimated 90% risk for individuals with normal blood pressure at age 55 (Vasan and others 2002)!

For people with prehypertension, exercise is one of the primary recommended lifestyle changes that can lower blood pressure to a normal level. In fact, even in individuals who have a normal blood pressure, regular aerobic exercise may lower resting blood pressure, although the extent of lowering is less than that seen in individuals with frank hypertension. In the *JNC 7*, physical activity compares favorably with the other recommended lifestyle changes when it comes to reducing resting blood pressure—table 10.2 gives examples of these changes and their relative effects. Also, physical activity may contribute to weight loss, which can reduce blood pressure even further.

So, although the distensibility or pliability of blood vessels decreases with age and makes hypertension more likely, hypertension is not necessarily an unavoidable consequence of aging, because regular physical activity is clearly effective in preventing hypertension and reducing its intensity. A single session of aerobic exercise creates a sustained hypotensive response that lasts up to 24 h (Hagberg and others 2000), and over time participation in regular physical activity can have long-term benefits. Postulated mechanisms for the reduction of both systolic and diastolic blood pressure that occurs with regular exercise training include the following:

- Reduced visceral fat
- Improved sodium elimination due to altered renal function
- Reduced plasma renin and catecholamine activity
- Reduced sympathetic and increased parasympathetic tone

Because physical activity decreases hypertension, it also reduces the potential contribution of hypertension to cardiovascular risk. In addition to lowering blood pressure, physical activity generally causes a regression of left ventricular hypertrophy, improves lipid profiles, and lowers the risk for type 2 diabetes. This combination of risk reduction has a substantial long-term benefit on cardiovascular risk. There are also notable benefits to other health parameters. Individuals with hypertension may not necessarily achieve normal blood pressure through exercise, but they will likely see improvements in

TABLE 10.1 Classification of Blood Pressure

Classification	Systolic blood pressure (mmHg)	Diastolic blood pressure (mmHg)
Normal	<120	And <80
Prehypertension	120-139	Or 80-89
Stage 1 hypertension	140-159	Or 90-99
Stage 2 hypertension	≥160	Or ≥100

TABLE 10.2 Relative Reduction in Resting Blood Pressure by Different Lifestyle Modifications

Modification	Approximate systolic blood pressure reduction
Weight reduction	5-20 mmHg for every 22 lb or 10 kg of weight lost
Adoption of DASH eating plan (see chapter 17)	8-14 mmHg
Dietary sodium restriction (<2.4 g/d of sodium or 6 g/d of sodium chloride)	2-8 mmHg
Physical activity (at least 30 min/d most days of the week)	4-9 mmHg
Moderation of alcohol consumption	2-4 mmHg

lean body mass, glycemic control, and inflammatory biomarkers.

Overall, a regular physical activity program lowers both systolic and diastolic blood pressure. However, during aerobic exercise, the normal physiological response is a gradually rising systolic blood pressure with a stable or slightly decreasing diastolic blood pressure. In addition, during resistance training and particularly during high-intensity efforts, blood pressure may rise significantly. These natural responses to exercise necessitate guidelines for people with hypertension so that they may receive the benefits of activity safely without increasing their risk for injury or disease.

The use of antihypertensive medications also necessitates physical activity guidelines. Exercise is recommended as an adjunct to antihypertensive medication as there appears to be an additive effect of physical activity with most antihypertensive medications. Beta-blockers, however, may interfere with the antihypertensive effect of exercise (Ades and others 1988). For hypertensive individuals who regularly participate in physical activity, beta-blockers also blunt the peak heart rate and blood pressure; the former effect requires using an alternative to target heart rate for measuring exercise intensity.

American College of Sports Medicine Guidelines

The ACSM published a position statement providing exercise guidelines for individuals with hypertension initially in 1994 and then released an update to these guidelines in 2004. The recommended frequency of physical activity is most days—most likely this frequency is given to help people with hypertension to take advantage of the postexercise hypotensive response. Resistance training guidelines are also included. Note that the AHA has also issued separate guidelines for resistance training. Individuals who exercise frequently and also need antihypertensive medications are recommended to take either angiotensin converting enzyme (ACE) inhibitors or calcium channel blockers. The ACSM guidelines state that exercise should be delayed if blood pressure is significantly elevated, but a specific parameter is not given.

Although the ACSM guidelines were updated in 2004, they still contain some gray areas. For example, the position statement recommends aerobic exercise on most, if not all, days of the week but does not fully delineate this recommendation. Also, the strength training section

GUIDELINE 10.1

Title: "Exercise and Hypertension"

Organization: ACSM

Year published: 2004

Purpose: To prevent, treat, and control hypertension

Location: *Medicine and Science in Sports and Exercise* 36(3): 533-553

Population: Individuals with hypertension and no CVD or renal complications

GUIDELINES

Component	Frequency	Intensity	Duration	Type
Aerobic exercise	Most, preferably all, days of the week	40%-<60% $\dot{V}O_2R$; moderate intensity	At least 30 min/d of continuous or accumulated physical activity	Primarily endurance physical activity

>continued

GUIDELINE 10.1 >*continued*

Component	Frequency	Intensity	Duration	Type
Resistance training	Should serve as an adjunct to an aerobic exercise program; guidelines follow 2000 AHA report (see guideline 10.10)			
Flexibility training	Not specified			

Special considerations include the following:

- Graded exercise testing may be warranted for men 45 y or women 55 y or older who are planning a vigorous exercise program (of at least 60% $\dot{V}O_2R$).
- If medication is needed, ACE inhibitors and calcium channel blockers are the drugs of choice.
- Individuals with markedly elevated blood pressures should wait to initiate exercise until after physician evaluation and initiation of pharmacological therapy.

GUIDELINE 10.2

Title: "Hypertension"

Organization: ACSM

Year published: 2009

Purpose: To increase aerobic capacity, muscle strength, endurance, and ability to perform ADLs; to decrease hypertensive and heart rate response to submaximal exercise; to decrease CAD risk factors; and to decrease submaximal myocardial $\dot{V}O_2$ demand

Location: *ACSM's Exercise Management for Persons with Chronic Diseases and Disabilities, Third Edition*

Population: Individuals with a history of hypertension

GUIDELINES

Component	Frequency	Intensity	Duration	Type
Aerobic exercise	3-7 d/wk	50%-80% HRmax; 40%-70% $\dot{V}O_2$max or HRR; RPE of 11-14 out of 20	30-60 min each session or energy expenditure of 700-2,000 kcal/wk	Large-muscle activities
Resistance training	2-3 d/wk	Low resistance	High number of repetitions (10-15)	Circuit training

Special considerations include the following:

- Resting blood pressure of either >200 mmHg systolic or >115 mmHg diastolic is a contraindication to exercise.
- The initial goal for caloric expenditure is 700 kcal/wk, ultimately progressing to 2,000 kcal/wk.

Adapted, from N.F. Gordon, 2009, Hypertension. In *ACSM's exercise management for persons with chronic diseases and disabilities*, 3rd ed., edited by J.L. Durstine et al. (Champaign, IL: Human Kinetics), 112.

simply acknowledges that the AHA guidelines are appropriate without providing any other stipulations.

The 2009 *ACSM's Exercise Management for Persons With Chronic Diseases and Disabilities, Third Edition*, was an update to earlier guidelines, which had been published prior to the updated 2004 position statement on hypertension. These 2009 guidelines provide a greater amount of detail although overall they are not significantly different from the 2004 recommendations. One point addressed in *ACSM's Exercise Management for Persons With Chronic Diseases and Disabilities* is that resting blood pressure seems to be lowered by moderate-intensity activity just as much as, if not more than, it is lowered by activity at higher intensities.

The eighth edition of *ACSM's Guidelines for Exercise Testing and Prescription*, published in 2010, also contains guidelines for individuals with hypertension. These are slightly different from those reco`mmended in the 2004 position statement, possibly due to the 2008 publication of the *Physical Activity Guidelines for Americans (PAGA)* as well as the 2007 publication of the joint ACSM and AHA guidelines that occurred in the interim.

International Guidelines

In 1999 the Canadian Hypertension Society released guidelines on using physical activity for the prevention and control of hypertension. Although the report states that its recommendations agree with those of the ACSM, among others, there are differences. The Canadian guidelines recommend moderate-intensity physical activity for 50 to 60 min 3 or 4 times per week. This recommendation was based off of a literature review that found that individuals who exercised in sessions of 50 to 60 min were more likely than people exercising in 30 to 45 min sessions to induce significant decreases in systolic or diastolic blood pressure (or both). The report noted that vigorous-intensity exercise was less effective than moderate-intensity exercise for producing antihypertensive outcomes. Like the ACSM, the Canadian group recommended that exercise be prescribed as an adjunctive therapy for people requiring pharmacological therapy for hypertension. The Canadian guide also noted that beta-blockers may not be the optimal drug for individuals with hypertension

GUIDELINE 10.3

Title: "Exercise Prescription for Other Clinical Populations"

Organization: ACSM

Year published: 2010

Purpose: To improve blood pressure at rest and submaximal exercise workloads

Location: *ACSM's Guidelines for Exercise Testing and Prescription, Eighth Edition*

Population: People with hypertension

GUIDELINES

Component	Frequency	Intensity	Duration	Type
Aerobic exercise	Most, preferably all, days of the week	40%-<60% $\dot{V}O_2R$	30-60 min of continuous or intermittent activity (in bouts of 10 min or more)	Walking, swimming, jogging, cycling
Resistance training	2-3 d/wk	60%-80% 1RM	At least 1 set of 8-12 repetitions for 8-10 exercises	Machine weights or free weights for major muscle groups

Special considerations include the following:

- Resting blood pressure of >200 mmHg systolic or >110 mmHg diastolic is a contraindication to exercise (maintain an exercise blood pressure of 220/105 mmHg or lower).
- Avoid the Valsalva maneuver during exercise.
- Severe or uncontrolled hypertension should be managed with medication before an exercise program is initiated.

Adapted, by permission, from ACSM, 2010, *ACSM's guidelines for exercise testing and prescription*, 8th ed. (Baltimore, MD: Lippincott, Williams, and Wilkins) 249-250.

who choose to exercise. This is in line with the ACSM recommendations (guideline 10.1), which state that frontline antihypertensive medications for people who exercise are ACE inhibitors and calcium channel blockers.

The British Hypertension Society (BHS) issued its guidelines for management of hypertension in 2004 in the *BHS IV* (the British equivalent of the *JNC 7*). Its recommendations for the primary prevention of hypertension include regular aerobic activity as outlined by the U.S. NHBPEP: at least 30 min of aerobic physical activity such as brisk walking on most days of the week. Along with the 30 min of activity, the guidelines also recommend maintaining a normal body weight as defined by a BMI between 20 and 25 kg/m^2. For people with established hypertension, the BHS endorses the recommendations as set forth in the U.S. NHLBI's DASH plan—that is, either moderate-intensity activity for 30 min most days of the week or (for fitter individuals) 20 min of vigorous activity 3 times a week. The BHS guidelines do not recommend isometric exercise due to its associated pressor effects. Also,

people with severe or poorly controlled hypertension should postpone heavy exercise until blood pressure under better control. The guidelines recommend exercising consistently due to the fact that the protective effect of physical activity on blood pressure is lost when exercise is discontinued. Lastly, the BHS paper states that although some forms of activity such as yoga or breathing exercises may result in short-term blood pressure reductions, it does not recommend them due to limited evidence and the resulting difficulty in developing a proper exercise prescription (Williams and others 2004).

The WHO and the International Society of Hypertension issued a joint statement on the management of hypertension in 2003. This statement supported the use of lifestyle interventions, specifically physical activity, as a valuable adjunct in the prevention and management of hypertension. Although the statement referenced Hagberg and others' 2000 article on the role of exercise training in the treatment of hypertension, it suggested no specific guidelines. Hagberg and others' 2000 review did go into more detail about

GUIDELINE 10.4

Title: "Recommendations on Physical Exercise Training"

Organization: Canadian Hypertension Society, Canadian Coalition for High Blood Pressure Prevention and Control, Laboratory Centre for Disease Control at Health Canada, and the Heart and Stroke Foundation of Canada

Year published: 1999

Purpose: To prevent and control hypertension

Location: *Canadian Medical Association Journal* 160(9 Suppl): S21-S28

Population: Otherwise healthy adults with and without hypertension

GUIDELINES

Component	Frequency	Intensity	Duration	Type
Aerobic exercise	3-4 times per week	Moderate intensity	50-60 min	Rhythmic movements with the lower limbs, such brisk walking or cycling

Special considerations include the following:

- Exercise should be prescribed as an adjunctive therapy for people who require pharmacological therapy for hypertension and especially for people who are not receiving beta-blockers.
- People who do not have hypertension should participate in regular exercise as it will decrease blood pressure and reduce the risk of CAD even though there is no direct evidence that it will prevent hypertension.

GUIDELINE 10.5

Title: *Guidelines for Management of Hypertension: Report of the Fourth Working Party of the British Hypertension Society, 2004—BHS IV*

Organization: BHS

Year published: 2004

Purpose: To manage, treat, and prevent hypertension in order to reduce the risk of CVD

Location: www.bhsoc.org/pdfs/BHS_IV_Guidelines.pdf

Population: People with hypertension

GUIDELINES

- Engage in regular aerobic activity such as brisk walking (for 30 min or more on most days of the week).
- Fit younger patients may engage in 3 vigorous training sessions per week, while older patients may engage in brisk walking for 20 min/d.
- People with hypertension should not participate in isometric exercise such as heavy weightlifting.
- Patients with severe or poorly controlled hypertension should postpone heavy physical exercise until appropriate drug therapy has been instituted and found to be effective.

---

the frequency, intensity, and duration of exercise and the quantitative effects of exercise on blood pressure, but again no specific guidelines were noted.

Guidelines for Coronary Artery Disease

Guidelines for CAD are included in this chapter because the physiological demands on the heart are closely tied to the rate–pressure product achieved during exertion, which is dependent on blood pressure. Currently, 16.8 million people in the United States are living with CAD (AHA 2009). According to the AHA, data from 2008 suggest that heart disease, including CAD, stroke, and hypertension, costs the United States more than $287 billion U.S. every year. Direct and indirect costs of CHD alone are estimated at more than $156 billion U.S. every year! Physical activity and exercise have been a cornerstone in the rehabilitation and prevention of CAD. As a result, in the United States there are perhaps more medically supervised cardiac rehabilitation exercise programs than there are programs for any other disease process. Categories of cardiac

conditions affecting exercise prescription are listed in the sidebar on p. 175.

Although just a few decades ago there was no consensus for the appropriate amount of physical activity for patients with CAD (and exercise was frowned upon as potentially harmful in worsening myocardial oxygen demand), today exercise as part of cardiac rehabilitation is a mainstay in the treatment and secondary prevention of heart disease. Mechanisms for cardiovascular benefits of physical activity are now well established and go beyond improving traditional risk factors such as hypertension, lipid profile, body composition, and glycemic control. Regular exercise reduces inflammatory markers and thrombotic factors as well as improves autonomic regulation.

Aerobic Exercise Guidelines

In 1994, the ACSM issued the position stand "Exercise for Patients With Coronary Artery Disease" in order to delineate the risks and benefits of exercise for CAD. In the setting of a properly designed exercise program, individuals with CAD can reap significant physical and emotional benefits. The aerobic activity recommendations seem somewhat shorter in duration when compared with guidelines for other individuals, but these

GUIDELINE 10.6

Title: "Exercise for Patients With Coronary Artery Disease"

Organization: ACSM

Year published: 1994

Purpose: To achieve optimal physical and emotional health while minimizing risk for cardiovascular complications

Location: *Medicine and Science in Sports and Exercise* 26(3): i-v

Population: Individuals with CAD

GUIDELINES

Component	Frequency	Intensity	Duration	Type
Aerobic exercise	At least 3 non-consecutive days a week and up to daily if desired	40%-85% of HRR or $\dot{V}O_2$max); 55%-90% of HRmax; an RPE of *moderate*; below ischemia-provoking levels	20-40 min sandwiched between 10 min of warm-up and 10 min of cool-down	Continuous large-muscle activities, including walking, jogging, bicycling, swimming, aerobics, rowing
Resistance training	As part of aerobic exercise, using circuit training	10-12 repetitions performed comfortably	10-12 separate exercises	Circuit training
Flexibility exercise	Before and after aerobic exercise	Low	10 min warm-up and 10 min cool-down	Not available

Special considerations include the following:

- Graded exercise testing is required before participation in physical activity.
- People with acute cardiac issues (such as exertional chest pain, uncontrolled hypertension or unstable angina) should defer exercise until problems are resolved.
- Reevaluation should be performed regularly, generally 2 to 3 mo after the exercise program begins and annually thereafter.
- High-risk patients may exercise under direct medical supervision.

1994 recommendations should be taken in context with other guidelines issued during that time. For example, the 1996 *Physical Activity and Health: A Report of the Surgeon General* recommended at least 30 min of aerobic activity on most (usually interpreted as 5) days of the week—a total of 150 min/wk. The ACSM guidelines for CAD are for at least 3 d of 20 to 40 min of activity—a total of 60 to 120 min/wk. Nevertheless, when compared with the recent *PAGA*—which suggests at least 150 min/wk and optimally 300 min/wk—the ACSM CAD guidelines seem somewhat conservative.

The 2009 *ACSM's Exercise Management for Persons With Chronic Diseases and Disabilities* has an entire section dedicated to individuals with various types of CVD. The severity of CVD ranges along a continuum from asymptomatic atherosclerosis picked up with computed tomography to substantial and symptomatic ischemia or angina representing a mismatch between myocardial oxygen supply and demand. People with a history of myocardial infarction or CAD critical enough to warrant revascularization via percutaneous transluminal coronary angioplasty (PTCA) or coronary artery bypass graft (CABG) surgery are considered in this 2009 ACSM text. Although there are some differences between these ACSM guidelines and those published in the 1994 position statement, the rationale for exercise training of individuals with CVD is the

Categorization of Cardiac Conditions for Aerobic Exercise Prescription

Cardiac Contraindications to Exercise

- Unstable angina
- Severe aortic stenosis
- Uncontrolled cardiac arrhythmia
- Decompensated congestive heart failure
- Acute myocarditis or infectious disease

High-Risk Cardiac Conditions Requiring Medical Supervision During Exercise

- Severely depressed left ventricular function
- Decreased systolic blood pressure with exercise
- History of sudden cardiac arrest
- Recent myocardial infarction complicated by congestive heart failure
- Marked exercise-induced ischemia
- Resting complex ventricular arrhythmia
- Ventricular arrhythmia appearing or increasing with exercise

GUIDELINE 10.7

Title: "Myocardial Infarction" and "Coronary Artery Bypass Graft Surgery and Percutaneous Transluminal Coronary Angioplasty"

Organization: ACSM

Year published: 2009

Purpose: To increase aerobic capacity, muscle strength, endurance, and ability to perform ADLs; to decrease hypertensive and heart rate response to submaximal exercise; to decrease CAD risk factors; and to decrease submaximal myocardial $\dot{V}O_2$max demand

Location: *ACSM's Exercise Management for Persons With Chronic Diseases and Disabilities, Third Edition*

Population: Individuals with a history of myocardial infarction, PTCA, or CABG

GUIDELINES

Component	Frequency	Intensity	Duration	Type
Aerobic exercise	At least 3 d/wk	40%-80% of $\dot{V}O_2$max or HRR; RPE of 11-15 on scale of 20	20-40 min each session (myocardial infarction); 20-60 min each session (PTCA or CABG surgery)	Large-muscle activities or arm or leg ergometry
Resistance training	2-3 d/wk	40%-50% maximal voluntary contraction (avoid Valsalva maneuver)	1-3 sets of 10-15 repetitions of 8-10 exercises	Circuit training

>continued

GUIDELINE 10.7 >*continued*

Component	Frequency	Intensity	Duration	Type
Flexibility exercise	2-3 d/wk	Not specified	Hold for 10-30 s	Static stretches of the upper and lower body

Special considerations include the following:

- Supervision is suggested for moderate- to high-risk patients, including patients with exercise-induced myocardial ischemia, ST segment depression, or angina pectoris and patients with an ejection fraction of less than 30%.
- Individuals who are less fit and have a functional capacity of 5 METs or less may work at the lowest end of the range.
- After CABG surgery, individuals should start with 1 to 2 lb (0.5-1 kg) weights and wait 12 wk before utilizing heavier weights.

Adapted, by permission, from B.A. Franklin, 2009, Myocardial infarction. In *ACSM's exercise management for persons with chronic diseases and disabilities*, 3rd ed., edited by J.L. Durstine et al. (Champaign, IL: Human Kinetics), 55.

GUIDELINE 10.8

Title: "Revascularization of the Heart"

Organization: None

Year published: 2009

Purpose: To improve cardiac performance, exercise capacity, and angina-free exercise tolerance

Location: *Clinical Exercise Physiology, Second Edition*

Population: Individuals with a history of PTCA with or without stenting or a history of CABG surgery

GUIDELINES

Component	Frequency	Intensity	Duration	Type
Aerobic exercise	Daily	70%-85% of HRmax or RPE of 11-15 out of 20 for asymptomatic individuals; below ischemic threshold or RPE of 11-15 out of 20 for symptomatic individuals	At least 30 min, may be intermittent or continuous	Treadmill walking, cycling, arm and leg exercise, rowing, and stair-climbing
Resistance training	2-3 times per week	Weight that is moderately difficult to lift during the last repetitions (and does not cause significant straining)	8-10 repetitions	Elastic bands, hand weights, free weights, and multiple-station machines
Flexibility exercise	Daily	Static stretching	5-10 min	Range-of-motion and flexibility exercises

Special considerations include the following:

- Groin soreness (due to transcatheter procedures) or incisional discomfort (due to CABG surgery) may require transient restriction of certain activities.
- After CABG surgery, patients should initially start upper-extremity exercise with very little resistance.

GUIDELINE 10.9

Title: "Exercise Prescription for Other Clinical Populations: Cardiac Disease"

Organization: ACSM

Year published: 2010

Purpose: To implement a safe and effective formal exercise and lifestyle physical activity program

Location: *ACSM's Guidelines for Exercise Testing and Prescription, Eighth Edition*

Population: Individuals with CVD

GUIDELINES

Component	Frequency	Intensity	Duration	Type
Cardiorespiratory exercise	At least 4-7 d/wk	40%-80% of HRR or $\dot{V}O_2R$; RPE of 11-16 out of 20	20-60 min total (may begin with 1-10 min sessions)	Rhythmic exercises working large-muscle groups
Resistance training	2-3 d/wk with at least 48 h separating training sessions for the same muscle group	60%-80% 1RM (may start with 30%-40% 1RM upper body and 50%-60% 1RM lower body); RPE of 11-13 out of 20	8-10 exercises, 2-4 sets, 8-12 repetitions per set	Include each major muscle group, using large groups before small groups; include multijoint exercises

Adapted, by permission, from ACSM, 2010, *ACSM's guidelines for exercise testing and prescription*, 8th ed., (Baltimore, MD: Lippincott, Williams, and Wilkins), 212-213.

same: Exercise induces physiological changes in body composition, lipid levels, blood pressure, heart rate, $\dot{V}O_2$max, and oxygen demand that protect against future adverse clinical outcomes.

The most recent edition of *ACSM's Guidelines for Exercise Testing and Prescription* includes a section on outpatient cardiac rehabilitation programs. These programs are recommended to help individuals with cardiac disease implement a formal exercise and lifestyle physical activity program. Guidelines are similar to those recommended for all healthy individuals, although there is an emphasis on safety, developing independence, and modifying the program if necessary.

Resistance Training Guidelines

The AHA issued specific guidelines for resistance exercise in 2000; these were later updated in 2007. These guidelines address the proper type of exercise both for individuals with and for individuals without CVD. The guidelines for healthy individuals are presented in chapter 2. For people with CVD, there are several modifications to the general resistance training recommendations. These reflect the fact that several types of CAD may predispose individuals to different risks. For example, individuals who have undergone CABG surgery may experience sternal wound complications and thus should not do upper-body exercises initially after their surgery. Nevertheless, a minimum amount of regular exercise is necessary to perform ADLs and to prevent adhesions. It is recommended that flexibility activities be initiated as early as 24 h after CABG surgery and 2 d after myocardial infarction.

These guidelines are important because although the prescription and benefits of aerobic training have been well known for awhile now, the prescription and benefits of resistance training—while known by professionals—have not been as widely embraced by either the public or the media. This holds true for resistance exercise for all individuals—not just people with hypertension or CAD. Perhaps this lack of awareness has to do with the push to promote at least a *basal*

GUIDELINE 10.10

Title: "Resistance Exercise in Individuals With and Without Cardiovascular Disease"

Organization: AHA Committee on Exercise, Cardiac Rehabilitation, and Prevention, Council on Clinical Cardiology

Year published: 2000 (updated in 2007)

Purpose: To improve muscular strength and endurance, cardiovascular function, metabolism, coronary risk factors, and psychosocial well-being

Location: *Circulation* 116: 572-584

Population: Adults with CVD

GUIDELINES

Group	Frequency	Intensity	Duration	Type
RESISTANCE TRAINING				
Individuals who have recently had myocardial infarction or CABG surgery	2-3 d/wk, starting 2-3 wk after event	To moderate fatigue; 12-13 out of 20 RPE	1 set of 10-15 repetitions	1-2 lb (0.5-1 kg) dumbbells or wrist weights, progressing by 1-2 lb (0.5-1 kg) increments every 1-3 wk
Individuals who have had myocardial infarction or CABG surgery and have completed the training recommended in the previous row for 4-6 wk	2-3 d/wk	Individuals at moderate risk should exercise to RPE of 15 or less out of 20; individuals at low risk may progress to volitional fatigue	10-15 repetitions	Start at low weight and progress by 2-5 lb/wk (1-2.3 kg/wk) for arms and 5-10 lb/wk (2.3-4.5 kg/wk) for legs
FLEXIBILITY TRAINING				
Individuals who have recently had myocardial infarction or CABG surgery	Daily, starting at 24 h after surgery or 2 d after infarction	Light to somewhat hard (RPE of 11-13 out of 20)	10-15 repetitions	Upper- and lower-extremity range-of-motion (flexibility) exercises

Special considerations include the following:

- Individuals may progress to using regular barbells or weight machines after 4 to 6 wk.
- Individuals who have had CABG surgery or sternotomy within the past 3 mo should avoid resistance training exercises that pull on the sternum.
- Pure isometric exercise is not recommended for patients with CVD.

level of activity and the assumption that individuals will become overwhelmed if presented with too many criteria and recommendations. Traditionally, at least, the media's interpretation of influential guidelines has emphasized the aerobic component while often leaving out the strength component. Physiological parameters that improve with resistance training include body composition and muscle mass, $\dot{V}O_2$max, ventilatory anaerobic threshold, and local muscular endurance. Of interest is a recent study on individuals with documented CAD that showed that an exercise program consisting of both aerobic and resistance training yielded greater gains in $\dot{V}O_2$max, muscular strength and endurance, and body composition (Marzolini and others 2008). What is notable from this study is that the subjects substituted sessions of resistance exercise for

GUIDELINE 10.11

Title: "Resistance Exercise Program"

Organization: American Association of Cardiovascular and Pulmonary Rehabilitation

Year published: 2004

Purpose: To safely participate in a resistance exercise program

Location: *Guidelines for Cardiac Rehabilitation and Secondary Prevention Programs, Fourth Edition*

Population: Adults with CVD

GUIDELINES

Frequency	Intensity	Sets and type of program
2-3 d/wk	RPE of 11-13 out of 20; initial load should allow 12-15 repetitions comfortably (30%-40% of 1RM for upper body or 50%-60% of 1RM for hips and legs); increase loads by 5% when 12-15 repetitions are comfortable	1 set of 6-8 exercises (major muscle groups); an additional set may be added, but additional gains are not proportionate; exercise large-muscle groups before small-muscle groups

Special considerations include the following:

- Lift weights with slow, controlled movements; emphasize complete extension of the limbs when lifting.
- Avoid straining.
- Exhale during the exertion phase of the lift and inhale when lowering the weight.
- Avoid sustained, tight gripping, which may evoke an excessive blood pressure response.
- Stop exercise if warning signs or symptoms occur, especially dizziness, dysrhythmia, unusual shortness of breath, or anginal discomfort.

Participation criteria include the following:

- Minimum of 5 wk elapsed after date of myocardial infarction or cardiac surgery, including 4 wk of consistent participation in a supervised cardiac rehabilitation endurance training program
- Minimum of 3 wk following transcatheter procedure (PTCA or other), including 2 wk of consistent participation in a supervised cardiac rehabilitation endurance training program
- No evidence of the following conditions:
  - Congestive heart failure
  - Uncontrolled dysrhythmia
  - Severe valvular disease
  - Uncontrolled hypertension (patients with moderate hypertension such as a systolic pressure >160 mmHg or a diastolic pressure >100 mmHg) should be referred for appropriate management, although these values are not absolute contraindications for participation in a resistance training program)
  - Unstable symptoms

Adapted, by permission, from AACVPR, 2004, *AACVPR guidelines for cardiac rehabilitation and secondary prevention programs*, 4th ed. (Champaign, IL: Human Kinetics), 119.

sessions of aerobic exercise; that is, the multiple improvements occurred despite a 28% reduction in the number of aerobic training sessions! The latest guidelines on resistance training as part of cardiac rehabilitation issued by the American Association of Cardiovascular and Pulmonary Rehabilitation were published in 2004. These guidelines emphasize ensuring that an individual can safely participate in a program before initiating that program. Individuals should be cleared by a medical director—after participating in a supervised cardiac rehabilitation program—before beginning a personal resistance exercise program. People who have undergone a percutaneous procedure such as angioplasty should wait a minimum of 3 wk before starting a resistance program. People who have either had a myocardial infarction or cardiac surgery should wait a minimum of 5 wk. This is longer than what is recommended in the more current 2007 AHA guidelines, perhaps reflecting the more recent belief that a quicker return to activity is not only safe but also beneficial. Contraindications to participation are included in the guidelines.

The ACE has issued a *Fit Facts* on exercising with heart disease. Because this publication addresses individuals with a diverse group of heart diseases, including CAD, hypertension,

stroke, and congestive heart failure, the recommendations are very generalized. These guidelines advise consulting with a physician before initiating an exercise program, not only to obtain medical clearance to participate in a program but also to obtain specific instructions on the features to be included in the exercise program. General guidelines include exercising 3 or 4 d/wk and starting with a warm-up and ending with a cool-down. Exercise duration may be as short as 5 to 10 min completed several times a day and as long as 60 min once the individual is appropriately conditioned. Lastly, exercising within a specified target zone of comfort and safety is emphasized.

GUIDELINE 10.12

Title: *Exercising With Heart Disease*

Organization: ACE

Year published: 2001

Purpose: To positively influence blood pressure, cholesterol, diabetes, and obesity risk

Location: www.acefitness.org/fitfacts/fitfacts_display.aspx?itemid=2600

Population: Individuals with a history of CAD, hypertension, stroke, or congestive heart failure

GUIDELINES

- Obtain medical clearance and guidelines from a physician before increasing activity level.
- Monitor exercise closely and stay within an individualized heart rate zone.
- Exercise at least 3 or 4 times per week.
- If necessary, start with 5 to 10 min sessions performed 2 or 3 times per week.
- Perform a gradual warm-up and cool-down of at least 10 min.
- Gradually increase total exercise duration to 60 min over a time frame of 1 to 6 mo.
- Stop exercising and call a physician if abnormal signs or symptoms occur, including chest pain, labored breathing, and extreme fatigue.

Adapted, by permission from American Council on Exercise Fit Facts, San Diego 2001. www.acefitness.org/fitfacts Accessed 10/10/2009.

SUMMARY

Hypertension and CAD are very common chronic diseases that have a significant overlap in prevalence and risk. Exercise is not only permissible but also beneficial for individuals with hypertension and CAD, providing both short-term and long-term effects. The guidelines covered in this chapter emphasize both aerobic and resistance training for individuals with hypertension and CAD, and specific details are given for their safe implementation. With the success of preventive medicine, many more individuals are surviving with CAD; physical activity is a mainstay not only in the rehabilitation from CAD but also in the prevention of recurrent CAD.

Arthritis and Osteoporosis

Both arthritis and osteoporosis are major public health problems, as 2.1 million Americans have been diagnosed with rheumatoid arthritis (RA), 27 million Americans with osteoarthritis (OA), and more than 10 million Americans with osteoporosis (PAGAC 2008). The prevalence of OA and osteoporosis increases with age; both are more common in women. RA is often diagnosed between the ages of 20 and 50, although it can manifest at any age. Individuals who are overweight are more likely to develop OA and may also experience more symptoms from OA due to the increased stress that added body weight puts on the joints. On the other hand, individuals who are underweight are actually at a higher risk for osteoporosis because they get less of the overall day-to-day impact on the bones that stimulates bone health.

Physical activity is one of the cornerstones in OA management. Exercise modulates symptoms of arthritis in multiple ways, by improving body weight, muscular strength, and flexibility and also by possibly modulating inflammatory cytokines. RA is an autoimmune disease in which the immune system attacks healthy joint tissue, causing joint injury and inflammation. Because much of the disability associated with RA is attributed to lack of fitness, it follows that regular physical activity should be an important part of the treatment for RA. In fact, the risks for CAD, diabetes, and osteoporosis are higher for individuals with RA, and physical activity can effect positive outcomes for each of these diseases. For people with osteoporosis, exercise is recommended not only as a bone-strengthening endeavor but also a means to reduce fall risk. In addition, people with osteoporosis who exercise have a lower risk for fracture—even if they do fall—when compared with nonexercisers. Thus physical activity, particularly activity that focuses on functional activities, balance, and flexibility, reduces the primary morbidity associated with osteoporosis: fracture.

Although RA, OA, and osteoporosis are separate entities, they are interrelated. Systemic factors such as genetics, diet, estrogen use, and bone density as well as local biomechanical factors such as muscle weakness, obesity, and joint laxity play a role in both arthritis and osteoporosis. For example, individuals with RA are at increased risk for osteoporosis and fracture, possibly due to pain and loss of joint function, direct bone loss, and glucocorticoid medications that are often prescribed for significant disease. Modifying some of these risk factors through physical activity could help to prevent pain and disability associated with OA, RA, and osteoporosis.

Rheumatoid Arthritis

Criteria for the diagnosis of RA were developed in 1958 and revised in 1987. An individual meeting four of the following criteria receives the diagnosis of RA (JHAC n.d.):

- Morning stiffness: Stiffness that occurs within and surrounding joints and lasts at least 1 h before maximal improvement
- Arthritis: Soft tissue swelling or fluid in a minimum of three joint areas (possibilities include right or left proximal interphalangeal, metacarpophalangeal, wrist, elbow, knee, ankle, and metatarsophalangeal joints)
- Arthritis of hand joints: Soft tissue swelling or fluid in a minimum of one wrist, metacarpophalangeal, or proximal interphalangeal joint
- Symmetric arthritis: Simultaneous involvement of the same joint areas bilaterally
- Rheumatoid nodules: Subcutaneous nodules observed by a physician
- Serum rheumatoid factor: Abnormal amounts of serum rheumatoid factor by any method for which the result has been positive in <5% of normal control subjects
- Radiographic changes: Radiographic changes typical of RA on posteroanterior hand or wrist radiographs

Goals for physical activity in people with RA include preserving or restoring range of motion and flexibility around affected joints, increasing muscle strength and endurance to build joint stability, and increasing aerobic capacity in order to both enhance psychological state and decrease the risk of CVD (Nieman 2000). Exercise appears to have an analgesic effect on joint pain, presumably through the release of beta-endorphins. Also, vigorous exercise can improve joint mobility, muscle strength, and aerobic capacity without causing joint damage.

The three components of the exercise prescription—aerobic exercise, strength training, and flexibility exercise—are all essential to a program for individuals with RA. However, individuals who are currently experiencing diffuse flare-ups (whether localized joint inflammation or uncontrolled systemic disease), should consider resting or at least significantly modifying their program to include only static strengthening and gentle range-of-motion exercises.

- **Aerobic exercise:** Both short-term and long-term aerobic exercise programs can benefit people with RA, increasing aerobic capacity by at least 20%, improving endurance, and enhancing mood. In general, low-impact activities such as walking, biking, swimming, rowing, and cross-country skiing are safe and present low risk for injury to individuals who are deconditioned and have poorly supported joints. Of course, activity selection should be based on the extent of RA joint involvement, underlying aerobic capacity, and patient preferences. A good alternative to continuous exercise is interval training utilizing sporadic exercise and rest sessions to improve aerobic capacity while limiting and dispersing joint loads. Various entities have recommended using heart rate, RPE, or the talk test to measure intensity. The concept of progression is very important for people with RA because some individuals who can perform very little activity in the beginning can work up gradually in frequency, intensity, and duration. As with general physical activity recommendations, accumulating aerobic exercise in short sessions is a viable option, particularly because it may enable an appropriate *total* volume of activity while preventing a significant degree of joint loading all at once. The ACSM recommends using large-muscle activities and gradually working up to 30 min sessions. Initial emphasis is on improving the duration of exercise, while the goals for improving $\dot{V}O_2$max, peak work, and endurance may be achieved over 4 to 6 mo (NCPAD 2007b).

- **Strength training:** Strength training in the form of dynamic weight-bearing exercise can significantly enhance muscle strength for people with RA (van den Ende and others 1996; Hakkinen and others 1999; Stenstrom 1994; Iversen and others 1999). By improving a person's muscle strength and ability to perform ADLs, a strengthening program can truly make a difference in quality of life. Isometric, isotonic, and isokinetic exercises all improve muscle strength (Iversen and others 1999).

- **Flexibility training:** A major morbidity of RA is pain and loss in joint range of motion and the accompanying decline in the ability to perform ADLs. Several studies have found that flexibility programs can preserve range of motion across several joints, including those in the trunk, shoulder, elbow, wrist, hip, knee, thumb, and ankle (Minor and others 1989; Stenstrom 1994; van den Ende and others 1996). The ACSM recommends stretching 1 or 2 times daily, even during acute flare-ups, and especially when other types of physical activity are contraindicated. Static stretching is safest, and 10 to 30 s can lengthen the tissues surrounding a joint enough to increase range of motion and flexibility. Ultimately, a person with RA should work up to performing several sets of 10 repetitions of stretching and range-of-motion exercises daily.

Osteoarthritis

OA is the most common type of arthritis and is a degenerative condition affecting one or more joints. Although individuals experiencing OA or RA share many common characteristics, the different pathophysiology inherent to the different diseases affects individuals differently. For example, RA is an inflammatory disease that may cause systemic effects and predispose individuals to CAD, while OA damage is generally localized to the involved joints and first appears as deficits in articular cartilage. Table 11.1 summarizes the similarities and differences between the two diseases.

As important as exercise is for individuals with RA, it is also a necessity for people with OA. Because OA becomes more prevalent as individuals age, the associated pain may contribute to disability and may decrease physical activity. People who have OA and do not exercise may be doing themselves harm not only by increasing their propensity for obesity, CVD, and diabetes (Ries and others 1995) but also by allowing their OA to worsen. Additionally, according to the first National Health and Nutrition Examination Survey (NHANES I) and the epidemiological follow-up, knee OA in women is associated with a decreased rate of survival (Lawrence and others 1990). Prolonged inactivity due to the pain of OA may decrease muscle strength. Additionally, fluid distention of the knee joint capsule inhibits quadriceps muscular contraction, which exacerbates losses of muscle strength. Well-conditioned muscles are necessary to provide joint stability and function and may decrease the impact load on the joint itself.

Exercise programs for people with OA focus on the three core components of any general exercise program (see chapter 3): aerobic or cardiorespiratory exercise, resistance training or muscle conditioning, and flexibility exercises. The last seems obvious when addressing a disease defined by joint stiffness, joint inflammation, and a general lack of mobility, but muscle conditioning and aerobic fitness are imperative as well. In the FAST Trial, older individuals with knee OA had significant improvements in symptoms with both aerobic and muscle-strengthening exercises (Ettinger and others 1997).

Exercise programs for individuals with OA may be performed as part of a home exercise

TABLE 11.1 Key Features of OA and RA

Disease	Features	Affected joints	Common features
OA	• Articular cartilage destruction • Localized degeneration • Osteophyte formation	Cervical and lumbar spine, hips, knees, hands (distal interphalangeal joints), thumbs	• Joint pain and stiffness • Anti-inflammatories and physical activity are mainstays of treatment • Increased metabolic cost of physical activity due to pain, stiffness, gait abnormalities, and biomechanical inefficiency • Multiple benefits of physical activity, including less joint swelling, pain, weakness, muscle atrophy, and fatigue
RA	• Systemic inflammatory process • Morning stiffness lasting >60 min • Possible rheumatoid nodules or positive rheumatoid factor	Cervical spine, shoulders, elbows, wrists, hands (metacarpophalangeal joints, proximal interphalangeal joints), knees, feet (metatarsophalangeal joints)	

routine or as part of a supervised program. Muscular conditioning incorporates training for muscular strength and endurance at functional speeds and patterns. Flexibility activities utilize low-intensity, controlled movements that do not increase symptoms or pain. Joints need a sufficient range of motion as well as an adequate elasticity of periarticular tissues in order to protect themselves from high-impact loads. By engaging in range-of-motion exercises, individuals can improve function and comfort in performing ADLs. Aerobic activity provides numerous health benefits (described throughout earlier chapters of this text); prevention of weight gain and obesity are among the more relevant factors for patients with OA.

In 2000, the American College of Rheumatology (ACR) issued the most recent guidelines for medical management of OA of the hip and knee. These recommendations were based on a review of the scientific evidence available at that time and were put together by an ad hoc committee, several members of which were authors of the previous ACR guidelines in 1995. Many of the recommendations were updated from those issued in 1995. The results of the FAST Trial had become available during this time frame and confirmed the beneficial effects of both strengthening the quadriceps and participating in aerobic exercise on OA of the knee. Another randomized controlled trial for individuals with OA of the hip and knee was published in 1998 and demonstrated the efficacy of an exercise program in improving pain, muscle strength, mobility, and coordination. Interestingly, although weight loss traditionally has been regarded as a fundamental part of managing OA, there have not been convincing data to support this recommendation. However, a randomized trial from Japan, published in 1998, showed that loss of body *fat* (not body weight) of an average of 8.6 lb (3.9 kg) improves symptoms of knee OA.

The ACR recommendations include both pharmacological and nonpharmacological interventions. Several of the nonpharmacological therapy recommendations, which are listed here, involve various types of physical activity:

- Weight loss (if overweight)
- Aerobic exercise programs
- Physical therapy
- Range-of-motion exercises
- Muscle-strengthening exercises
- Occupational therapy
- Patient education
- Self-management programs (such as the Arthritis Foundation Self-Help Program)
- Personalized social support through telephone contact
- Assistive devices for ambulation
- Patellar taping
- Appropriate footwear
- Lateral-wedged insoles (for genu varum)
- Bracing
- Joint protection and energy conservation
- Assistive devices for ADLs

Another point made in the ACR recommendations is that pharmacological therapies should be used in conjunction with nonpharmacological measures, because that is how they are the most effective. The importance of various types of exercise, among the other nonpharmacological interventions, cannot be overstressed; since the publishing of the ACR guidelines in 2000, data have come to light relating an increased risk for thrombotic cardiovascular events with COX-2 selective nonsteroidal anti-inflammatory drugs. These drugs are among the first line of agents prescribed for OA, and thus it seems even more important to maximize relief of OA by nonpharmacological means.

There are multiple recommendations for exercises that influence various measured outcomes in individuals with OA. Some apply to individuals with a common diagnosis of OA, while others have been specifically studied in people with OA of one joint or compartment. A combination of manual therapy and regular exercise is recommended for OA, as this combination is clinically proven to reduce pain. In general, people with chronic pain and stiffness in weight-bearing joints can improve aerobic capacity and participate in physical activity by regularly jogging in water. An even easier activity is walking. A walking program is one of the simplest ways to start exercising, and its significant benefits include the following:

- Improved pain
- Improved functional status
- Increased stride length
- Decreased disability in getting out of bed
- Decreased disability when bathing
- Improved aerobic capacity
- Higher energy level
- Decreased medication use

General Arthritis Guidelines

The ACSM, in addition to publishing general guidelines for exercise prescription, includes a specific exercise prescription for individuals with arthritis in the eighth edition of *ACSM's Guidelines for Exercise Testing and Prescription*. Exercise and physical activity are a core component of the treatment plan for arthritis and can help alleviate symptoms of the disease as well as reduce other comorbid risk factors often present in individuals with arthritis. When pain or inflammation is present, the exercise prescription should be modified or limited. Also, a gradual warm-up may be helpful in minimizing pain. Exercise that produces weight loss may be particularly beneficial, as it decreases the load on the joints. Water exercise may also be helpful in decreasing joint loading.

The ACSM presents some slightly different guidelines in the 2009 *ACSM's Exercise Management for Persons With Chronic Diseases and Disabilities*. Whether addressing OA, RA, gout, lupus, or ankylosing spondylitis, the goals and benefits of physical activity are similar. The most salient benefit of exercise is that it reduces the effects

---

GUIDELINE 11.1

Title: "Exercise Prescription for Other Clinical Populations: Arthritis"

Organization: ACSM

Year published: 2010

Purpose: To reduce pain, maintain muscle stretch around affected joints, reduce joint stiffness, prevent functional decline, and improve mental health and quality of life

Location: *ACSM's Guidelines for Exercise Testing and Prescription, Eighth Edition*

Population: Individuals with arthritis

GUIDELINES

Component	Frequency	Intensity	Duration	Type
Aerobic exercise	3-5 d/wk	40%-<60% $\dot{V}O_2R$	Start with 5-10 min to accumulate 20-30 min/d (total of 150 min/wk)	Walking, swimming, cycling (low joint stress)
Resistance training	2-3 d/wk	40%-60% 1RM (start with 10% of max, progress at 10% per week as tolerated)	10-15 repetitions per exercise	Voluntary isometric contractions progressing to dynamic training for major muscle groups
Flexibility training	At least daily	Not specified		Stretching or range-of-motion exercises for all major muscle groups

Special considerations include the following:

- An adequate warm-up and cool-down of 5 to 10 min are critical.
- Exercise at the time of day when pain is least severe or pain medications are working.

Adapted, by permission, from ACSM, 2010, *ACSM's guidelines for exercise testing and prescription*, 8th ed., (Baltimore, MD: Lippincott, Williams, and Wilkins), 227.

of inactivity, including loss of flexibility, muscle atrophy, weakness, osteoporosis, pain, depression, and fatigue. Emphasis on a gradually progressive, low- to moderate-intensity program enhances compliance and participation. People who follow the exercise guidelines should expect improved aerobic capacity, endurance, strength, and flexibility; decreased joint swelling and pain; and reduced anxiety and depression. Preparticipation exercise testing may be performed at a modified level of intensity, depending on muscle strength and joint pain. Avoiding further pain during exercise is important to encourage continued physical activity. By utilizing low-impact activities, strengthening muscles, improving joint range

of motion, and accumulating exercise through several daily sessions if necessary, individuals with arthritis can maximize the benefits of exercise while protecting the joints.

Just as the ACSM includes arthritis in its text on exercise programs for chronic disease and disability, the NCPAD also considers arthritis a disabling condition. In fact, the NCPAD has published guidelines for both RA and OA. These primarily reflect those guidelines issued by the ACSM, with some additional information presented from other authors and sources.

Another source of recommendations (which does not differentiate between OA and other forms of arthritis) is the *Exercise and Arthritis* fact

---

GUIDELINE 11.2

Title: "Arthritis: Exercise Programming"

Organization: ACSM

Year published: 2009

Purpose: To improve cardiovascular status, muscular fitness, flexibility, and general health status

Location: *ACSM's Exercise Management for Persons With Chronic Diseases and Disabilities, Third Edition*

Population: Individuals with arthritis, including OA and RA

GUIDELINES

Component	Frequency	Intensity	Duration	Type
Aerobic exercise	3-5 d/wk	60%-80% of HRmax or 40%-60% of $\dot{V}O_2$max or RPE of 11-16 out of 20	30 min each session (may start with 5 min sessions and build up)	Large-muscle activities (walking, cycling, rowing, dancing, swimming, water aerobics)
Resistance training	2-3 d/wk	Less than pain tolerance	10-12 repetitions (may start with 2-3 repetitions and build up)	Circuit training with free weights, machines, isometric exercises, elastic bands
Flexibility exercise	Not specified	Not specified	Not specified	Not specified
Functional or neuromuscular training	Not specified	Not specified	Not specified	Exercises that improve gait and balance

Special considerations include the following:

- Progression should occur with increases in duration rather than intensity.
- High-repetition, high-resistance, and high-impact exercise is not recommended.
- Avoid overstretching unstable joints.
- Avoid medial or lateral forces.

Adapted, by permission, from M.A. Minor, 2009, Arthritis. In *ACSM's exercise management for persons with chronic diseases and disabilities*, 3rd ed., edited by J.L. Durstine et al. (Champaign, IL: Human Kinetics), 264.

sheet published online by the ACR. These recommendations provide more detail with respect to the FITT principle and emphasize a volume of activity that more closely represents activity volumes prescribed for the general population by the ACSM (at the time). According to these ACR guidelines, aerobic exercise should be performed for 30 to 60 min/d, light resistance exercise should be performed on 2 or 3 d/wk, and each session should be preceded and followed by stretching.

The American Council on Exercise also has guidelines for individuals with arthritis. These are listed in guideline 11.4. These are not differentiated between those with OA and those with RA; guidelines are the same.

The Arthritis Foundation offers specific recommendations for individuals with arthritis, although these are not separated into guidelines for different types of arthritis. One suggestion is to seek physician approval before initiating an exercise program. Also, like the ACR recommendations, these guidelines suggest that physical therapists may be helpful not only to assess specific exercises but also to advise regarding joint protection.

Osteoarthritis Guidelines

The ACR has issued guidelines for OA, as discussed earlier. These guidelines, published in 2000, do not provide many specifics. However, there are patient education handouts and suggestions for individuals with arthritis on the ACR Web site, www.rheumatology.org. On the Web site, exercises are classified as therapeutic

GUIDELINE 11.3

Title: *Exercise and Arthritis*

Organization: ACR

Year published: 2006

Purpose: To realize less pain, more energy, improved sleep, and better day-to-day function

Location: www.rheumatology.org/public/factsheets/diseases_and_conditions/exercise.asp

Population: Individuals with OA and RA

GUIDELINES
Start and finish each session with stretching, choosing exercises that minimize the stress on the most painful joints.

Component	Frequency	Intensity	Duration	Type
Aerobic exercise	5 d/wk	Moderate	30-60 min/d (may be accumulated)	Walking, aerobic dance, aquatic exercise, bicycling, stationary biking, treadmill or elliptical trainer exercise, golfing, gardening
Resistance training	2-3 times per week	Amount of weight needed to challenge the muscles without causing increased joint pain	1 set of 8-10 exercises, with 8-12 repetitions or 10-15 repetitions for older individuals	Exercises using gravity, weights, elastic bands, or weight machines
Flexibility exercise	A minimum of 3 d/wk	Not specified	5-10 times per day	Stretching, yoga, range-of-motion exercises
Body awareness training	Not specified	Not specified	Not specified	Tai chi and yoga

GUIDELINE 11.4

Title: *Exercise and Arthritis*

Organization: ACE

Year published: 2009

Purpose: To improve functional capacity to help reduce pain and fatigue associated with ADLs and to improve physical fitness

Location: www.acefitness.org/fitfacts/fitfacts_display.aspx?itemid=2593

Population: Individuals with OA and RA

GUIDELINES

Start and finish each session with stretching, choosing exercises that minimize the stress on the most painful joints.

Component	Frequency	Intensity	Duration	Type
Aerobic exercise	3-5 times per week	Not specified	Start with as little as 2 min at a time, 3 times a day; progress to 20 min sessions per day	Walking, swimming, or bicycling
Resistance training	2-3 times per week, with at least 1 d off between sessions	Not specified	Not specified	Isometric or isotonic exercises utilizing weights, elastic tubing, or exercise bands
Flexibility exercise	At least once per day	Full range of motion, never to the point of pain or discomfort	Hold stretches for 30 s	Not specified

Adapted, by permission from American Council on Exercise Fit Facts, San Diego 2008. www.acefitness.org/fitfacts Accessed 10/10/2009.

and rehabilitative, recreational and fitness, or competitive and elite, and specific exercises are recommended according to the FITT principle. This information is considerably more detailed than that presented in the 2000 ACR guidelines.

The ACR guidelines stress that exercise is a mainstay of arthritis treatment and that the overload principle is important for patients with OA. First the arthritis-related functional impairments should be addressed, and then a generalized exercise program can be started. Flexibility exercises should improve range of motion, decrease joint stiffness, and prevent contractures. Static stretching is generally included as part of the warm-up. Strength training can improve muscle strength and thus decrease joint loading. Submaximal resistance with a higher number of repetitions will achieve this goal.

The National Institute of Arthritis and Musculoskeletal and Skin Diseases (NIAMS), which is a division of the U.S. NIH, addresses exercise for both RA and OA. However, the information available stresses the importance of exercise but does not provide a specific exercise prescription. For RA, NIAMS emphasizes exercise as an ongoing foundation for joint health but suggests to avoid exercise during flare-ups. For OA, NIAMS recommends an exercise program composed of four components: (1) strengthening exercises with weights or exercise bands, (2) aerobic activities to promote cardiorespiratory fitness, (3) range-of-motion activities to improve plasticity of the joints, and (4) agility exercises to help patients maintain daily living skills. Exercise is also encouraged as a means to losing weight, which is paramount for individuals with OA.

GUIDELINE 11.5

Title: *Top Three Types of Exercise*

Organization: Arthritis Foundation

Year published: 2007

Purpose: To decrease fatigue, strengthen muscles and bones, increase flexibility and stamina, and improve general sense of well-being

Location: www.arthritis.org/types-exercise.php

Population: Individuals with OA and RA

GUIDELINES

Start and finish each session with stretching, choosing exercises that minimize the stress on the most painful joints.

Component	Frequency	Intensity	Duration	Type
Aerobic exercise	3-4 d/wk	In target heart rate zone	30 min, starting with as little as 5 min and progressing	Walking, dancing, swimming, and bicycling
Resistance training	Every other day	Not specified	Not specified	Isometric and isotonic exercises
Flexibility exercise	Every day	Not specified	15 min	Tai chi and yoga

GUIDELINE 11.6

Title: *Recommendations for the Medical Management of Osteoarthritis of the Hip and Knee*

Organization: ACR

Year published: 2000

Purpose: To control pain and improve function and health-related quality of life while avoiding toxic effects of therapy

Location: www.rheumatology.org/publications/guidelines/oa-mgmt

Population: Individuals with OA of the hip and knee

GUIDELINES

- Participate in regular exercise, including aerobic, muscle-strengthening, and range-of-motion exercises.
- Utilize physical therapy for assessment and instruction in an exercise program.
- Strengthen quadriceps weaknesses.
- Lose excess weight.

From "Recommendations for the Medical Management of Osteoarthritis of the Hip and Knee, American College of Rheumatology Subcommittee on Osteoarthritis Guidelines." *Arthritis & Rheumatism.* 43(9), 2000, pp. 1905-1907. Copyright 2007 American College of Rheumatology. Reprinted with permission of John Wiley & Sons, Inc.

GUIDELINE 11.7

Title: *Handout on Health: Osteoarthritis*

Organization: NIAMS

Year published: 2002 (updated 2006)

Purpose: To result in better function and to reduce pain

Location: www.niams.nih.gov/Health_Info/Osteoarthritis/default.asp

Population: Individuals with OA and RA

GUIDELINES

An exercise program should include

- strengthening exercises with weights or exercise bands,
- aerobic activities to promote cardiorespiratory fitness,
- range-of-motion activities, and
- agility exercises.

OA increases in both prevalence and severity with aging. In fact, among all individuals aged 65 and over, about half are affected by OA (American Geriatrics Society Panel on Exercise and Osteoarthritis 2001). This increases to 85% for adults aged 75 and older (Verbrugge 1995)! For this reason, the American Geriatrics Society (AGS) has issued clinical practice guidelines for older adults with OA pain. These were derived from a review of the literature and a consensus among a panel of experts from geriatrics, internal medicine, orthopedics, physical therapy and rehabilitation, exercise physiology, nursing, and pharmacy. The guidelines focus on reducing the physical impairments of OA and the burden of comorbidities in an effort to improve quality of life. Some of the risk factors for development of OA are amenable to treatment, and their modification can actually reduce the symptoms of OA in patients who already have the disease. These risk factors include obesity, muscle weakness, joint laxity, reduced proprioception, and altered biomechanics.

One point that the AGS makes is that graded exercise testing is not necessarily required before prescribing increased physical activity, despite the fact that there is an increased prevalence of CVD among older adults. Because more serious cardiovascular events occur during high-intensity training and the AGS recommends a more moderate intensity, individuals without significant cardiovascular risk factors do not need a physician-supervised exercise test. Addressing functional impairment is the first priority in the AGS guidelines; this is followed by a generalized fitness program designed to improve health and functional capacity.

The AGS has also issued guidelines for exercise prescription and OA. Patients with joint pain not only improve their symptoms with regular activity but also prevent further declines. The goal is to reduce the risk of injury while maximizing the benefits obtained by exercise.

In 1999, the NIH held a multidisciplinary conference, bringing together experts to address OA and the prevention of its onset, progression, and disability. Although the resulting report, which was published in 2000, advised that exercise programs should include training for strength and endurance, it did not provide a specific exercise prescription. It simply emphasized that low-intensity, controlled movements that do not increase pain are likely to help regain or maintain motion and flexibility in order to alleviate some of the joint stiffness and muscle deconditioning associated with OA.

The Ottawa Panel is a group of health professionals that has developed sets of evidence-based guidelines for many clinical entities related to physical health, exercise, and orthopedics. In

GUIDELINE 11.8

Title: "Exercise Prescription for Older Adults With Osteoarthritis Pain: Consensus Practice Recommendations"

Organization: AGS

Year published: 2001

Purpose: To combat declines in health and functional capacity caused by chronic diseases such as OA

Location: *Journal of the American Geriatrics Society* 49: 808-823

Population: Older individuals with OA

GUIDELINES

Exercise should start with 5 to 10 min of low-intensity range-of-motion exercises and end with 5 min of static muscle stretching.

Component	Frequency	Intensity	Duration	Type
Aerobic exercise	3-5 d/wk	Low to moderate: 40%-60% of HRmax or $\dot{V}O_2$max; RPE of 12-14 = 60%-65% $\dot{V}O_2$max; talk test	Accumulation of 20-30 min/d	Bicycling, swimming, walking, dancing, tai chi, treadmill exercise, rowing machine exercise
Resistance training (isometric)	Daily	Low to moderate: 40%-60% of maximal voluntary contraction	1-10 submaximal contractions involving key muscle groups; hold for 1-6 s	Isometric exercises
Resistance training isotonic)	2-3 sessions per week	Low: 40% 1RM Moderate: 40%-60% 1RM High: >60% 1RM	10-15 repetitions 8-10 repetitions 6-8 repetitions	Isotonic exercises
Flexibility exercise (initial)	Once every day	Stretch to subjective sensation of resistance	5-15 s each stretch; 1 stretch per key muscle group	Static stretching
Flexibility exercise (goal)	3-5 d/wk	Stretch to full range of motion	20-30 s each stretch; 3-5 stretches per key muscle group	Static stretching

Special considerations include the following:

- Aquatic exercise may reduce joint loading and provide resistance.
- During flare-ups, stretching and resistance exercise should be modified.
- Exercise should be performed when pain and stiffness are minimal or should be preceded by moist heat.
- Submaximal exercise is emphasized; working to fatigue increases the chance for injury.

The following are key muscle groups for stretching and strengthening in the AGS exercise program:

- Head and neck extensors and flexors
- Shoulder forward flexors and extensors, abductors and adductors, external and internal rotators, scapular retractors and depressors
- Elbow extensors and flexors
- Forearm and wrist pronators, supinators, extensors, flexors
- Finger flexors and extensors, thumb adductors and abductors
- Trunk flexors, extensors, and rotators
- Hip flexors and extensors, abductors and adductors
- Knees extensors and flexors
- Ankle dorsiflexors, plantar flexors, inverters, evertors, toe flexors and extensors

GUIDELINE 11.9

Title: "Osteoarthritis: New Insights. Part 2: Treatment Approaches"

Organization: NIH

Year published: 2000

Purpose: To prevent osteoarthritis onset, progression, and disability

Location: *Ann Intern Med* 133(9): 727-737

Population: Individuals with OA

GUIDELINE

Include low-intensity, controlled movements for strength and endurance.

GUIDELINE 11.10

Title: "Ottawa Panel Evidence-Based Clinical Practice Guidelines for Therapeutic Exercises and Manual Therapy in the Management of Osteoarthritis"

Organization: Ottawa Panel

Year published: 2005

Purpose: To use exercise to manage osteoarthritis

Location: *Phys Ther* 85(9):907-971

Population: Individuals with OA and RA

GUIDELINES

Condition	Type of exercise
Knee OA	Walking program Lower-extremity strengthening Isometric lower-extremity strengthening
Knee OA rest pain	Concentric and concentric–eccentric resistance training of the knee
Knee OA pain, functional status, energy level, flexion range of motion	Home strengthening program
Knee OA rest pain, range of motion	Progression lower-extremity strengthening programs
Knee OA pain and functional status	Whole-body functional exercises
Knee OA night pain and ability to use stairs	Generalized lower-extremity exercise program including muscle force, flexibility, and mobility and coordination exercises

2005, the panel published a set of criteria grading the strength of evidence pertaining to OA interventions. This group also issued guidelines for therapeutic exercises and manual therapy in the management of OA. Many of the exercises recommended in the guidelines can be incorporated into a home-based or supervised exercise program. These guidelines are different from many in this book, in that they are graded according to the evidence supporting them. The evidence-based grading system is beyond the scope of this text, and only the guidelines that have the strongest level and strength of evidence are included here.

Recommendations are to engage in general physical activity, including fitness and aerobic exercises, strengthening exercises, and a combination of exercises. Some types of exercise are recommended for one type of OA (e.g., bilateral knee OA) but not another (e.g., hand OA). Other types of exercise may be recommended for a certain outcome, such as improved functional status or range of motion but not pain relief or joint stiffness.

Lastly, the *Physical Activity Guidelines for Americans (PAGA)* includes some recommendations for individuals with OA, although there is not a separate section dedicated to the disease. The comprehensive *Physical Activity Guidelines Advisory Committee Report*, which was issued a few months before *PAGA* was made public, had reviewed and identified the scientific evidence for physical activity and health relating to OA. The conclusion was that physical activity provides therapeutic benefits to people with OA, physical activity does *not* make the condition worse, and physical activity in people with OA can lower the risk for other chronic diseases including heart disease, type 2 diabetes, and cancer. The specific recommendations are similar to those for all adults, with the modification that people with OA should do activities that are low impact, do not cause pain, and have a low risk for joint injury.

Osteoporosis

Osteoporosis, literally "porous bone," is estimated to affect 10 million Americans, while another 34 million have low bone mass that puts them at future risk for osteoporosis (NOF 2008). Osteoporosis is a cause of major morbidity in older Americans—particularly women—because of the significantly increased risk of fractures in the spine, hip, and wrist seen in individuals with osteoporosis. The disease is a major public health problem and becomes increasingly prevalent as people age. In the population aged 50 and older, osteoporosis affects 1 in 3 women and 1 in 5 men (Melton 1992; Kanis 2000). In 2005, osteoporosis-related fractures were responsible for an estimated $19 billion U.S. in costs, a figure that is expected to rise to more than $25 billion U.S. by 2025 (NOF 2008). While many people associate osteoporosis with falls and hip fractures, the hip is not even the most common site for osteoporosis-associated fractures. According to the NOF, in 2005 there were more than 2 million fractures attributed to osteoporosis, including those of the vertebrae, wrist, hip, pelvis, and other sites (see figure 11.1).

GUIDELINE 11.11

Title: *Physical Activity Guidelines for Americans*

Organization: USDHHS

Year published: 2008

Purpose: To improve joint pain, physical function, quality of life, and mental health as well as reduce the risk for heart disease and type 2 diabetes

Location: www.health.gov/paguidelines

Population: Adults with OA

GUIDELINES
- Individuals should match the type and amount of physical activity to their abilities and the severity of their condition.
- Most people should do moderate-intensity activity for 150 min/wk and may choose to be active 3 to 5 d/wk in episodes lasting 30 to 60 min.
- Low-impact activities posing low risks for the joints, such as swimming, walking, and strength training, are good activities.
- Individuals may participate in vigorous-intensity activity, if tolerated.

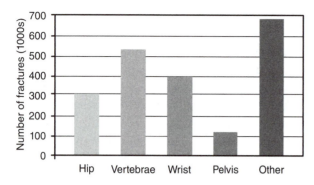

Figure 11.1 Sites of osteoporosis-related fractures occurring in 2005.

Data from nof.org 2008

Risks of Decreased Bone Density

Standards for the WHO definition of osteoporosis utilize T-scores to measure bone density; a T-score is the number of standard deviations above or below that of young, white adult women. A T-score of less than –2.5 is consistent with osteoporosis, while osteopenia is a T-score of –1.0 to –2.5. Bone mineral density naturally declines with age, so older individuals are consequently at higher risk for osteoporosis. Many of the guidelines covered in previous chapters—particularly those for older adults—aim to prevent osteoporosis by recommending types of exercise known to improve bone density (or at least diminish age-related bone loss) and enhance balance.

Osteopenia is a decreased bone density—or preosteoporosis—without a definitive increase in fracture risk. However, individuals with bone density that measures as normal or at the lower end of normal may still be at risk for fractures. Also, individuals with nontraumatic fractures as a result of diminished bone density (not necessarily in the osteoporosis range) are considered to be at a heightened risk for future fracture. A new fracture risk prediction tool, FRAX, is now available from the WHO. FRAX helps to estimate the 10 y fracture risk (www.shef.ac.uk/FRAX), which may be helpful in determining who all may benefit from the exercise prescriptions that follow in this chapter.

Benefits of Physical Activity

A major objective of several general physical activity recommendations is to prevent osteoporosis, but for people who already have diminished bone mass, preserving functional capacity is another equally important target. Individuals with superior functional capacity (whatever their bone mineral density) have less of a risk of falling and have improved outcomes if they do fall.

Falls and resultant fractures are a major cause of morbidity and mortality in the United States. Falls are the leading cause of injury deaths and the most common cause of injuries and trauma-related hospital admissions among adults aged 65 or more. Between 360,000 and 480,000 fall-related fractures occur annually among older Americans (CDC 2007). According to the IOF, 90% of hip fractures result from falls. One-third of people over the age of 65 fall annually, and 10% to 15% of these falls result in fractures. Physical activity and fitness not only improve bone mineral density (see figure 11.2) but also reduce the risk of hip fracture and fall-related injuries. Specific activities, such as back-strengthening exercises, can reduce the risk for vertebral fractures and kyphosis. Because exercise is one of the major elements associated with preservation of bone mass, exercise has been increasingly promoted as a way of reducing risk of osteoporosis-related fracture. Most of the studies evaluating physical activity and fracture risk have been cohort, case control, or cross-sectional studies. Although these types of studies prevent making a conclusion about causality, these studies do suggest an improved fracture risk for people participating in physical activity. Both weight-bearing exercise and resistance training have been shown to benefit bone density, although generally this finding has been borne

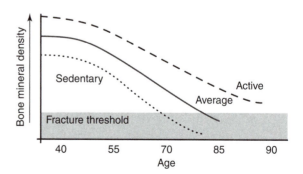

Figure 11.2 Changes in bone mineral density over time.

Reprinted, by permission, from IOF, 2005, *Move it or lose it* (Nyon, Switzerland: IOF).

out only in measures of lumbar spine density (Kelley 1998; Kelley and others 2001; Martyn-St James and Carroll 2006a, 2006b). Individuals who already have a decreased bone mass can benefit from a specific exercise program involving muscle strengthening and fall risk reduction.

Weight-bearing exercise not only prevents the decreased bone mineral density that often accompanies aging but also improves functional ability and decreases fall risk in individuals who already have osteoporosis. At least for postmenopausal women, the age-related decline in bone mineral density can be halted or even reversed with therapeutic exercise (IOF 2007). Tai chi stands out in that people who practice tai chi have been found to have a 47% decrease in fall rate and a 25% reduction in hip fracture rate when compared with people who do not practice tai chi (IOF 2007). A unique characteristic of physical activity guidelines for osteoporosis is the focus on not only improving the disease state itself but also enhancing other physiological adaptations that reduce risks for complications. Adaptations in bone strength may be site specific, and thus guidelines recommend exercising the trunk, upper-body, and lower-body muscles, particularly the extensor groups.

The disease may necessitate modification of standard aerobic and resistance exercises. People with severe kyphosis may have mechanical limitations in respiratory muscle function as well as a shift in the center of gravity; these individuals should avoid treadmill exercise or anything else that might be unsafe based on limited forward vision. Stationary equipment or walking with support is a better alternative. Forward flexion and twisting of the spine both increase the risk for vertebral fractures and are not recommended.

Guidelines for Prevention of Osteoporosis

In 2004, the ACSM's 1995 position stand on "Osteoporosis and Exercise" was updated and titled "Physical Activity and Bone Health." The latter includes recommendations to optimize bone health even before diminished bone density occurs. For example, the position stand sug-

gests an exercise prescription to help children and adolescents augment bone mineral accrual (see guideline 5.30). Maintaining bone health throughout adulthood is also emphasized. The guidelines point out that even the most frail elderly individuals benefit from exercises to maintain bone mass, improve balance, and prevent falls. Although walking, which is the most popular form of physical activity (particularly among the elderly who have the highest risk of osteoporosis, falls, and fractures), is weight bearing, evidence suggests that it provides only modest effects on the preservation of bone mass. Thus the guidelines for maintaining bone mass in adulthood recommend activities with higher-intensity loading forces such as jogging, jumping, and resistance training. Of note, the ACSM position stand does *not* contain details for physical activity in people already diagnosed with osteoporosis.

The most recent edition of *ACSM's Guidelines for Exercise Testing and Prescription* includes a specific recommendation for people with osteoporosis as well as for people at risk for osteoporosis. The two sets of guidelines are similar, although people *with* osteoporosis should utilize slightly lower bone-loading forces in order to avoid injury. On the other hand, moderate- to high-intensity bone-loading forces during resistance training are recommended to help individuals who are at risk preserve their bone health. Also, the type of activity differs between the two groups. While both populations should participate in weight-bearing and resistance activities, people without osteoporosis (but at risk for it!) should also include higher-impact activities such as tennis, intermittent jogging, and jumping in addition to walking and climbing and descending the stairs, which are recommended for both groups.

In 2000, the NIH held a conference to focus on the prevention, diagnosis, and treatment of osteoporosis. Although a specific exercise prescription was not made, qualitative recommendations for higher-impact and resistance exercise were presented.

The ACE has published guidelines for the *prevention* of osteoporosis. These guidelines were adapted from the 2004 *Bone Health and Osteoporosis: A Report of the Surgeon General* and recommend physical activity starting in early

GUIDELINE 11.12

Title: "Physical Activity and Bone Health"

Organization: ACSM

Year published: 2004

Purpose: To augment bone mineral accrual in children and adolescents and to preserve bone health during adulthood

Location: *Medicine and Science in Sports and Exercise* 36(11): 1985

Population: Children, adolescents, and adults

GUIDELINES

Component	Frequency	Intensity	Duration	Type
CHILDREN AND ADOLESCENTS				
Aerobic exercise	At least 3 d/wk	High	10-20 min (2 times per day or more may be effective)	Impact activities such as gymnastics, plyometrics, and jumping
Resistance training	At least 3 d/wk	Moderate (<60% 1RM)	10-20 min (2 times per day or more may be effective)	Not specified
ADULTS				
Aerobic exercise	3-5 times per week	Moderate to high	30-60 min/d, combining weight-bearing endurance activities, activities that involve jumping, and resistance exercise	Weight-bearing endurance activities such as tennis, stair-climbing, jogging, and jumping
Resistance training	2-3 times per week	Moderate to high bone-loading forces	30-60 min/d, combining weight-bearing endurance activities, activities that involve jumping, and resistance exercise	Weightlifting exercises targeting all major muscle groups

Special considerations include the following:

- While exercise may help prevent age-related declines in bone mass, it may not prevent menopause-related loss in women.

- Exercise programs for elderly women and men should also include activities to improve balance and prevent falls.

Adapted, by permission, from W.M. Kohrt, et al., 2004, "Physical Activity and Bone Health," *Medicine & Science in Sports & Exercise* 36(11):1985.

infancy as well as make suggestions regarding calcium and vitamin D intake. The guidelines emphasize bone-specific, aerobic weight-bearing exercise such as walking or cross-country skiing on 3 d/wk for most adults. To help improve upper-body strength and stimulate different bones, upper-body muscle strength and endurance training on 2 d/wk is also recommended.

The 2004 *Bone Health and Osteoporosis: A Report of the Surgeon General* recommends meeting the basic guidelines for physical activity (at least 30 min/d for adults and 60 min/d for children) as well as participating in specific strength and weight-bearing activities in order to maintain and build bone mass throughout life. For people who already have osteoporosis or are at a high risk for

GUIDELINE 11.13

Title: "Exercise Prescription for Other Clinical Populations: Osteoporosis"

Organization: ACSM

Year published: 2010

Purpose: To preserve bone health (individuals at risk for osteoporosis) and prevent disease progression (individuals with osteoporosis)

Location: *ACSM's Guidelines for Exercise Testing and Prescription, Eighth Edition*

Population: Individuals with or at risk for osteoporosis

GUIDELINES

Component	Frequency	Intensity	Duration	Type
INDIVIDUALS AT RISK				
Aerobic exercise	3-5 d/wk	Not specified	30-60 min daily of combined resistance and aerobic activities	Weight-bearing activities (tennis, climbing and descending stairs, walking with intermittent jogging) and jumping (volleyball and basketball)
Resistance training	2-3 d/wk	Moderate (60%-80% 1RM for 8-12 repetitions) or high (80%-90% 1RM for 5-6 repetitions)	30-60 min daily of combined resistance and aerobic activities	Weightlifting
INDIVIDUALS WITH OSTEOPOROSIS				
Aerobic exercise	3-5 times per week	Moderate (40%-<60% $\dot{V}O_2R$ or HRR)	30-60 min daily of combined resistance and aerobic activities	Weight-bearing activities such as walking and climbing and descending stairs
Resistance training	2-3 times per week	60%-80% 1RM for 8-12 repetitions	30-60 min daily of combined resistance and aerobic activities	Weightlifting

Special considerations include the following:

- Exercise should not cause or exacerbate pain.
- Avoid explosive or high-impact loading, twisting, bending, or compression of the spine.
- Older men and women at increased risk for falling should also include balance activities for fall prevention.

Adapted, by permission, from ACSM, 2010, *ACSM's guidelines for exercise testing and prescription*, 8th ed., (Baltimore, MD: Lippincott, Williams, and Wilkins), 257-258.

fracture, an extensive fall-prevention program is recommended.

Although the NOF does not have specific guidelines relating to physical activity for those who already *have* osteoporosis or fractures due to low bone mineral density, the NOF Web site does detail an appropriate physical activity program to prevent osteoporosis or losses in bone mass and density. Exercises include weight-bearing impact and nonimpact activities. Each type of activity improves health through different means. Weight-bearing exercises can be high or low impact—the higher the impact, the better the stimulus for bone formation.

GUIDELINE 11.14

Title: *Osteoporosis Prevention, Diagnosis, and Therapy*

Organization: NIH

Year published: 2000

Purpose: To stimulate bone mineral accrual and improve muscular strength and balance

Location: www.consensus.nih.gov/2000/2000Osteoporosis111html.htm

Population: Individuals with osteoporosis

GUIDELINE

Engage in regular exercise with a focus on high-impact and resistance training exercises.

GUIDELINE 11.15

Title: *Reduce Your Risk for Osteoporosis Now*

Organization: ACE

Year published: 2009

Purpose: To prevent bone loss and encourage bone growth

Location: www.acefitness.org/fitfacts/fitfacts_display.aspx?itemid=2609

Population: All individuals

GUIDELINES

Age group	Physical activity
Infants 0-12 mo	Interactive play
Children 1-18 y	Moderate to vigorous activity at least 60 min/d, emphasizing weight-bearing activity
Adults 18+ y	Moderate activity lasting at least 30 min on most, preferably all, days of the week, emphasizing weight-bearing activity

Special consideration:

- Fall-prevention programs should be considered for elderly patients who are frail or for patients with spine fracture.

Adapted, by permission from American Council on Exercise Fit Facts, San Diego 2001. www.acefitness.org/fitfacts Accessed 10/10/2009.

High-impact exercises are appropriate for individuals who have normal bone density and are not frail. Stair-climbing, high-impact aerobics, running or jogging, tennis, and hiking are all considered high-impact activities.

People who cannot perform high-impact activities can perform low-impact exercises, which are safer but still help to build bones. These activities include low-impact aerobics, working out on elliptical trainers or stair steppers, and walking indoors or outdoors. Resistance training also stimulates bone formation and can be done with body weight, free weights, weight machines, or elastic bands. Nonimpact activities are also an important part of the NOF recommendations, as these help people to maintain their ability to function in everyday activities. These include balance, posture, and functional exercises. People who have problems in one of these areas can enhance their abilities by repeating specific activities.

GUIDELINE 11.16

Title: *Bone Health and Osteoporosis: A Report of the Surgeon General*

Organization: Office of the Surgeon General

Year published: 2004

Purpose: To maintain and build bone mass throughout life

Location: www.surgeongeneral.gov/library/bonehealth

Population: All individuals

GUIDELINES

- Adults should participate in at least 30 min/d of physical activity.
- Children should participate in at least 60 min/d of physical activity.
- Strength and weight-bearing activities should be included if possible.
- Older adults at high risk for fracture should participate in an extensive fall-prevention program; this can incorporate physical activity, in 30 to 45 min sessions 3 times a week for at least 3 mo, with enough intensity to improve muscle strength.
- Load-bearing physical activities such as jumping should be undertaken for 5 to 10 min daily; people who can tolerate impact activities should incorporate 50 3 in. (7.6 cm) jumps into their activity each day.
- Adults should use all muscle groups in a progressive program of weight training; people who cannot tolerate higher-impact physical activity should engage in jogging or stair-climbing.

GUIDELINE 11.17

Title: *Prevention: Exercise for Healthy Bones*

Organization: NOF

Year published: 2008 (copyright)

Purpose: To build bones, improve balance and posture, increase muscle strength, and decrease the risk of falls and broken bones

Location: www.nof.org/prevention/exercise.htm

Population: Healthy individuals

GUIDELINES

Type	Frequency	Intensity	Duration	Examples
Weight-bearing and impact exercises	Most days of the week	Not specified	30 min total, may be broken into 3 sessions of 10 min	Dancing, aerobics, hiking, running, jumping rope, playing tennis, walking, working out on an elliptical trainer

>continued

GUIDELINE 11.17 >*continued*

Type	Frequency	Intensity	Duration	Examples
Resistance and strengthening exercises	2-3 d/wk	Challenging enough to allow 8-10 repetitions per set, with 30-60 s rest between each set	1 set of 8-12 exercises, with 8-10 repetitions per set (adults who are frail may do 10-15 repetitions at a lighter weight)	Exercises working the major muscle groups using weights, machines, body weight, or elastic bands
Balance, posture, and functional exercises	Every day	Not specified	Not specified	Depending on needs, tai chi, yoga, or Pilates

GUIDELINE 11.18

Title: *Physical Activity Guidelines Advisory Committee Report*

Organization: PAGAC

Year published: 2008

Purpose: To increase bone mass or to prevent excessive loss of bone mass

Location: www.health.gov/paguidelines/committeereport.aspx

Population: Adults

GUIDELINES

Adults should strive to achieve the following, which appear to be related to a reduced fracture risk:

- At least 9 to 14.9 MET-h/wk of physical activity
- More than 4 h/wk of walking
- At least 1,290 kcal/wk of physical activity
- More than 1 h/wk of physical activity

For example, a person can practice rising out of a chair and assuming a standing position until tired. While the NOF recommends both weight-bearing and resistance exercises, it also urges people to consult with a physician before starting a program so that they can determine their fracture risk.

The PAGAC, whose report was the basis for *PAGA*, reviewed the available scientific evidence and concluded that the 2004 ACSM position stand on bone health was probably appropriate. As a result, its recommendations for increasing bone density or at least preventing excessive bone loss were similar to those issued by the ACSM. *PAGA*, issued 4 mo after the PAGAC report, did not include a specific guideline for osteoporosis. Nevertheless, the PAGAC report noted epidemiological studies quantifying the amount of physical activity associated with a reduced fracture risk. This amount is outlined in guideline 11.18, as it may be necessary for preventing fracture. Lastly, the PAGAC report commented that we are unlikely to ever see a randomized controlled trial with exercise as an intervention and osteoporotic fracture as an outcome due to the sheer size and length of time required for such a study. Therefore, the following recommendations are based on epidemiological (cross-sectional, prospective and retrospective cohort, and case-control) studies, which provide the best information currently available.

Guidelines for Individuals With Osteoporosis

The ACSM issued an exercise management plan for individuals with osteoporosis in the third edition of *ACSM's Exercise Management for Persons With Chronic Diseases and Disabilities*. Aerobic, weight-bearing activities are recommended in a similar fashion to those recommended for otherwise healthy individuals who do not have osteoporosis. Both resistance training and flexibility training are recommended as part of the weekly program. The resistance training prescription for individuals with diminished bone density consists of higher weight and fewer repetitions. This type of plan is felt to provide a greater stimulus for bone mass accumulation. The prescription also includes functional exercises as a way of preventing falls, as individuals who are adept at performing the daily activities that can trigger a fall are less likely to lose their balance.

Canadian guidelines for managing osteoporosis are slightly different from those issued for Americans. The Osteoporosis Society of Canada issued guidelines in 2002; these guidelines include pharmacological therapy, dietary changes, and lifestyle recommendations—the last of which include physical activity recommendations. The guidelines do not provide a specific exercise prescription

GUIDELINE 11.19

Title: "Osteoporosis"

Organization: ACSM

Year published: 2009

Purpose: To maximize mobility and prevent further falls, to improve or maintain work capacity, and to improve strength, posture, and range of motion

Location: *ACSM's Exercise Management for Persons With Chronic Diseases and Disabilities, Third Edition*

Population: Individuals with osteoporosis and osteopenia

GUIDELINES

Component	Frequency	Intensity	Duration	Type
Aerobic exercise	3-5 d/wk	40%-70% of HRmax or METs	20-30 min each session	Large-muscle activities (walking, cycling, swimming, walking through water)
Resistance training	2-3 d/wk	75% of 1RM, 8-12 repetitions or 2 sets of 8-10 repetitions	20-40 min	Calisthenics, cuff weights, machines, and dumbbells
Flexibility training	5-7 d/wk	Maintain stretch below point of discomfort	Not specified	Stretching and chair exercises
Functional exercises	3-5 d/wk (balance); 2-3 d/wk (sit to stand)	Not specified	Not specified	Brisk walking, balance exercises, chair sit to stand

Special considerations include the following:

- People with osteopenia should incorporate impact loading while people with osteoporosis should not.
- Avoid forward flexion and twisting.
- Enhanced bone mass occurs with higher intensity and fewer repetitions of resistance training.

Adapted from S.S. Smith, C.E. Wang, and S.A. Bloomfield, 2009, Osteoporosis. In *ACSM's exercise management for persons with chronic diseases and disabilities*, 3rd ed., edited by J.L. Durstine et al. (Champaign, IL: Human Kinetics), 276.

GUIDELINE 11.20

Title: "Clinical Practice Guidelines for the Diagnosis and Management of Osteoporosis"

Organization: Osteoporosis Society of Canada

Year published: 2002

Purpose: To manage and treat osteoporosis

Location: *CMAJ* 167(10 suppl)

Population: Individuals with osteoporosis

GUIDELINES

The guidelines do not provide a specific exercise prescription but do suggest including both weight-bearing exercises and those that improve balance and strength.

- Children, particularly those entering and passing through puberty, should be encouraged to participate in impact exercises or sports (mainly field and court sports)
- Throughout life, both men and women should be encouraged to participate in exercise, particularly in weight-bearing exercises, which include impact as a component
- For older men and women at risk of falling or who have fallen, tailored programs that are based on individual assessment, contain exercises to improve strength and balance and, where necessary, are multidisciplinary in nature should be made available

but do suggest at least 30 min of activity performed at least 3 times a week.

The IOF created its *Move It or Lose It* report in conjunction with World Osteoporosis Day 2005 to encourage regular activity in patients with osteoporosis. In this report, generalized aerobic fitness is a second priority to exercises that focus on balance, coordination, and hip and trunk stabilization. Exercise guidelines for improving bone health and *preventing* osteoporosis are included along with the guidelines addressing individuals who already have osteoporosis. Preventive measures are similar to those recommended by other organizations. Weight-bearing and high-impact exercises stimulate bone formation, while lower-impact exercises strengthen muscles and prevent falls. The IOF suggests to start slowly and progress gradually in order to reduce the risk of stress fractures or joint damage. In addition, it stresses the importance of maintaining exercise frequency. The IOF states that "two short exer-

cise sessions separated by 8 hours is better than one long one." (Minne and Pfeifer 2005, p. 10) Certain activities and postures should be avoided, as these strain areas that are prone to fracture in people with osteoporosis. Postural and balance exercises are encouraged to protect from falls and reduce the chance of fracture. Additionally, the IOF report addresses dietary recommendations for providing an appropriate environment for bone building, including adequate calcium and vitamin D intake and avoiding smoking and excessive alcohol consumption.

The following guidelines are the *Move It or Lose It* recommendations for individuals who already have osteoporosis or already have a fracture. As in other guidelines, consulting with a health care professional before starting physical activity is recommended to reduce the risk of injury. Fall prevention is important, as is avoiding inappropriate exercises that might worsen a fracture.

GUIDELINE 11.21

Title: *Invest in Your Bones. Move It or Lose It*

Organization: IOF

Year published: 2005

Purpose: To build and maintain strong bones, prevent falls and fractures, and speed rehabilitation

Location: www.iofbonehealth.org/download/osteofound/filemanager/publications/pdf/move_it_or_lose_it_en.pdf

Population: Individuals with osteoporosis

GUIDELINES

Things to do	Things to avoid
Start a basic strengthening program Be aware of falling (enroll in a fall-prevention program if possible)	Jarring or twisting movements Abrupt and sudden or high-impact movements Abdominal curl-ups Bending forward from the waist Heavy lifting

Special considerations include the following:

- A doctor, nurse, or physiotherapist should be consulted before beginning an exercise program.
- If a fracture occurs, exercise should aim to relieve pain and help regain range of motion and independence. Exercise in warm water is the first thing to attempt.

Adapted, with permission, from Minne HW, Pfeifer M. Invest in your bones. Move it or lose it. International Osteoporosis Foundation, 2005. www.iofbonehealth.org Accessed on 10/10/2009.

SUMMARY

Arthritis and osteoporosis are common bone diseases that affect many individuals. Physical activity has a role in the treatment of both, because of the beneficial physiological modifications it creates in the muscles, joints, and bones. Adapted programs may be necessary for individuals with various related conditions, and guidelines take these into account so that people with disease can achieve benefit without causing further harm. There is a particular emphasis on resistance training for these conditions, as this component of physical activity does seem to enhance balance, build muscle strength, and reduce joint stresses. As the aging population with OA and osteoporosis continues to grow, these guidelines will become increasingly important in reducing the associated morbidities of chronic pain, falls, and fractures.

Diabetes

Diabetes is a devastating disease that results in higher-than-normal blood glucose levels either through inadequate insulin secretion (type 1 diabetes) or impaired response to elevated circulating levels of insulin (type 2 diabetes). Complications of diabetes include CAD, peripheral arterial disease, stroke, kidney failure, blindness, and neuropathy and may lead to circulatory problems, amputation, and premature death. More than 21 million Americans have been diagnosed with diabetes, the majority of which is type 2 diabetes (ADA 2008a). Nearly as many Americans are classified as prediabetic, which the ADA defines as having a fasting glucose level greater than 100 mg/dl. Worldwide, diabetes affects 246 million people, and each year 3.8 million deaths are attributable to diabetes, with a significantly higher number of diabetes deaths attributed to CVD; these CVD deaths likely are exacerbated by diabetes-related hypertension and lipid disorders (International Diabetes Federation). In fact, CVD is the leading cause of death among people with diabetes, accounting for 50% of deaths (IDF) as well as a significant amount of disability. As the obesity epidemic continues to grow, the diabetes epidemic does also. As described in chapter 8, the incidence of type 2 diabetes is clearly related to obesity, weight gain, and sedentary lifestyle. People with type 2 diabe-

tes currently make up nearly 90% to 95% of all people with diabetes, and this number continues to increase. Incidence increases with age, and at least 20% of patients over the age of 65 y have diabetes (ADA 2008a); this number is expected to continue to increase as the older population increases. The ADA estimates that by the year 2030, the number of American adults with type 1 or type 2 diabetes will be more than 30 million, though only about 1 million of those cases will be type 1 diabetes.

As opposed to type 1 diabetes, which is caused by autoimmune destruction of the insulin-producing beta cells of the pancreas, type 2 diabetes is caused primarily by insulin resistance, which is associated with elevated insulin concentrations. The definition of diabetes has changed over time, and the current classification is based on the knowledge that the physiological consequences of insulin resistance or impaired glucose tolerance (IGT) significantly increase both cardiovascular risk factors and CVD itself—even at glucose levels once thought to be only at the higher end of the normal range. Diagnostic criteria for diabetes are presented in the following sidebar. Because of the mechanistic differences between type 1 and type 2 diabetes, the guidelines for the two conditions are considered separately within this chapter.

Diagnosis of Diabetes

A diagnosis of diabetes requires any of the following:

- Fasting plasma glucose level of ≥126 mg/dl (7.0 mM)
- 2 h plasma glucose level in an oral glucose tolerance test of ≥200 mg/dl (11.1 mM)
- Casual plasma glucose level of ≥200 mg/dl *with* symptoms of hyperglycemia, such as polyuria, polydipsia, and unexplained weight loss

Notes:

- In the absence of unequivocal hyperglycemia, these criteria should be confirmed by repeat testing on a different day.
- Impaired fasting glucose (IFG) is a fasting plasma glucose value between 100 and 125 mg/dl.
- The use of hemoglobin A1c (HbA1c) as a diagnostic test for diabetes is not recommended because of a lack of standardized methodologies.

Benefits of Physical Activity for Preventing and Managing Diabetes

The ADA recommends daily activity not only to prevent diabetes but also to prevent complications in people who have diabetes. The ACSM endorses physical activity and exercise for people with diabetes to improve glucose tolerance and insulin sensitivity as well as manage weight, improve well-being, and reduce other cardiovascular risk factors. Physical activity is a cornerstone in the management of diabetes, along with diet and nutrition. Goals for physical activity in people with type 2 diabetes include weight control, glycemic control, improved insulin sensitivity, reduced CVD, reduced cancer rates, and lower overall mortality.

Evidence-based clinical trials have consistently shown that intensive management and interventions can improve long-term diabetes outcomes, but it is estimated that 70% of patients with diabetes still die prematurely from heart disease (Haffner and others 1998). CVD is the leading cause of death in individuals with diabetes, and stroke causes an additional 15% of deaths. In fact, many would consider type 2 diabetes to be a CAD equivalent. It has been postulated that an individual with diabetes but *without* CAD

has a greater risk for subsequent cardiac events than individuals with known CAD have! Regular physical activity is associated with a substantially reduced risk for cardiovascular events (Wei and others 2000; Ford and DeStefano 1991), particularly in women (Hu and others 2001). In one study, women with diabetes who spent at least 4 h/wk in moderate or vigorous exercise had a 40% lower risk for CVD than women who did not exercise had. This study concluded that there is a dose–response relationship between leisure-time physical activity and reduction in cardiovascular risk (Hu and others 2001). Another study evaluated the association between walking and mortality (both overall and cardiovascular related) and found that individuals walking at least 2 h/wk had a 39% lower all-cause mortality rate and a 34% lower cardiovascular mortality rate. This study also discovered a dose–response relationship between activity and health outcomes, observing the lowest mortality rates among individuals who walked 3 to 4 h/wk (Gregg and others 2003). Another study looking at cardiorespiratory fitness (Church 2005) concluded that lower levels of fitness increase the risk of death due to CVD independently of BMI.

Both physical activity and exercise training have favorable effects on other factors in people with diabetes, such as lipid profiles, glycemic control, blood pressure, and insulin resistance. In people with type 2 diabetes and obesity, exercise as an adjunct to diet yields greater and more sustained weight loss. In one study of 16 patients with well-controlled type 2 diabetes, aerobic exercise improved body fat percentage and waist-to-hip ratio, although body weight did not significantly change (Lehmann and others 1995). In addition to inducing weight loss, exercise alters glucose metabolism by increasing both insulin-mediated and non-insulin-mediated glucose disposal. The increased insulin-mediated glucose uptake from a single exercise session lasts for more than 24 h (Annuzzi and others 1991). Both vigorous and nonvigorous activities improve insulin sensitivity (Yamanouchi and others 1995; Mayer-Davis and others 1998).

In a meta-analysis of studies evaluating the effect of exercise on HbA1c, there was a significant improvement in glycemic control in the groups who exercised (Boule' and others 2001). The paper also evaluated the effect of combined diet and exercise on HbA1c, and the effect was similar to that seen with exercise alone. The size of the effect seen with regular physical activity and exercise was a 0.66% reduction; this magnitude is similar to the magnitude of difference found in people with intensive glycemic control in the United Kingdom Prospective Diabetes Study (UKPDS). Subjects with that degree of HbA1c lowering had a 32% reduction in diabetes-related clinical end points and a 42% reduction in diabetes-related deaths. It is possible that this extent of improvement in glycemic control may afford even further reductions in CVD than those observed in the UKPDS because exercise in general is correlated with a multitude of other protective benefits when compared with insulin or oral medications. There also seems to be a dose–response relationship between activity intensity and outcomes: People who exercise at a higher intensity experience greater improvement in HbA1c and fitness (Boule' and others 2003).

As previously noted, many individuals with diabetes also possess other cardiovascular risk factors, including hypertension, obesity, and a sedentary lifestyle. The combination of two or more of these confers a particularly higher risk for CVD. Although many trials of exercise training in people with hypertension and diabetes focus on glycemic control or improved blood pressure, there are other mechanisms by which regular exercise reduces cardiovascular risk. The most robust evidence for a benefit from exercise are observations on improvements in endothelial vasodilator function and left ventricular diastolic function. Data also suggest that exercise improves arterial wall stiffness and systemic inflammation as well as reduces left ventricular mass (Stewart 2002). The reduction in both total and abdominal fat that occurs with exercise mediates improved insulin sensitivity and blood pressure and may improve endothelial vasodilator function.

In addition to the health benefits of physical activity for people with diabetes—including improved blood pressure, reduced risk for CAD, and decreased HbA1c—there are substantial economic benefits as well. The 3 y cost of managing diabetes and coexisting CHD and hypertension is 300% higher than it is for managing diabetes alone! In absolute numbers, this translates to $46,000 U.S. versus $14,000 U.S. (Gilmer and others 2005). According to another estimate, if all the members of the American Academy of Family Physicians (AAFP) reduced the median HbA1c levels of their patient population with diabetes by 2%, the savings would be more than $2.5 billion U.S. annually (Unger 2007). Clearly, by promoting physical activity to people with diabetes, health professionals can reduce both the economic and the CVD burden in this population.

Potential Concerns Regarding Physical Activity

Because of the potential nerve damage and metabolic derangements associated with diabetes, there are specific concerns regarding physical activity. Blood sugar levels may be higher or lower than normal, depending on a person's medication, degree of disease control, and activity level, and many people with diabetes require closer monitoring during physical activity to avoid fluctuations in blood glucose. Also, hydration status may both cause and result from changes in blood glucose, and so it requires attention as well. Nerve damage may occur both in peripheral nerves and in the autonomic nervous system, necessitating caution in individuals with either complication. Lastly, people with other organ damage such as eye or kidney damage may require modification of activity.

Hypoglycemia and Hyperglycemia

During exercise, glucose is a primary source of energy. It is derived from skeletal muscle and the liver and is mediated by glucagon and catecholamines. Hypoglycemia is the most common problem for people with diabetes who also exercise. Patients may or may not be aware of hypoglycemia; the signs of hypoglycemia are presented shortly. In patients with poorly controlled diabetes, rapid drops in blood glucose may precipitate signs or symptoms of hypoglycemia even if the absolute glucose level is normal or elevated. Physical activity may cause hypoglycemia in individuals who take insulin or other oral hypoglycemic agents because glucose mobilization is attenuated by high levels of insulin. Either the medication dose or the preactivity carbohydrate consumption may have to be altered in order to prevent this problem. In individuals who do not take insulin or insulin secretagogues (e.g., sulfonylurea medications), exercise-induced hypoglycemia is uncommon (ADA/ACSM, 1997), and preactivity measures are not routinely recommended. Signs of hypoglycemia include the following:

Hypoglycemia (Plasma glucose level of <80 mg/dl)

- Altered mentation
- Blurred vision
- Loss of consciousness
- Slurred speech
- Irritability
- Seizure
- Somnolence
- Strokelike symptoms

People with type 1 diabetes with inadequate insulin levels may experience an excessive release of counterregulatory hormones that further increases glucose levels and sets off a cascade of worsening hyperglycemia, ketone production, and eventually ketoacidosis, so hyperglycemia may be more of a risk for people with type 1 diabetes and poor glycemic control. Hyperglycemia may worsen thermoregulation through polyuria-induced dehydration. In the presence of ketosis, exercise can worsen hyperglycemia. Signs of hyperglycemia include the following:

Hyperglycemia (Plasma glucose level of >300 mg/dl)

- Nausea
- Vomiting
- Blurred vision
- Frequent urination
- Weakness
- Excessive thirst
- Irregular respiration
- Dehydration

Exercise and physical activity both increase glucose uptake. This effect persists after exercise and can last for up to 48 h after exercise (McDonald 1987). For this reason, plasma glucose levels should be monitored after exercise and carbohydrate should be consumed as needed. The ACSM recommends monitoring blood glucose to prevent both hypoglycemia and hyperglycemia during exercise, especially if the exercising individual is taking insulin or oral hypoglycemic agents that increase insulin production. The specific recommendations for monitoring glucose levels during activity are as follows:

- Monitor blood glucose before exercise and following exercise, especially when beginning or modifying the exercise program.
- Be aware of late-onset hypoglycemia, which can occur up to 48 h following exercise, especially when beginning or modifying the exercise program.
- Avoid physical activity if fasting glucose is >250 mg/dl and ketosis is present, and use caution if glucose is >300 mg/dl and no ketosis is present.
- Before exercise, adjust carbohydrate intake or insulin injections based on blood glucose and exercise intensity in order to prevent hypoglycemia associated with exercise; ingest 20 to 30 g of additional carbohydrate if the preexercise blood glucose level is <100 mg/dl.
- To lower the risk of hypoglycemia associated with exercise, avoid injecting insulin

into exercising limbs; an abdominal injection site is preferred.

- When exercising late in the evening, be aware that an increased consumption of carbohydrate may be required to minimize the risk of nocturnal hypoglycemia.

- If taking insulin or oral hypoglycemic agents, know that intense resistance exercise often produces an acute hyperglycemic effect, whereas there is an increased risk for postexercise hypoglycemia in the hours following basic resistance training.

Autonomic Neuropathy and Other Complications

Autonomic neuropathy is another problem for many people with diabetes. This type of neuropathy may prevent individuals from being aware of hypoglycemia or even angina. Individuals with autonomic neuropathy may have a blunted systolic blood pressure response to exercise, postural hypotension, blunted oxygen uptake kinetics, and decreased sweating. They may also have an unpredictable delivery of carbohydrate due to gastroparesis and thus may be predisposed to hypoglycemia. Close monitoring of signs or symptoms of hypoglycemia, hypotension, or hypertension during exercise is important in these individuals because the autonomic neuropathy may preclude their own awareness. Likewise, poor thermoregulation necessitates caution and proper precautions in both hot and cold environments. Because of the blunted physiological response, traditional heart rate or blood pressure recommendations for exercise may not be accurate, and thus the ACSM recommends that people with autonomic neuropathy should use RPE to regulate exercise intensity. Individuals with autonomic neuropathy in general have a higher risk for CVD than other people with diabetes have (Wackers 2004; Valensi and others 2001). For this reason, individuals with diabetic autonomic neuropathy are recommended to undergo a cardiac examination before beginning a physical activity program at a new or higher intensity than that to which they are accustomed (ADA 2008b).

Peripheral neuropathy may develop as a result of long-term poor glycemic control. This type of neuropathy may be particularly problematic for people starting a physical activity program; the physical activity is essential for improving glycemic control and yet the exerciser must take precautions to avoid further damage to the nerve or denervated area. Peripheral neuropathy occurs most often in the feet; proper care must be taken both before and after physical activity in order to prevent foot ulcerations or infection. Shoes that fit well, cushion the feet, and redistribute pressure may prevent callus formation. Individuals with bony deformities such as hammertoes, bunions, or prominent metatarsal heads may need special accommodations to prevent skin breakdown. Individuals whose peripheral neuropathy compromises their proprioception and balance may need to limit weight-bearing exercise in order to lessen the chance of falling. Swimming, bicycling, or arm ergometry are good activity alternatives for these individuals.

Retinopathy is the leading cause of blindness in people with diabetes, and its incidence is inversely correlated to both glycemic and blood pressure control. Vigorous exercise may increase the likelihood of retinal detachment and vitreous hemorrhage. For this reason, activities that significantly elevate the blood pressure should be avoided. Specific guidelines for people with retinopathy are listed in table 12.1.

Nephropathy is another complication of diabetes, and although physical activity may acutely increase urinary protein excretion, this outcome does not seem to be correlated with an increased risk for progression of diabetic kidney disease. Thus, the ADA does not recommend any specific restriction to physical activity programs for individuals with diabetic kidney disease.

Guidelines for Individuals With Type 2 Diabetes

The ADA recommends regular physical activity as an essential part of managing type 2 diabetes. Specific guidelines were created with the goal of CVD risk reduction as well as weight control and glycemic control. Stress testing before the initiation of an exercise program may be recommended, depending on the previous exercise history of the individual. Despite the fact that

TABLE 12.1 Activity Modifications for Individuals With Diabetic Retinopathy

Type of retinopathy	Modification
Proliferative	Avoid strenuous activities, Valsalva maneuvers, or pounding or jarring activities.
Moderate nonproliferative	Avoid activities that dramatically elevate the blood pressure.
Severe nonproliferative	Avoid exercise that increases systolic blood pressure to more than 170 mmHg.

GUIDELINE 12.1

Title: "Physical Activity/Exercise and Diabetes"

Organization: ADA

Year published: 2008

Purpose: To favorably affect health outcomes of patients with diabetes

Location: Executive Summary: Standards of Medical Care in Diabetes – 2008, *Diabetes Care*, 31(S1): S5-S11, 2008; Standards of Medical Care in Diabetes – 2008, *Diabetes Care*, 31(S1): S12-S54, 2008; and Physical Activity/Exercise and Type 2 Diabetes. A consensus statement from the American Diabetes Association. *Diabetes Care*, 29(6): 1433-1438, 2006.

Population: Individuals with type 2 diabetes

GUIDELINES

Type	Frequency	Intensity	Duration
Aerobic exercise	Distributed over at least 3 d/wk; no more than 2 consecutive days without physical activity	Moderate: 50%-70% HRmax or 40%-60% $\dot{V}O_2$max	At least 150 min/wk
		Vigorous: >60% $\dot{V}O_2$max or >70% HRmax	At least 90 min/wk
Resistance exercise	3 times a week	At a weight that cannot be lifted more than 8-10 times	Progressing to 3 sets of 8-10 repetitions targeting all major muscle groups

Special considerations include the following:

- Individuals may do moderate-intensity exercise, vigorous-intensity exercise, or both. People who do both are expected to have greater CAD risk reduction.
- For long-term maintenance of major weight loss of 30 lb (13.6 kg) or more, 7 h of weekly activity (moderate or vigorous) is recommended.
- Individuals on insulin or insulin secretagogues who have a preexercise glucose level of <100 mg/dl should ingest additional carbohydrate to prevent hypoglycemia.
- After laser photocoagulation for retinopathy, individuals should wait 3 to 6 mo before initiating or resuming resistance exercise.
- People with severe peripheral neuropathy should do non-weight-bearing activities such as swimming, bicycling, or arm exercises.
- People with autonomic neuropathy should undergo cardiac testing (possibly thallium scintigraphy) before increasing their physical activity.
- People with microalbuminuria and proteinuria should complete a stress test before beginning a physical activity program more intense than ADLs.

diabetes significantly increases the risk for CAD, the current consensus from the ADA is that *routine* screening for CAD before recommending an exercise program is not necessary (Bax and others 2007). The ADA guidelines also detail the timing of medications with exercise. For people with diabetes, the combination of diet, exercise, and behavior modification is important to achieve and sustain both glucose and weight control. Resistance training is as important as aerobic exercise is, as both tend to improve insulin resistance equally (Ivy 1997). There is an additive benefit of combined aerobic and resistance exercise in adults with type 2 diabetes (Sigal and others 2007). The most recent ADA guidelines are similar to the ACSM and AHA 2007 guidelines (see guideline 2.1) in that an individual can fulfill the appropriate volume of activity through either a greater amount of moderate-intensity work or a lesser amount of vigorous-intensity work. Of course, a combination of the two is also acceptable.

Diabetes is addressed in the 2008 *Physical Activity Guidelines for Americans (PAGA)* as part of the section on physical activity for people with chronic medical conditions. In the extensive scientific review undertaken to produce this report, the PAGAC evaluated all the available data before reaching the conclusion stated in the

PAGA recommendations: Physical activity not only provides therapeutic benefits for people with type 2 diabetes but also facilitates a healthy body weight and reduces the risk for other diseases. The minimum recommendation of 150 min/wk of moderate-intensity activity may lower the risk for heart disease, while the preferred recommendation of 300 min/wk is expected to impart a more significant benefit. There are concerns, however, for people with diabetes who participate in physical activity: (1) It may be necessary to work with a health care provider to adapt the exercise program in order to ensure that physical activity is safe, (2) blood glucose levels may need to be monitored, and (3) foot injuries should be avoided.

The ACSM has issued several guidelines for people with diabetes. The ACSM position stand on exercise and type 2 diabetes was issued in 1997 and updated in 2000. Also, *ACSM's Exercise Management for Persons With Chronic Diseases and Disabilities, Third Edition* has a section on exercise programming for people with diabetes. Lastly, *ACSM's Guidelines for Exercise Testing and Prescription, Eighth Edition* has a section with details for reducing risk and optimizing benefits of exercise among people with diabetes. These recommendations are similar to those from the ACSM position stand, as both emphasize the end

GUIDELINE 12.2

Title: *Physical Activity Guidelines for Americans*

Organization: USDHHS

Year published: 2008

Purpose: To promote a healthy body weight and reduce the risk of other diseases

Location: www.health.gov/paguidelines

Population: Adults with type 2 diabetes

GUIDELINES

- Participate in a minimum of 150 min/wk of moderate-intensity physical activity.
- For greater health benefits, increase to 300 min/wk of moderate-intensity activity.
- Work with a health care provider to adapt physical activity so that it is appropriate for the chronic condition.
- Be careful to monitor blood glucose.
- Avoid injury to the feet.

result of a total weekly caloric expenditure of at least 1,000 kcal. There are minor differences, though, as the position stand generally advocates a slightly *easier* absolute intensity, while *ACSM's Guidelines for Exercise Testing and Prescription* gives a specific recommendation for daily time spent in aerobic activity (20-60 min). Each has slightly different goals as well, although the overall theme is the same—regular (preferably daily) activity is important for improving fitness and diabetes status while lowering the risk for chronic disease. Intensity ranges from lower-intensity training that corresponds to a goal of simply maintaining cardiorespiratory fitness to higher-intensity exercise that corresponds

to improved body fat and reduced medication requirements.

The ACE guidelines are similar to those seen in the seventh edition of *ACSM's Guidelines for Exercise Testing and Prescription.* These stress regular (and preferably daily) moderate-intensity activity and recommend an energy expenditure of a minimum of 1,000 and up to 2,000 kcal/wk. For people with type 1 diabetes, the exercise is somewhat lighter—although the ACE is one of the few entities with specific recommendations for people with type 1 diabetes, the aerobic component is the only one specified directly. The ACE guidelines for type 1 diabetes are presented in the following section.

GUIDELINE 12.3

Title: "American College of Sports Medicine Position Stand: Exercise and Type 2 Diabetes"

Organization: ACSM

Year published: 2010

Purpose: To develop and maintain cardiorespiratory fitness, body composition, and muscular strength and endurance

Location: *ACSM's Guidelines for Exercise Testing and Prescription, Eighth Edition*

Population: Individuals with type 2 diabetes

GUIDELINES

Component	Frequency	Intensity	Duration	Type
Aerobic exercise	3-7 d/wk	RPE of 10-12 out of 20; 40%-70% $\dot{V}O_2max$	That which will expend a minimum of 1,000 kcal/wk (start with 10-15 min sessions and increase up to 30 min)	Walking or non-weight-bearing activities
Resistance exercise	At least 2 d/wk	10-15 repetitions, to near fatigue	A minimum of 1 set of 8-10 exercises	Major muscle groups

Special considerations include the following:

- If weight loss is a goal, greater amounts of caloric expenditure (≥2,000 kcal/wk), including daily exercise, may be required.
- People with diabetic complications may require modifications such as reducing lifting intensity, gripping, isometric contractions, or exercising to the point of exhaustion.
- All people with type 2 diabetes should be screened before exercise participation and receive appropriate monitoring and supervision.
- Glucose monitoring before and after exercise sessions is recommended; modification of medication administration may be needed.

Adapted, by permission, from ACSM, 2010, *ACSM's guidelines for exercise testing and prescription*, 8th ed., (Baltimore, MD: Lippincott, Williams, and Wilkins), 234-235.

GUIDELINE 12.4

Title: "Exercise Prescription for Other Clinical Populations: Diabetes Mellitus"

Organization: ACSM

Year published: 2010

Purpose: To improve glucose control and insulin sensitivity; decrease HbA1c; and improve lipid profiles, blood pressure, weight, physical work capacity, and well-being

Location: *ACSM's Guidelines for Exercise Testing and Prescription, Eighth Edition*

Population: Individuals with diabetes

GUIDELINES

Component	Frequency	Intensity	Duration	Type
Aerobic exercise	3-7 d/wk (PS); 3-7 d/wk (GETP)	RPE of 10-12 out of 20 or 40%-70% $\dot{V}O_2$max (PS); 50%-80% $\dot{V}O_2$R or HRR or RPE of 12-16 out of 20 (GETP)	Enough to expend a minimum of 1,000 kcal/wk, starting with 10-15 min sessions and increasing up to 30 min (PS); 20-60 min (GETP)	Walking or non-weight-bearing activities
Resistance exercise	At least 2 d/wk (PS); 2-3 d/wk (GETP)	10-15 repetitions to near fatigue (PS); 60%-80% 1RM, 8-12 repetitions (GETP)	2-3 sets of 8-10 exercises (GETP); 1 set of 8-10 exercises (PS)	Exercises working the major muscle groups

Special considerations include the following:

- Patients with type 2 diabetes should strive to accumulate a minimum of 1,000 kcal/wk of physical activity. (PS and GETP)

- If weight loss is a goal, greater amounts of caloric expenditure (≥2,000 kcal/wk), including daily exercise, may be required. (PS and GETP)

- Modifications such as reducing lifting intensity, gripping, isometric contractions, or exercising to the point of exhaustion may need to be made for people with diabetic complications. (PS/GETP)

- All people with type 2 diabetes should be screened before participation and receive appropriate monitoring and supervision. (PS)

- Glucose monitoring before and after exercise sessions is recommended; modification of medication administration may be needed. (PS)

PS = Based on Albright et al. 2000. GETP = Adapted, by permission, from ACSM, 2010, *ACSM guidelines for exercise testing and prescription*, 8th ed. (Baltimore, MD: Lippincott, Williams, and Wilkins), 234.

Guidelines for Individuals With Type 1 Diabetes

Type 1 diabetes is a result of autoimmune destruction of the beta cells in the pancreas. This type of diabetes usually manifests at a much younger age; 75% of all cases of type 1 diabetes are diagnosed in individuals younger than 18 y. Despite the young diagnosis, people who keep their diabetes well controlled often live complication free for several decades. The risks and benefits of a regular physical activity program differ slightly between people with type 1 diabetes and people with type 2 diabetes because of the differences in the pathophysiology of the disease and the underlying health concerns. CVD is again the leading cause of death, and the AHA categorizes children with type 1 diabetes in the highest tier for cardiovascular risk. As in people with type 2 diabetes, complications and risk for death are

GUIDELINE 12.5

Title: "Diabetes: Exercise Programming"

Organization: ACSM

Year published: 2009

Purpose: To improve blood glucose control, lower medication requirement, improve insulivity, reduce body fat, reduce CVD risk, and reduce stress (for people with diabetes) and to prevent type 2 diabetes (for people with IGT, gestational diabetes, or a family history of type 2 diabetes)

Location: *ACSM's Exercise Management for Persons With Chronic Diseases and Disabilities, Third Edition*

Population: Individuals with type 1 or type 2 diabetes, gestational diabetes, IGT, or IFG

GUIDELINES

Component	Frequency	Intensity	Duration	Type
Aerobic exercise	4-7 d/wk	50%-80% HRmax, 50%-85% $\dot{V}O_2$max or RPE	20-60 min each session	Large-muscle activities
Resistance training	3 d/wk	Low resistance of 40%-50% 1RM (high resistance OK for athletes with well-controlled diabetes)	2-3 sets of 10-15 repetitions	Exercises using free weights, machines, isokinetic machines
Flexibility exercise	2-3 sessions per week	Maintain stretch below point of discomfort	20-60 s each stretch, starting at only 6-10 s each stretch	Stretching, yoga
Functional training	Intermittent			ADLs, activity-specific exercise

Special considerations include the following:

- High-intensity intervals may be acceptable for healthy athletes in good diabetic control.
- Snack or insulin dosage change may be needed 30 to 60 min before exercise.
- Blood glucose should be monitored before and after exercise.
- If complications are present or if diabetes has been present for a long time, lower-intensity activity (40%-70% $\dot{V}O_2$max) may be advisable.
- People with autonomic neuropathy or on beta-blockers may have an altered heart rate response and should use RPE to monitor intensity.

Adapted, by permission, from W.G. Hornsby, Jr., 2009, Diabetes. In *ACSM's exercise management for persons with chronic diseases and disabilities*, 3rd ed., edited by J.L. Durstine et al. (Champaign, IL: Human Kinetics), 188.

tied to the degree of glycemic control. Current standards reflect a goal of maintaining glucose levels at or near normal, although in people with type 1 diabetes, the risk of hypoglycemia is higher.

All levels of exercise, including leisure activities, recreational sports, and professional sports, can be performed by people with type 1 diabetes who do not have complications and are in relatively good glycemic control. Close monitoring of blood glucose levels and appropriate carbohy-drate consumption or insulin administration are important elements of a safe exercise program. Many individuals with type 1 diabetes now use an insulin pump, which gives them the flexibility to make appropriate insulin dose adjustments for various levels of activity. People who exercise routinely can become aware of their own personal metabolic response to exercise and use this information to improve their glycemic control during activity. Many individuals with type 1 diabetes

GUIDELINE 12.6

Title: *Exercise and Type 2 Diabetes*

Organization: ACE

Year published: 2001

Purpose: To prevent and reverse the course of diabetes, improve insulin response, and decrease the risk of CVD

Location: www.acefitness.org/fitfacts/fitfacts_display.aspx?itemid=2608

Population: Individuals with type 2 diabetes

GUIDELINES

Component	Frequency	Intensity	Duration	Type
Aerobic exercise	3-4 d/wk minimum; daily is recommended	Moderate	20-60 min each session	Walking, water aerobics, and cycling
Resistance training	At least 2 d/wk	Low resistance, low intensity	1 set of 10-15 repetitions	Not specified
Flexibility exercise	At least 2-3 d/wk	Stretch to the point of tightness but not pain	15-30 s each stretch, 2-4 times per stretch	Stretching

Special considerations include the following:

- The ultimate goal is to expend a minimum of 1,000 kcal/wk for health benefits or 2,000 kcal/wk for weight loss.
- Monitor glucose before and after exercise.
- Wear an ID bracelet indicating type 2 diabetes.
- Exercise with a partner whenever possible.
- Check with a physician before beginning a physical activity program, particularly if eye, kidney, or heart complications are present.

Adapted, by permission from American Council on Exercise Fit Facts, San Diego 2001. www.acefitness.org/fitfacts Accessed 10/10/2009.

are diagnosed in childhood or adolescence—a time that is crucial to establishing a foundation for lifetime physical activity, which is even more important in the population with type 1 diabetes. One Australian study found that the epidemiology of participation in regular physical activity parallels that of children and adolescents without diabetes: Children spend more time in physical activity than adolescents do, and girls (particularly adolescents) tend to be less active than boys are (Valerio and others 2003). This study also found that children and adolescents with type 1 diabetes spend less time in physical activity than their peers without diabetes do, underscoring the need to promote physical activity to all individuals in this age group in light of the considerable benefits.

Although studies discussed in the earlier section on type 2 diabetes indicated that exercise training significantly improved glycemic control as measured by HbA1c, this effect has not been seen in people with type 1 diabetes (Roberts and others 2002). Nevertheless, the benefits of regular physical activity and exercise in reducing atherosclerosis are evident through improvements in lipid profile, blood pressure, and cardiorespiratory fitness.

In 2006 to 2007, the International Society for Pediatric and Adolescent Diabetes (ISPAD) issued clinical practice consensus guidelines regarding exercise and diabetes for youths. While there is no specific guideline regarding the minimum amount of activity recommended for youths, there are several stipulations about timing of exercise and modifications to carbohydrate intake and insulin administration as well as cautionary advice regarding postexercise hypoglycemia. The importance of planning for exercise—particularly of durations longer than those to which the individual is accustomed—as well as keeping track of carbohydrate ingested, insulin administered, and blood glucose levels, is highlighted. Exercise beyond that to which an individual is accustomed may require additional carbohydrate to maintain blood glucose level. An alternative is to reduce insulin administration in order to match glucose availability and utilization. Useful information regarding exercise and insulin administration modification is presented in table 12.2. For example, insulin administration may be reduced by nearly 100% for exercise lasting more than 60 min at 75% of $\dot{V}O_2$max. An important caution is to monitor for late hypoglycemia, which may occur several hours after exercise that is higher in duration or intensity than usual. Prolonged activity may increase insulin sensitivity as well as delay the replenishment of liver and muscle glycogen stores, thus increasing the risk for hypoglycemic events even in the day after strenuous exercise. Highlights of the ISPAD recommendations for exercise in children and young people with diabetes, as well as the ACE recommendations for exercise in type 1 diabetes, are presented in the following guidelines. Notably, the ISPAD recommendations do not specify whether they refer to people with type 1 or type 2 diabetes, although the majority of children with diabetes have type 1. This is slowly changing, however, with the increasing levels of childhood obesity and insulin resistance, and greater numbers of children and adolescents are being diagnosed with type 2 diabetes all the time.

TABLE 12.2 Recommendations for Percent Reduction in Premeal Insulin Bolus

Intensity/duration of exercise	Reduction in pre-meal insulin bolus
Low (25% $\dot{V}O_2$max)/30 minutes	25%
Low/60 minutes	50%
Moderate (50% $\dot{V}O_2$max)/30 minutes	50%
Moderate/60 minutes	75%
Heavy (75% $\dot{V}O_2$max)/30 minutes	75%
Heavy/60 minutes	-

Adapted from Robertson 2009.

GUIDELINE 12.7

Title: "Exercise in Children and Adolescents With Diabetes"

Organization: ISPAD

Year published: 2008

Purpose: To address blood glucose regulation during various forms of sport and exercise

Location: *Pediatric Diabetes* 9: 65-77

Population: Children and adolescents with diabetes (type not specified)

GUIDELINES

- Daily physical activities should be a part of the normal routine for both health benefits and consistency in blood glucose management.
- Regular and accustomed exercise is easier to manage, but adjustments may be necessary for sporadic extra physical activity.
- Progressively increase the intensity and duration of activity.
- Keep notes on the following:
  - Timing and intensity of physical activity
  - Carbohydrate intake
  - Blood glucose response before, during, and after activity
- Depending on the insulin administration schedule, preexercise and basal rates should be reduced before, during, and after exercise.
- If physical activity is not foreseen, insulin dose during and after exercise should be decreased.
- Ingest slowly absorbing carbohydrate in the few hours preceding exercise.
- Avoid physical exercise at the time of peak action of insulin.
- Always have peers or adults nearby while physically active.
- Glucose tablets or quick-acting carbohydrate should be available and adults should be aware of treatment procedures for hypoglycemia.
- Reduction in insulin dose should be greater for higher-intensity or longer-duration exercise, reaching 100% reduction at an intensity of 75% $\dot{V}O_2$max lasting for 60 min.
- Inject insulin at a site that will not be heavily involved in muscular activity.
- Insulin pumps should be removed for up to 2 h for contact sports. Dosage should be reduced to 50% of basal rate 90 min before activity.
- Strenuous exercise should be avoided if preexercise blood glucose levels are >250 mg/dl and accompanied by ketonuria or ketonemia; these individuals should take a bolus dose of 5% of the total daily insulin requirement and postpone exercise until ketones have cleared.
- Measure blood glucose before going to bed on the evening after major activity. Caution should be taken when bedtime glucose is <7.0 mM (~125 mg) due to risk for postexercise nocturnal hypoglycemia.
- Glucometer and test strips should be kept at 59 to 86 °F (15-30 °C) for accuracy.

GUIDELINE 12.8

Title: *Exercise and Type 1 Diabetes*

Organization: ACE

Year published: 2001

Purpose: To reduce the onset of complications, maintain body weight, lower blood pressure, and favorably alter lipids and lipoproteins

Location: www.acefitness.org/fitfacts/fitfacts_display.aspx?itemid=2607

Population: Individuals with type 1 diabetes

GUIDELINES

Component	Frequency	Intensity	Duration	Type
Aerobic exercise	4-5 d/wk at a minimum	Low to moderate	30-40 min each session	Not specified
Resistance training	Not specified	Not specified	Not specified	Not specified
Flexibility exercise	Not specified	Not specified	Not specified	Not specified

Special considerations include the following:

- Monitor glucose before and after exercise and carry high-carbohydrate food in case energy is needed.
- Wear an ID bracelet indicating type 1 diabetes.
- Exercise with a partner whenever possible.
- Check with a physician to get intensity recommendations for the activity program, particularly if eye, kidney, or heart complications are present.

Adapted, by permission from American Council on Exercise Fit Facts, San Diego 2001. www.acefitness.org/fitfacts Accessed 10/10/2009.

SUMMARY

Diabetes is a common disease that continues to increase in prevalence both nationally and worldwide. Exercise and physical activity are cornerstones in the treatment and prevention of diabetes and its complications. Guidelines focus on achieving a level of activity to improve glycemic control, reduce CVD, and maintain weight. In general, physical activity is very safe and highly recommended, although people with diabetes may have to modify their programs in order to minimize any problems associated with the risks of physical activity.

Neuromuscular Disorders

Until recently, the primary focus of physical activity for individuals with disabilities centered on rehabilitation, and these individuals often felt isolated or segregated. In the past decade, there has been a paradigm shift toward integration and inclusion and toward exercise participation not only for rehabilitation but also for improved fitness for play, sport, and the physical demands of employment (Seaman 1999). Both the NCPAD and United Cerebral Palsy (UCP) have exercise guidelines for people with disabilities. The recent *Physical Activity Guidelines for Americans (PAGA)* issued by the USDHHS also has a specific section relating to activity for individuals with disabilities. A diverse group of conditions are classified as disability; *PAGA* includes (but is not limited to) cerebral palsy, multiple sclerosis, muscular dystrophy, Parkinson's disease, spinal cord injury, stroke, Alzheimer's disease, limb amputations, and mental illness. This chapter focuses on people with disabilities due to neuromuscular disorders, including cerebral palsy, multiple sclerosis, spinal cord injury, muscular dystrophy, Parkinson's disease, and stroke-related neuromuscular deficits.

Rates of physical activity participation among individuals with disabilities are lower than those of individuals without disability. According to data from 2005, inactivity rates for people with disabilities were twice those of people without disability (28% versus 15%; CDC 2006). Also, there is a higher incidence of cardiovascular risk factors, including obesity and smoking, among people with disabilities, underscoring the importance of a physical activity program for this population. Regular physical activity has the potential to slow or reverse some of the health and functional deficits associated with various neuromuscular disorders, but there are barriers to participation, including accessibility, environment, and information.

Benefits of Physical Activity

The benefits of an exercise program for people with neuromuscular disability include improvements in muscular strength and endurance, cardiorespiratory fitness, weight control, self-perception, lipoprotein profile, energy, and body composition. A benefit unique to people with disabilities is an improved economy of movement, which is particularly important because both spasticity and inefficient movement patterns can be significant obstacles in functioning for daily activities of living. The goals of exercising with neuromuscular disorders include increasing stamina, balance, and strength without exacerbating ongoing injuries or conditions. Flexibility training with an emphasis on range-of-motion exercises is stressed. Improving cardiorespiratory fitness can be done safely and can enhance quality of life. Focusing on correcting imbalances and strengthening weaker muscles while stretching tighter ones improves well-being.

The NCPAD provides exercise guidelines for several different diseases, including those that cause neuromuscular dysfunction. The NCPAD is a good resource not only for these guidelines but also for pathophysiology and scientific evidence regarding exercise benefits for the various conditions (see www.ncpad.org).

Gradually increasing exercise in duration, frequency, and intensity improves fitness while reducing the risk for injury. Exercising with a partner is recommended for both social and safety issues. Due to diversity in baseline functional status, history of physical activity, and type of disorder, the exercise prescription should be individualized for a person with disability. Nevertheless, most individuals with neuromuscular disabilities should strive for regular physical activity to enhance outcomes for health, function, and secondary conditions.

- **Musculoskeletal health:** Both aerobic exercise and resistance training increase muscle strength. Flexibility, which is important for many with neuromuscular diseases, can be enhanced through regular flexibility training. Bone density can be improved, and age-related declines can be slowed through regular exercise. Bone health is particularly crucial because weakened balance, a common feature of neuromuscular disease, places individuals at an already heightened risk for fracture. Interventions that improve both balance (and consequently fall risk) and bone density are ideal and may significantly decrease morbidity.

- **Functional status:** Several measures of functional status can be improved through regular physical activity, including ADLs, walking speed and distance, and quality of life. Even people with spinal cord injury who use a wheelchair benefit from increased propulsion speed through exercise.

- **Secondary conditions:** *Healthy People 2010* defines secondary conditions as "physical, medical, cognitive, emotional or psychosocial consequences to which persons with disabilities are more susceptible by virtue of an underlying impairment, including adverse outcomes in health, wellness, participation and quality of

life" (USDHHS 2000). Physical activity helps reduce these secondary conditions by improving quality of life and reducing chronic diseases such as hypertension, heart disease, diabetes, and cancer.

General Recommendations From *Physical Activity Guidelines for Americans*

As noted earlier, the 2008 *PAGA* provided the first set of comprehensive national guidelines to specifically address people with disabilities. The guidelines state that it is acceptable for individuals with disabilities who can follow the guidelines for all adults to do so. For those who cannot follow the general adult guidelines, there are adaptations that can make physical activity safe, appropriate, and beneficial. Because there is a large amount of dissimilarity among individuals with different disabilities, consultation with a health care provider or exercise specialist is recommended so that the physical activity program can be modified as needed. The *PAGA* recommendations are presented here and are followed by separate sections on guidelines for various disabilities as recommended by various entities.

Guidelines for Cerebral Palsy

Cerebral palsy (CP) is a nonprogressive brain disorder interfering with normal brain development that occurs before or right around the time of birth. About 1.5 to 5 of every 1,000 births in the United States involve CP. The disorder may result from faulty embryonic development, faulty blood supply, or genetic or chromosomal abnormality or may occur during or shortly after birth due to hypoxia, head trauma, infection, toxic injury, or other insult to the brain. Manifestations of CP are varied and range from severe spastic or athetoid tetraplegia to minimally affected monoplegia or athetosis. This continuum ranges from those who are unable to function independently to those who can ambulate, participate in competitive sports, and live independently.

Individuals with CP who exercise can expect augmented physical fitness and work capacity,

GUIDELINE 13.1

Title: *Physical Activity Guidelines for Americans*

Organization: USDHHS

Year published: 2008

Purpose: To lower the risk for all-cause mortality, CHD, stroke, hypertension, and type 2 diabetes

Location: www.health.gov/paguidelines

Population: Adults with disabilities

GUIDELINES

- Adults with disabilities, who are able to, should get at least 150 min/wk of moderate-intensity aerobic activity or 75 min/wk of vigorous-intensity aerobic activity or an equivalent combination of moderate- and vigorous-intensity activity.
- Aerobic activity should be performed in episodes lasting at least 10 min and should be spread throughout the week.
- Adults with disabilities, who are able to, should also perform muscle-strengthening activities that are moderate or high intensity and involve all the major muscle groups on 2 or more days a week, as these activities provide additional health benefits.
- When adults with disabilities are not able to meet the suggested guidelines, they should engage in regular physical activity according to their abilities and should avoid inactivity.
- Adults with disabilities should consult their health care providers about the amounts and types of physical activity that are appropriate for their abilities.

although the improvements are generally less than those seen in people without disability. There is a reduced mechanical efficiency due to spasticity and increased muscle tone. Although fatigue, spasticity, and incoordination may increase after a strenuous exercise session, this increase is temporary. Over the long term, regular physical activity and exercise can attenuate spasticity and not only improve function and independence but also reduce reliance on antispasmodic medications. Regular exercise has many benefits for people with CP, including the following:

- Improved peak oxygen uptake
- Higher ventilatory threshold
- Increased work rate for a given heart rate
- Improved range of motion
- Improved coordination and skill of movements
- Increased muscular strength and endurance
- Growth of skeletal muscle

- Improved sense of wellness and body image
- Improved capacity to perform ADLs
- Less severe spasticity and athetosis
- Decreased use of medications

Additionally, because of the heightened risk for CAD and its risk factors in sedentary individuals with disability, it is important to consider the reduction in CAD risk. Such a reduction can be achieved by regular physical activity.

Aerobic exercise is as important for individuals with CP as it is for individuals without CP. Because CP decreases movement efficiency, caloric expenditure for a person with CP is likely to be higher than that achieved by a person without CP performing at a similar workload. Muscular training, for strength and endurance, is an important component of the activity program. Depending on the degree of impairment from CP, an individual may use weight machines or free weights. For people who have very limited strength, gravity-reduced exercise training is a viable way of utilizing gravity to aid movements.

Gravity-reduced exercise training may also help improve abnormal muscle tone in spastic muscles. Safety is a significant concern because stability and range of motion are often limited by CP. Using a spotter or straps can increase safety and stability. Individuals who use wheelchairs or other assistive devices definitely benefit from increased muscular strength, as the energy expenditure required just to ambulate or navigate around architectural barriers may be 15 times that required by a person without disability! Flexibility is the third key element in the CP exercise program. Increased flexibility may help people with contractures be able to perform ADLs independently.

The ACSM delineates guidelines for people with CP in the third edition of *ACSM's Exercise Management for Persons With Chronic Diseases*

and Disabilities and the eighth (and most recent) edition of *ACSM's Guidelines for Exercise Testing and Prescription.* Regular aerobic exercise through some form of nonambulatory ergometry is recommended, while walking is utilized as a way to improve endurance. Because individuals with CP may be easily fatigued, an increased frequency of shorter training sessions often is preferred. Both flexibility and resistance training are encouraged. Resistance exercises that strengthen weaker muscle groups may help with balancing tone between muscle groups while reducing spasticity.

The NCPAD has a significant compilation of resources and recommendations for various types of disabilities. Although the NCPAD does provide some information on the disease process

GUIDELINE 13.2

Title: "Cerebral Palsy"

Organization: ACSM

Year published: 2009

Purpose: To improve sense of wellness and body image, to enhance capacity to perform ADLs, to lessen severity of symptoms such as spasticity and athetosis, and to reduce the need for antispasmodic medications

Location: *ACSM's Exercise Management for Persons With Chronic Diseases and Disabilities, Third Edition*

Population: Individuals with CP

GUIDELINES

Component	Frequency	Intensity	Duration	Type
Aerobic exercise	3-5 d/wk	40%-85% $\dot{V}O_2$max or HRR	20-40 min each session	Exercise on Schwinn Airdyne, arm or leg ergometer
Endurance exercise	1-2 sessions per week	Not specified	6-15 min	Walking (ambulatory), wheelchair pushes
Resistance training	2 d/wk	Resistance as tolerated	3 sets of 8-12 repetitions	Exercises with free weights, machines
Flexibility exercise	Before and after aerobic and endurance exercise	Maintain stretch below point of discomfort	60 s each stretch	Stretching

Special considerations include the following:

- Duration is more important than intensity for all exercise modes.
- Hands or feet should be strapped to pedals during arm and leg ergometry.
- Spotters or supervising personnel should be present to prevent falls during treadmill use.

GUIDELINE 13.3

Title: "Exercise Prescription for Other Clinical Populations: Cerebral Palsy"

Organization: ACSM

Year published: 2010

Purpose: To increase fitness levels

Location: *ACSM's Guidelines for Exercise Testing and Prescription, Eighth Edition*

Population: Individuals with CP

GUIDELINES

- General ACSM guidelines for healthy individuals (guideline 3.1) apply with some individualized modifications.
- Individuals with more significant disease should start aerobic exercise with frequent, short bouts (at 40%-50% $\dot{V}O_2R$) and progress to 20 min bouts (at 50%-85% $\dot{V}O_2R$).
- While the type of exercise depends on the degree of disease, tricycles or hand cycles are recommended to facilitate balance, allow for a wide range of power output, and minimize injury.
- Introduce new skills early in the session, before onset of fatigue.
- Flexibility training should always be included with resistance training. Stretch muscles slowly to their limits throughout the workout in order to maintain their length.
- Good positioning of joints, head, and trunk is important, although strapping is not recommended.

Adapted, by permission, from ACSM, 2010, *ACSM's guidelines for exercise testing and prescription*, 8th ed. (Baltimore, MD: Lippincott, Williams, and Wilkins), 239-240.

GUIDELINE 13.4

Title: *Exercise/Fitness: Resistance Training for Persons with Physical Disabilities: Resistance Training Guidelines for Cerebral Palsy*

Organization: NCPAD

Year published: 2005

Purpose: To safely strengthen muscles, reduce physical impairment, and improve posture and balance

Location: www.ncpad.org/exercise/fact_sheet.php?sheet=107&section=813

Population: Individuals with CP

GUIDELINES

- Focus on strengthening the hip abductors in order to balance the strong pull of the hip adductors.
- Flexibility and range-of-motion exercises should receive the same amount of attention resistance training receives.
- If there is complete paralysis on one side of the body, substitute resistance training with flexibility training.
- Standing exercises should be performed only if the individual's static and dynamic balance is appropriate.
- Cuff weights and machines are more appropriate than free weights or elastic bands.
- Active assistive exercise is often needed to perform smooth motions.

and basic exercise principles of CP, there is not a specific set of guidelines addressing the FITT principle. The published guidelines focus on safety and what to expect with a training program. Specific recommendations relate to the variability of clinical presentations of CP and how to approach these different situations. An emphasis is made on including flexibility training as part of the program. Although NCPAD guidelines for CP are lacking in detail when compared with NCPAD guidelines for other disabilities, the NCPAD CP guidelines are included here for the sake of completion.

UCP, a national organization focusing on progress not only on CP but also on all disabilities, has issued exercise principles for persons with CP and neuromuscular disorders. Goals include strengthening muscles and bones and improving flexibility as well as heart and lung function. Exercise sessions should be performed at least 3 to 5 times per week. These should be sandwiched between an ample warm-up and an ample cooldown. Suggested types of physical activity are given; each provides one or more of the desired benefits. These are summarized in Guideline 13.5 Table B.

The UCP guidelines also include warning signs that should prompt termination of exercise and contact with a physician. Most of the following signs result from overexertion, which may worsen spasticity for people with CP (UCP 2008c):

- Lightheadedness
- Chest pain
- Difficulty breathing
- Excessive fatigue
- Nausea
- Moderate to severe joint or muscle pain

Guidelines
for Parkinson's Disease

Parkinson's disease is a movement disorder characterized by tremors, slowness of movement, freezing or akinetic episodes, and decreased mobility due to muscle rigidity. These are all caused by reduced levels of dopamine in the substantia nigra. Symptoms generally manifest when people reach their mid-50s and grow increasingly prevalent with aging. Parkinson's disease affects approximately 3% of Americans aged 65 y and older (Ferrini and Ferrini 2000). Exercise is important to individuals with Parkinson's disease for many reasons. It improves range of motion, balance, and function (Schenkman and others 1998) as well as grip strength, fine motor coordination, arm tremor, and gait (Palmer 1986). One of the most debilitating manifestations of Parkinson's disease is a shuffling gait that is often punctuated with freezing episodes in the middle of the gait cycle. A 1999 study found that the motor disability associated with Parkinson's disease significantly improved when individuals participated in an activity program for 1 h twice a week over 14 wk (Reuter and others 1999).

Exercise programs are a valuable component of the overall treatment plan for Parkinson's disease; individuals treated with combined exercise and medication have a lower level of disability than those treated with drug therapy alone (Formisano and others 1992). Exercise is essential to decrease respiratory complications such as pneumonia and CVD as well as depressed function of the respiratory muscles. Exercise can also improve spinal flexibility, which in turn facilitates coordinated movements and makes activities that are difficult for many individuals with Parkinson's disease significantly easier.

Many individuals with Parkinson's disease experience dysfunction of the autonomic nervous system that alters heart rate and blood pressure responses to exercise, sweating patterns, and thermal regulation. These can all be concerns during physical activity. Other considerations include working out with supervision because of increased fall risk (due to loss of balance and shuffling or freezing-induced gait abnormalities) and reduced exercise efficiency due to rigidity. Lastly, it is safer for individuals with Parkinson's disease to utilize weight machines that can provide support that may help with balance and prevent falls.

Medications used to treat Parkinson's disease aim at correcting dopamine, norepinephrine, epinephrine, and acetylcholine imbalances within the nervous system. A concern of physical activity is the effect that exercise may have on the

GUIDELINE 13.5

Title: *Sports and Leisure: Exercise & Fitness: Exercise Principles and Guidelines for Persons with Cerebral Palsy and Neuromuscular Disorders: Components of an Exercise Session*

Organization: UCP

Year published: 2008

Purpose: To increase and maintain heart and lung efficiency, to improve flexibility and range of motion, to strengthen muscles and increase endurance, and to improve and maintain bone structure and strength

Location: www.ucp.org/ucp_channeldoc.cfm/1/15/15/15-15/643

Population: Individuals with CP

GUIDELINES

Type	Activity	Duration	Frequency	Intensity
Warm-up	Gentle activity such as slow walking or slow wheeling to warm up the muscles and joints to be exercised, followed by stretching that includes muscles and joints of primary concern	2-5 min of each type	Each session	Not specified
Primary exercise	Aerobic exercise, resistance exercise, swimming, aquatics, cycling	20-30 min (may start with several 5-10 min sessions per day)	At least 3-5 times a week, goal is 7 d/wk	Moderate
Resistance exercise	Exercises using weight machines, free weights, plastic tubing, medicine balls, circuit training	5-8 repetitions for strength or 8-12 repetitions for endurance		To complete the set of repetitions
Flexibility exercise	Stretching, yoga, Pilates	Hold stretches and progress slowly; spend more time on tight muscle groups	Before and after every cardiovascular and strength workout	Thorough stretching without pain
Cool-down	Slow walking or slow wheeling followed by stretching that includes joints of primary concern	3-5 min of activity and 2-5 min of stretching		

Special considerations include the following:

- Exercising with a partner is recommended.
- Duration should be increased progressively before intensity is increased.
- Performance goals should be assessed at least every 6 mo.
- A physician should be consulted before initiation of an exercise program.

>continued

GUIDELINE 13.5 *>continued*

Suggested Exercises for Individuals With CP

Type	Benefit
Arm cycling	HL, M, FR
Chair aerobics	HL, M, FR
Dancing	HL, B, FR
Exercise band activities	M, B
Jogging	HL, B, M
Leg cycling	HL, M , FR
Rowing	HL, M, FR
Stair-climbing	HL, M, B
Swimming	HL, M, FR
Walking	HL, B
Water exercise	HL, B, FR
Weight training	M, B
Wheeling	HL, M
Yoga and tai chi	FR, B

B = bone structure and strength; FR = flexibility and range of motion; HL = heart and lung efficiency; M = strength and endurance.

Adapted with permission, from UCP's Exercise Principles and Guidelines for Persons with Cerebral Palsy and Neuromuscular Disorders, www.ucp.org

absorption and metabolism of levodopa, the most effective drug used to treat the symptoms of Parkinson's disease. It is recommended that individuals taking levodopa should ingest the prescribed dose, rest quietly for 60 min, and then proceed with regular vigorous exercise (Goetz and others 1993). While levodopa and levodopa plus carbidopa may cause transient tachycardia, they may also produce exercise bradycardia. Because the therapeutic effects of the drugs used to treat Parkinson's disease depend on their absorption, plasma levels, and brain levels, clinical motor function (or dysfunction) may vary widely depending on medication timing. To complicate things further, the same exercise may affect the same individual differently and produce different autonomic and vascular responses depending on the medication plasma level. All of these phenomena need to be considered when exercising; keeping both the timing of medication ingestion and physical activity constant helps to reduce variability.

Both the ACSM and NCPAD have guidelines for individuals with Parkinson's Disease. Although these differ slightly, there is a common goal which is to improve or maintain function through regular exercises which can diminish the progressive decline in function often experienced by individuals with this disease.

Guidelines for Muscular Dystrophy

There are several types of muscular dystrophy (MD), which together form a group of hereditary disorders of skeletal muscle that result in progressive loss of muscle strength. Types include Becker, Duchenne, and Emery-Dreifuss, which are generally classified as X-linked recessive genetic disorders, and facioscapulohumeral, limb girdle, and myotonic. While X-linked MD generally presents with symptoms during the first few years of life, some types of MD, such as myotonic MD, may not manifest until adulthood. Variability in muscular involvement causes a wide array of clinical pictures. An individual with facioscapulohumeral MD, for example, might not be able to lift the arms overhead but could have a strong handgrip. Involvement of cardiac muscle is common and can present with cardiomyopathy, cardiac conduction

GUIDELINE 13.6

Title: *Disability/Condition: Parkinson's Disease and Exercise*

Organization: NCPAD

Year published: 2009

Purpose: To decrease disability

Location: www.ncpad.org/disability/fact_sheet.php?sheet=59&view=all#4

Population: Individuals with Parkinson's disease

GUIDELINES

Component	Frequency	Intensity	Duration	Type
Aerobic exercise	3 times per week at a minimum	60%-75% of HRmax	30 min or more per session	Walking, recumbent stationary biking, or swimming
Resistance training	2-3 times per week	Resistance as tolerated	1 set of 8-12 repetitions done for 8-10 exercises	Machine exercises using major muscle groups
Flexibility exercise	After 5 min warm-up and 5 min cool-down; also on off days or at home	Not specified	20-30 s for each stretch	Static stretching for all major muscle groups

Special considerations include the following:

- Baseline cardiorespiratory fitness testing is recommended, using a recumbent stationary bicycle and measuring $\dot{V}O_2$max.
- Baseline strength should be measured using 1RM or 10RM for all major muscle groups.
- Active range-of-motion testing should also be performed (see table that follows) to obtain baseline flexibility.

Flexibility Range-of-Motion Testing for Parkinson's Disease

Assessment	Type
Cervical region	Flexion and extension
Vertebral region	Flexion and extension
Hip	Internal and external rotation, flexion and extension, straight-leg raise
Shoulder	Flexion and extension, rotation and elbow flexion
Posture	Assessment to evaluate for vertebral deviation such as kyphosis, lordosis, or scoliosis
Gait	Notes on general gait patterns and timing

The following are recommended flexibility exercises for Parkinson's disease:

- Chin tuck
- Head tilt
- Head turn
- Trunk bend (forward and backward and side to side)
- Trunk twist
- Hip stretch
- Hamstrings stretch
- Shoulder stretch
- Shoulder raise
- Facial mobility (grimaces, closing eyes tightly, etc.)
- Arm stretch

GUIDELINE 13.7

Title: "Parkinson's Disease"

Organization: ACSM

Year published: 2009

Purpose: To improve function

Location: *ACSM's Exercise Management for Persons With Chronic Diseases and Disabilities, Third Edition*

Population: Individuals with Parkinson's disease

GUIDELINES

Component	Frequency	Intensity	Duration	Type
Aerobic exercise	3 d/wk	60%-80% HRmax	60 min or less each session	Rowing or leg and arm ergometry
Endurance training	4-6 sessions per *day*	Speed dependent on individual	20-30 min at a time	Short walking bouts
Strength training	3 sessions per week	Light resistance	1 set of 8-12 repetitions	Exercises on weight machines
Flexibility exercise	Not specified	Not specified	20 s for each stretch; 30 min total	Stretching
Functional training	Not specified	Not specified	15-20 min	Transfer exercises, ADLs, postural changes
Neuromuscular training	Not specified	Not specified	15-20 min	Gait and balance activities, orofacial activities

Special considerations include the following:

- Duration and intensity should receive equal priority.
- Duration should be increased slowly (every 4-5 wk).
- Time of day for both exercise and medication should be kept as constant as possible.

Adapted, by permission, from E.J. Protas, R.K. Stanley, and J. Janovik, 2009, Parkinson's Disease. In *ACSM's exercise management for persons with chronic diseases and disabilities*, 3rd ed., edited by J.L. Durstine et al. (Champaign, IL: Human Kinetics), 353, 355, 356.

defects, incomplete atrial filling, and dropped beats. For this reason, people with MD should obtain a resting ECG and echocardiogram as well as a cardiology consult before starting a physical activity program.

Individuals with MD can expect mild improvements in strength and ADLs after engaging in moderate resistance exercise and stretching. Improvements in bone mass are another important benefit of a regular exercise program, particularly in light of the fact that individuals with MD have lower bone mass to begin with and are often treated with steroids, which causes loss of bone mass. Exercise can attenuate the loss

of both bone and muscle mass that is associated with corticosteroid use. MD causes progressive loss of muscle function and is often accompanied by a feeling of helplessness and despair. Participation in physical activity is one of the few ways individuals with MD can feel a sense of control, independence, strength, and self-direction. Both the NCPAD and ACSM have issued exercise guidelines for people with MD.

Duchenne MD is an X-linked skeletal muscle disorder and is the most common muscular dystrophy in the United States. Both heart and skeletal muscle are affected, and muscle weakness and wasting generally begin as early as 2 y of age. Loss of ambulation and respiratory difficulty occur due to severe contractures and muscle weakness, although an exercise program may help attenuate the natural progress of the disease.

Flexibility training is the cornerstone of the exercise program, as it helps to delay and moderate contractures. Because of the risk of muscle damage, ballistic stretching and certain types of resistance training should be avoided. These include concentric–eccentric contractions, in which muscle damage occurs with a change in directional velocity. Therefore, isokinetic exercise consisting of only eccentric or concentric contractions is preferred. This way, strength may be improved in order to assist in the performance of ADLs such as climbing stairs, rising from a chair,

and walking. Scoliosis is a common consequence of atrophy in the spinal musculature; strengthening the postural muscles may reduce scoliosis and ultimately some of the associated respiratory compromise. Lastly, obesity is a major problem for individuals with Duchenne MD, primarily because the ability to ambulate is often lost and energy expenditure simply does not match caloric intake. A regular exercise program may help to prevent excessive weight gain. Although there are some human studies suggesting that exercise can improve muscular strength for people with Duchenne MD, details with respect to intensity, duration, frequency, and mode are not fully clear. Nevertheless, the NCPAD has issued exercise guidelines for Duchenne MD.

GUIDELINE 13.8

Title: *Disability/Condition: Muscular Dystrophy*

Organization: NCPAD

Year published: 2007

Purpose: To maintain and improve muscular strength, slow rate of weakness and contracture development, maintain respiratory capacity, and strengthen postural muscles

Location: www.ncpad.org/disability/fact_sheet.php?sheet=73&view=all

Population: Individuals with MD, including Duchenne, Becker, facioscapulohumeral, and myotonic types

GUIDELINES

Component	Frequency	Intensity	Duration	Type
Aerobic exercise	Brief sessions several times a day	RPE of 12-14 out of 20	Various durations (focus should be on duration rather than intensity)	Walking, cycling, swimming, arm or leg ergometry
Resistance training	3 times per week	Low, utilize light weights or no resistance	1 set of 5-10 repetitions, work up to 3 sets of 8-12 repetitions	Multijoint exercises utilizing dynamometer, exercise bands, and weight machines
Flexibility exercise	Before and after each exercise program; 3-4 times *daily*	Avoid over-stretching or hypermobility	10-30 s for each stretch, repeated 3 times	Main muscle groups, including arms, wrists, fingers, shoulders, legs, and hips

Special considerations include the following:

- Preparticipation clearance by a physician is recommended.
- Fatigue or exercise-related cramps should prompt rest and a decrease in intensity.
- Consider nutritional counseling along with the exercise program to prevent weight gain.
- Walk on a level surface.

GUIDELINE 13.9

Title: "Muscular Dystrophy"

Organization: ACSM

Year published: 2009

Purpose: To gain muscle strength and muscular endurance, maintain or increase aerobic capacity, decrease cardiac risk factors, increase range of motion, prevent contractures, and maintain and enhance proficiency in ADLs

Location: *ACSM's Exercise Management for Persons With Chronic Diseases and Disabilities, Third Edition*

Population: Individuals with MD of all types

GUIDELINES

Component	Frequency	Intensity	Duration	Type
Aerobic exercise	4-6 d/wk	50%-80% of HRR	At least 20 min each session (to fatigue)	Cycling, walking on treadmill, exercising on elliptical trainer, rowing, arm ergometry
Strength training	3 d/wk	Light, start at 50% 1RM, gradually work up to 75% 1RM	1-3 sets of 10-12 repetitions	Isometric exercises
Flexibility exercise	Daily	Maintain stretch below point of discomfort	20 s for each stretch	Stretching
Functional training	Daily as tolerated	Not specified	Not specified	Activity-specific tasks such as wheelchair propulsion

Special considerations include the following:

- Avoid strenuous exercise in high heat and humidity.
- ECG and echocardiogram should be performed on all individuals with Becker, Duchenne, Emery-Driefuss, and limb girdle MD before the exercise program is begun.
- Consider bone density testing for individuals managed with steroids desiring to participate in contact sports.

Adapted, by permission, from M.A. Tarnopolsky 2009, Muscular dystrophy. In *ACSM's exercise management for persons with chronic diseases and disabilities*, 3rd ed., edited by J.L. Durstine et al. (Champaign, IL: Human Kinetics), 310.

GUIDELINE 13.10

Title: *Disability/Condition: Duchenne Muscular Dystrophy and Exercise*

Organization: NCPAD

Year published: 2007

Purpose: To improve performance of daily activities, reduce the progression of contractures, decrease the formation of scoliosis, and prevent obesity

Location: www.ncpad.org/disability/fact_sheet.php?sheet=142&view=all

Population: Individuals with Duchenne MD

GUIDELINES

Component	Frequency	Intensity	Duration	Type
Aerobic exercise	1-7 d/wk, depending on exercise tolerance	Spontaneous walking speed (low)	Varies based on exercise tolerance, 1-20 min	Walking, cycling, or swimming
Resistance training	1-5 times per week	Low	1-3 sets of 5-15 repetitions of 8-10 exercises	Isokinetic, concentric, or eccentric exercises only, using extensors of the hip, back, elbow, hand, and wrist and shoulder flexors and dorsiflexors
Flexibility exercise	3-4 times daily	Low	10-30 s for each stretch, repeat 3 times	Passive and active stretches of the hip, knee, shoulder, wrist, fingers, and plantar flexors

Special considerations include the following:

- Flexibility is the cornerstone of the Duchenne MD exercise program.
- Walking should be done on a level surface to avoid muscle damage.
- Ballistic activities and movements consisting of changes in directional velocity are contraindicated.

Guidelines for Multiple Sclerosis

Multiple sclerosis (MS) is a demyelinating central nervous system disease characterized by neurological deficits. One constant of MS is the variability in clinical presentation. MS has different symptoms, and many individuals experience a varying set of symptoms—or none at all—at various times during the course of the disease. Some of the more common symptoms of MS are the following:

- Paresis or paralysis
- Ataxia
- Weakness
- Tremor
- Heat sensitivity
- Fatigue
- Temperature sensitivity
- Sensory or balance loss
- Spasm
- Bladder control loss
- Visual loss
- Cardiovascular dysautonomia

The disease can be relapsing remitting, chronic progressive, or chronic relapsing, and roughly equal numbers of individuals with MS fit into each of these categories. Although exercise itself does not have any effect on the prognosis or progression of MS, it does provide short-term gains that may have a significant influence on quality of life, fitness, and comorbidity from other diseases. For example, aerobic exercise improves $\dot{V}O_2$max, physical work capacity, body composition, and lipid profile. Much of the disability occurring in people with MS is related to secondary deterioration from a sedentary lifestyle (Mulcare 2003).

However, exercise can also have negative effects on people with MS; it is important to consider the current clinical situation of the individual when planning a physical activity program. Before embarking on a program, people with MS should undergo graded exercise testing. Functional assessment for stretch, flexibility, and gait analysis is also recommended in order to determine which muscle groups are involved in the disease process. There are some factors that limit the improvement individuals with MS may receive from exercise. These include the baseline neurological dysfunction as well as cardiovascular dysautonomia, which during exercise translates into a blunted—if any—blood pressure response. For this reason, it is important to monitor physiological response to exercise. Also, because people with MS demonstrate a relatively lower threshold to fatigue, weakness, pain, and spasticity during exercise, it is imperative that intensity be very slowly and gradually increased to avoid overexertion. To adjust to the relapsing and remitting course of the disease, the physical activity program should be able to be modified or even suspended, depending on the severity of the exacerbation.

MS is associated with a preponderance of neurological conditions affecting the lower extremities, including foot drop, spasticity, and proprioception loss, and thus a seated activity utilizing back support is recommended for exercisers who have MS. Options include upright or recumbent leg ergometers or combined leg and arm ergometers. Utilizing mechanisms to stabilize extremities, such as foot or upper-extremity straps, may also help individuals properly use equipment, avoid injury, and avoid falling. Also, many individuals with MS are prone to heat sensitivity and not only need to maintain adequate hydration but also need to work out in a cool environment or at least with a good cooling device.

GUIDELINE 13.11

Title: *Disability/Condition: Multiple Sclerosis and Exercise*

Organization: NCPAD

Year published: 2009

Purpose: To increase daily activity energy expenditure, improve cardiovascular function, enhance general muscle strength, equalize agonist and antagonist strength, increase joint range of motion, improve balance, and counteract effects of spasticity

Location: www.ncpad.org/disability/fact_sheet.php?sheet=186&view=all

Population: Individuals with MS

GUIDELINES

Component	Frequency	Intensity	Duration	Type
Aerobic exercise	3 sessions per week	60%-75% of HRmax or 50%-65% of $\dot{V}O_2$max	A single 30 min session or 3 10 min sessions	Walking, cycling, chair exercise, or water aerobics
Resistance training	Perform on days not used for aerobic training	Not specified	Not specified	Exercises using elastic bands, free weights, pulley weights, and machines
Flexibility exercise	Daily	Not specified	For short durations, preferably more than once per day if fatigue occurs	Passive and active range-of-motion exercises, yoga, and tai chi
Physical activity	Daily	Low	30 min of accumulated physical activity	ADLs, built-in inconveniences, leisure activities, hobbies

Special considerations include the following:

- Energy conservation and cooling should be employed.
- Free weights should not be used with extremities that have sensory deficits.
- Flexibility exercises should be performed in a seated or lying position when possible.
- Preparticipation graded exercise testing should be performed.

Using the Schwinn Airdyne or working out in water that is 82 °F (27.8 °C) or cooler are good options. Finally, the medications used to treat MS may affect the patient's ability to exercise, and the potential side effects of the medication may also affect their response to the activity. Prednisone can exacerbate muscle weakness, cause hypertension and diabetes, and accelerate bone density loss. Baclofen is commonly used to decrease spasticity and spasms, but, like prednisone, it may increase muscle weakness as well (Petajan and White 1999).

Guidelines for Spinal Cord Injury and Disability

In the United States, there are approximately 75,000 individuals with spinal cord injury. Trauma is the major cause of spinal cord injury and can result from motor vehicle collisions, gunshot or stab wounds, other accidents, and falls. Tumor, infection, or surgical complications are other sources of injury. There are also congenital causes; in the United States about 1 out of every 1,000 live births—4,000 new cases

GUIDELINE 13.12

Title: "Multiple Sclerosis"

Organization: ACSM

Year published: 2009

Purpose: To increase or maintain cardiovascular function, increase functional capacity, and increase or maintain range of motion

Location: *ACSM's Exercise Management for Persons with Chronic Diseases and Disabilities, Third Edition*

Population: Individuals with multiple sclerosis

GUIDELINES

Component	Frequency	Intensity	Duration	Type
Aerobic exercise	3-5 d/wk	60%-85% of HRmax, 50%-70% of $\dot{V}O_2$max or RPE	30 min each session	Cycling, walking, or swimming
Resistance training	2-3 d/wk, on days separate from endurance exercise	50%-70% MVC	1-2 sets of 8-15 repetitions	Exercises using free weights, machines, isokinetic machines
Flexibility exercise	Daily	Not specified	Hold for minimum of 30-60 s for 2 repetitions	Positional stretching
Warm-up and cool-down	Before and after each session	Not specified	5-10 min	Not specified

Special considerations include the following:

- Reverse pedaling with resistance improves hip and ankle flexor strength for walking.
- Harness or partial support of body weight facilitates longer exercise duration and reduces fear while walking.
- Void before and during exercise to reduce incontinence.
- Non-weight-bearing exercise should be reserved for those individuals with poor balance, lower-extremity sensory loss, or significant orthopedic issues.
- Depression is common and may affect adherence.

Adapted, by permission, from K. Jackson and J.A. Mulcare 2009, Multiple sclerosis In *ACSM's exercise management for persons with chronic diseases and disabilities*, 3rd ed., edited by J.L. Durstine et al. (Champaign, IL: Human Kinetics), 324.

annually—results in spina bifida, a neural tube defect that causes incomplete development and closure of the vertebral arch. Spina bifida affects the lumbosacral neural segments, causing low-level loss of function. For all types of spinal cord disabilities, the loss of motor and sensory function depends on the level where the spinal cord is injured. Higher levels of cord injury may involve all four extremities and is called *tetraplegia* (formerly *quadriplegia*); bilateral lower-extremity involvement is termed *paraplegia*. In some cases the spinal cord is not fully impaired, or completely penetrated, and thus injury is only partial.

There are many reasons for people with spinal cord injury to participate in physical activity. Functional independence can be maintained and improved via a program involving cardiorespiratory conditioning and resistance training. Flexibility training is important to avoid painful contractures, immobility, and pressure sores and may help with balance. Resistance training specifically helps to decrease the propensity for osteoporosis, which is a significantly higher risk in individuals with spinal cord injury because of the baseline decrease in weight bearing.

Physical activity programs generally are safe for people with spinal cord injury, but supervision may be required, particularly when a person is initiating a program or performing exercises with a risk for injury or falling. Individuals who have a spinal cord injury at or above the sixth thoracic level may have autonomic dysreflexia, in which the autonomic response to a stimulus may cause abnormal changes in heart rate and blood pressure in spite of the individual's inability to respond to or even identify the stimulus. Both the individual and any supervising personnel need to be aware of the signs and symptoms of autonomic dysreflexia. Also, people with a high-level spinal cord injury have a lower HRmax; subtracting 20 to 40 beats from 220 *before* subtracting age is appropriate to adjust for the change in the autonomic nervous system that results from spinal cord injury (Shephard 1990). People with tetraplegia may not be able to sustain an exercise heart rate higher than 120 to 130 beats/min for similar reasons; in

---

GUIDELINE 13.13

Title: *Disability/Condition: Spinal Cord Injury and Exercise*

Organization: NCPAD

Year published: 2007

Purpose: To maintain functional capacity

Location: www.ncpad.org/disability/fact_sheet.php?sheet=130&view=all

Population: Individuals with spinal cord injury

GUIDELINES

Component	Frequency	Intensity	Duration	Type
Aerobic exercise	Not specified	Not specified	Not specified	Circuit training, interval training
Resistance training	Not specified	70%-75% of maximal lift	2-3 sets of 10-12 repetitions	Lifting through the full range of motion; multistation weight machines such as Uppertone Gym
Flexibility exercise	Not specified	Not specified	Not specified	Modified chair stretching or range of motion while lying on a firm surface

Special considerations include the following:

- Blood pressure may be quite low and orthostatic hypotension is common; these issues can be alleviated by maintaining hydration, wearing compression stockings, and getting up slowly.
- Breathing exercises may help sustain lung capacity.

GUIDELINE 13.14

Title: "Spinal Cord Disabilities: Paraplegia and Tetraplegia"

Organization: ACSM

Year published: 2009

Purpose: To increase active muscle mass and strength, maximize overall strength for functional independence, improve efficiency of manual wheelchair propulsion, and avoid joint contractures

Location: *ACSM's Exercise Management for Persons With Chronic Diseases and Disabilities, Third Edition*

Population: Individuals with spinal cord disability, either acquired, traumatic, or congenital

GUIDELINES

Component	Frequency	Intensity	Duration	Type
Aerobic exercise	3-5 d/wk	40%-90% $\dot{V}O_2R$	20-60 min each session	Exercises using arm ergometer, wheelchair ergometer, wheelchair treadmill, free wheeling, arm cycling, seated aerobics, swimming, wheelchair sports, and electrical stimulation leg cycle ergometry
Resistance training	2-4 d/wk	Not specified	2-3 sets of 8-12 repetitions	Exercises using free wrist weights, machines, and dumbbells
Flexibility exercise	Not specified	Not specified	Not specified	Stretching

Special considerations include the following:

- Exercise modes should be varied week to week to prevent overuse injury.
- People with tetraplegia should use an environmentally controlled, thermoneutral gym.
- Empty bladder or urinary collection device before exercise to prevent autonomic dysreflexia.
- People with hypotension should use supine postures and wear support stockings and an abdominal binder to maintain blood pressure.

Different modes of aerobic conditioning for individuals with spinal cord injury include the following:

- Arm ergometry
- Arm cycling
- Electrical-stimulation leg cycle ergometry with or without arm ergometry
- Free wheeling
- Seated aerobics
- Swimming
- Wheelchair ergometry
- Wheelchair sports
- Wheelchair treadmill exercise

Adapted, by permission, from S.F. Figoni, 2009, Spinal cord disabilities. In *ACSM's exercise management for persons with chronic diseases and disabilities*, 3rd ed., edited by J.L. Durstine et al. (Champaign, IL: Human Kinetics), 302.

these individuals, the RPE is a better indicator of exercise intensity (Seaman 1999). One other common effect of spinal cord injury is hypotension, and individuals may not mount the usual blood pressure response to exercise. As a result of sensory loss and some immobility, individuals with spinal cord injury are also at risk for skin breakdown, particularly over bony areas such as the hips, tailbone, and heels. Regular activity with frequent shifts in weight can help prevent this.

Guidelines for Stroke and Brain Injury

According to the AHA, approximately every 45 s a person in the United States experiences a stroke. Although stroke is the third leading cause of death in the United States, the disability experienced by stroke survivors is also a very serious consequence. Risk factors for stroke parallel those for CAD and include older age,

GUIDELINE 13.15

Title: "Exercise Prescription for Other Clinical Populations: Spinal Cord Injury"

Organization: ACSM

Year published: 2010

Purpose: To increase fitness

Location: *ACSM's Guidelines for Exercise Testing and Prescription, Eighth Edition*

Population: Individuals with spinal cord injury

GUIDELINES

- Initially, aerobic exercise should consist of short bouts of 5 to 10 min of 40% to 50% $\dot{V}O_2R$ (moderate) alternating with 5 min of active recovery. This may progress to 10 to 20 min of 85% to 90% $\dot{V}O_2maxR$ (vigorous).

- Strength training should be performed from both seated position in the wheelchair and non-wheelchair exercise and should include all trunk-stabilizing muscles.

- Avoid repetitive strain injuries by working to lengthen the anterior shoulder muscles and strengthen the upper back and posterior shoulder muscles.

- Avoid triggers for heat intolerance, including dehydration, glycogen depletion, sleep deprivation, alcohol, infection, and lack of acclimatization.

- Be wary of autonomic dysreflexia and avoid exercise if resting systolic blood pressure is $\geq$180 mmHg.

Reprinted, by permission, from ACSM, 2010, *ACSM's guidelines for exercise testing and prescription*, 8th ed. (Baltimore, MD: Lippincott, Williams, and Wilkins), 243-244.

obesity, diabetes, physical inactivity, hypertension, dyslipidemia, and tobacco use. Due to the increasing population of older individuals in the United States, along with the increasing numbers of people who are obese, have diabetes, or have heart failure, it is likely that the incidence of stroke will continue to rise. More individuals survive strokes than die from them; the AHA reports that in 2003 an estimated 4,700,000 stroke survivors lived in the United States.

Preventing recurrent stroke is a primary goal of exercise and physical activity programs for stroke survivors; nearly one-third of all strokes are recurrent! The other leading cause of mortality in stroke survivors is cardiac disease, thus underscoring the need for regular exercise to prevent mortality.

The range of disability in stroke survivors is quite large. While some individuals develop hemiplegia, sensory loss, or spasticity, others acquire difficulty with speech, memory, or mood. Luckily, several studies have shown multiple benefits from rehabilitative efforts in stroke survivors. In addition to improving strength, endurance, and functional status, regular physical activity can

reduce CVD risk factors—presumably decreasing the risk for the leading causes of death in individuals who have experienced stroke. Like other individuals with physical disabilities, stroke survivors often have an increased metabolic requirement to perform activities. For example, ambulatory individuals with a history of stroke may be able to perform at only 50% of the peak oxygen consumption or 70% of the peak power output that is achieved by age- and sex-matched controls without a history of stroke (Gordon 2004). Goals of physical activity for stroke survivors include optimizing functional motor performance, increasing aerobic capacity, increasing skill and efficiency in self-care and other activities, preventing complications of prolonged inactivity, and decreasing recurrent stroke and cardiovascular events.

Before initiating a physical activity program, a person with a history of stroke should be evaluated for safety or contraindication to participation. Both the AHA and ACSM recommend that patients with previous stroke undergo graded exercise testing before beginning an exercise program in order to evaluate for silent cardiac

GUIDELINE 13.16

Title: "Physical Activity and Exercise Recommendations for Stroke Survivors"

Organization: AHA Council on Clinical Cardiology, Subcommittee on Exercise, Cardiac Rehabilitation, and Prevention; Council on Cardiovascular Nursing; Council on Nutrition, Physical Activity, and Metabolism; Stroke Council

Year published: 2004

Purpose: To improve rehabilitation after stroke

Location: *Circulation* 109(16): 2031-2041

Population: Survivors of stroke

GUIDELINES

Component	Frequency	Intensity	Duration	Type
Aerobic exercise	3-7 d/wk	40%-70% $\dot{V}O_2$max or HRR; 50%-80% HRmax; RPE of 11-14 out of 20	20-60 min each session (or multiple 10 min sessions)	Large-muscle activities (walking or using treadmill, stationary cycling, combined arm and leg ergometry, arm ergometry, seated stepping)
Strength training	2-3 d/wk	Not specified	1-3 sets of 10-15 repetitions of 8-10 exercises involving the major muscle groups	Circuit training; exercises using weight machines, free weights, isometric exercise
Flexibility exercise	2-3 d/wk (before or after aerobic or strength training)	Not specified	Hold each stretch for 10-30 s	Stretching
Neuromuscular training	2-3 d/wk (same day as strength activities)	Not specified	Not specified	Coordination and balance activities

Special considerations include the following:

- Initially, intermittent training protocols may be needed due to deconditioning.
- Consider both depression and fatigue as barriers to participating in a program.

ischemia. Because stroke is considered a CAD equivalent, people who have had stroke are at higher risk for cardiovascular events. This risk, however, is significantly higher in individuals who become deconditioned from physical inactivity. The benefit of habitual physical activity in reducing the overall risk of sudden cardiac death is greater than the risk of a cardiac event during exercise and should be considered an essential lifestyle modification for stroke survivors. Many studies have shown that the risk of sudden cardiac death *during* exercise is higher in individuals who began exercise after being habitually sedentary (Kohl and others 1992; Thompson and others 1994), so it is important to encourage people to participate in regular activity at an appropriate level.

Another important factor to consider in stroke survivors is depression, which may affect up to 68% of individuals poststroke (Gordon 2004). Depression is a primary barrier to *any* type of poststroke therapy and certainly reduces participation in a physical activity program. Fatigue, both neurological and physiological, is also common and may prevent individuals from fully participating in exercise. Stroke-related fatigue seems to respond well to low-intensity aerobic exercise, which improves cardiac function as well as reduces energy demands in patients with a hemiparetic gait (Macko and others 1997).

GUIDELINE 13.17

Title: "Stroke and Brain Injury"

Organization: ACSM

Year published: 2009

Purpose: To decrease the risk of CVD, prevent contractures, increase range of motion of involved extremities, and increase independence and safety in ADLs

Location: *ACSM's Exercise Management for Persons With Chronic Diseases and Disabilities, Third Edition*

Population: Individuals with stroke and brain injury

GUIDELINES

Component	Frequency	Intensity	Duration	Type
Aerobic exercise	3-5 d/wk	40%-70% $\dot{V}O_2$max	20-60 min each session (or multiple 10 min sessions)	Upper and lower body ergometry, cycle ergometry, arm ergometry, seated stepping, treadmill exercise
Strength training	2 d/wk	Not specified	3 sets of 8-12 repetitions	Exercises using weight machines or free weights; isometric exercises
Flexibility exercise	2 d/wk (before or after aerobic or strength activities)	Not pecified	Not specified	Stretching
Neuromuscular training	2 d/wk (same day as strength activities)	Not specified	Not specified	Coordination and balance activities

Special considerations include the following:

- Individuals who need conditioning should start with an aerobic intensity at 40% to 50% $\dot{V}O_2$max and work up.
- Duration of aerobic activity should be gradually increased to an amount expending 300 kcal.
- Duration should be increased slowly (every 4-5 wk).
- Time of day for both exercise and medication should be kept as constant as possible.

SUMMARY

Many individuals with disabilities benefit from physical activity just as much as or even more than individuals without disabilities. Various organizations have issued guidelines for exercise and physical activity that are specific to different disabilities. Goals of physical activity in these varied groups are to maintain flexibility, preserve physical function, and prevent the onset of other chronic diseases such as diabetes, cancer, and CAD.

Asthma

Both children and adults with asthma can benefit from exercise. Although all individuals with asthma need to be prepared for flare-ups and people with exercise-induced asthma may develop airflow obstruction upon exertion, in general, exercise should not be curtailed due to fears of an asthma exacerbation. In fact, the ACSM recommends exercise as part of the asthma management plan (Clark 2003). Because asthma varies widely in terms of allergic and environmental triggers and severity and people with asthma vary widely in baseline physical fitness, guidelines for people with asthma should also vary as needed. Symptoms of asthma range from shortness of breath to chest tightness and pain to cough and wheezing. The goals of physical activity for individuals with asthma are to manage their symptoms so that they can improve (1) exercise tolerance so they can reap the benefits of physical activity, (2) musculoskeletal conditioning so they can participate in daily functions, and (3) fitness as judged by usual physical fitness criteria.

Exercise-Induced Asthma

Exercise-induced asthma (EIA) or exercise-induced bronchoconstriction (EIB—the terms are interchangeable) is a distinct entity that can be treated with anti-inflammatories and preexercise bronchodilators. Individuals with EIA experience transient airway obstruction 5 to 15 min after the onset of physical exertion. The criteria for EIA include either a fall in forced expiratory volume in 1 s (FEV_1) of 15% from baseline or a fall in peak flow of 20% from baseline. EIA should be assessed with a standard exercise test in which the person exercises at >75% HRmax for at least 8 min; airflow obstruction is measured 6 to 8 min after cessation of exercise.

The way in which exercise induces bronchospasm is not completely understood, although it may involve a combination of respiratory heat loss, water loss, and vascular events. Some individuals experience additional airway obstruction as late as 4 to 6 h after the initial episode of obstruction, although this late asthmatic response usually is mild. A potent stimulus to EIA is an increased minute ventilation or overbreathing; activity guidelines for people with EIA address this by suggesting a lengthy warm-up and cool-down. Many individuals with asthma—from 50% to 100%—also experience EIA (Cypcar and Lemanske 1994), but not all individuals with EIA have additional triggers for bronchoconstriction outside of exercise.

There is some controversy as to whether bronchospasm is a natural phenomenon that results from the noxious insult of exposure to large volumes of cold or dry air. As many as 75% of elite athletes who perform in cold environments demonstrate EIB, whereas 50% to 100% of all people with asthma (Cypcar and Lemanske 1994) and 10% to 15% of the general population experience EIB.

Interestingly, some of these athletes have profound or severe symptoms of EIB while in the field but do not have symptoms strong enough to meet the EIB criteria while in a laboratory. Clearly, both external environment and overbreathing due to exercise intensity contribute to the underlying reactivity and influence the degree of impairment.

General Guidelines for People With Asthma

The most recent (eighth) edition of *ACSM's Guidelines for Exercise Testing and Prescription* includes a section on individuals with asthma and other pulmonary diseases. The guidelines suggest using an inhaled bronchodilator (2-4 puffs) 15 min before exercise in order to reduce the risk of bronchoconstriction. A gradual warm-up of low-intensity exercise can also reduce the chance of bronchoconstriction. For individuals with well-controlled asthma, the ACSM text recommends an exercise prescription that is essentially the same as that for individuals without asthma (guidelines 3.1 and 5.8 for adults and

children, respectively), but the text does not have any further recommendations for people whose asthma is not well controlled. However, *ACSM's Exercise Management for Persons With Chronic Diseases and Disabilities* addresses exercise in people with asthma and provides guidelines that vary depending on the severity of the disease. For people who have mild asthma (defined by the American Thoracic Society as 60%-80% of predicted FEV_1) or EIA controlled by medications, the standard ACSM recommendations for exercise in sedentary individuals without asthma are appropriate. For people with moderate or severe asthma, different recommendations apply. Because of the variability in triggers and in the current respiratory status of even a single person, the goals for physical activity may need to be adjusted periodically. Individuals with well-controlled asthma or EIA should experience a normal adaptation to exercise training. When EIA limits exercise, the limited ventilation may prohibit improved fitness but may still allow improved endurance. Likewise, individuals with moderate to severe asthma who experience ventilatory limitation (and breathlessness) may also benefit from activity and should be encouraged

GUIDELINE 14.1

Title: "Exercise Prescription for Other Clinical Populations: Asthma"

Organization: ACSM

Year published: 2010

Purpose: To reduce stress on the pulmonary system during exercise by promoting adaptations in the musculoskeletal and cardiorespiratory systems

Location: *ACSM's Guidelines for Exercise Testing and Prescription, Eighth Edition*

Population: Individuals with well-controlled asthma

GUIDELINES

- Avoid environmental triggers such as cold, dry, dusty air and inhaled pollutants.
- Limit exercise until symptoms from an acute exacerbation have subsided.
- Utilize an inhaled bronchodilator (such as albuterol, 2-4 puffs) 15 min before beginning to exercise.
- Gradually warm up with low-intensity exercise.
- At least 3 to 5 d/wk, participate in 20 to 60 min of continuous or intermittent moderate-intensity aerobic exercise (walking or stationary cycling is recommended).
- Include resistance and flexibility training as desired (see guideline 3.1 or 5.8 for recommendations).

Reprinted, by permission, from ACSM, 2010, *ACSM's guidelines for exercise testing and prescription*, 8th ed. (Baltimore, MD: Lippincott, Williams, and Wilkins), 261-262 and 264.

GUIDELINE 14.2

Title: "Asthma"

Organization: ACSM

Year published: 2009

Purpose: To perform exercise at a higher intensity, to correct physical deconditioning, to reduce lactic acidosis, and to reduce ventilatory requirement during exercise

Location: *ACSM's Exercise Management for Persons With Chronic Diseases and Disabilities, Third Edition*

Population: Individuals with *mild asthma* (FEV_1 of 60%-80% of predicted value) or EIA controlled by medications

GUIDELINES

Component	Frequency	Intensity	Duration	Type
Aerobic exercise	3-5 d/wk	40%-85% HRR or $\dot{V}O_2R$ or 64%-94% HRmax	20-60 min of continuous or intermittent activity (in 10 min bouts)	Large-muscle activities, including walking, cycling, swimming
Resistance training	2-3 non-consecutive days per week	8-12 repetitions to volitional fatigue	8-10 separate exercises	Activities using bands, machines, or free weights to work the major muscles of the hips, thighs, legs, back, chest, shoulders, arms, and abdomen
Flexibility exercise	At least 2-3 d/wk; ideally 5-7 d/wk	To the end of the range of motion at a point of tightness without inducing discomfort	Hold stretches for 15-30 s; 2-4 repetitions for each stretch	Static stretching using the major musculo-tendinous units
Neuromuscular training	Daily	As tolerated	5 min	Walk drills, balance drills, and breathing exercises

Special considerations include the following:

- Several minutes of warm-up and cool-down may reduce the likelihood of EIA.
- Many people with chronic asthma have worse symptoms in the early morning, so avoiding exercise at this time may prevent flare-ups.
- Extremes in temperature or humidity may cause flare-ups.
- A beta-selective sympathomimetic agonist is recommended approximately 10 min before starting to exercise.
- A 6 wk introductory period to monitor exercise intensity is recommended.

Adapted, by permission, from C.J. Clark and L.M. Cochrane, 2009, Asthma. In *ACSM's exercise management for persons with chronic diseases and disabilities*, 3rd ed., edited by J.L. Durstine et al. (Champaign, IL: Human Kinetics), 149.

to work at a relatively high proportion of their maximum exercise tolerance. Individuals who are able to improve in objective symptoms such as FEV_1 may advance to a higher-intensity program.

The ACE has published *Fit Facts* sheets on exercise and asthma. In these guidelines, there is an emphasis on the need to obtain thorough medical evaluation and permission before starting a physical activity program. Although higher-intensity exercise may be acceptable for some people with asthma, starting slowly, warming up gradually, and exercising near the lower end of the

GUIDELINE 14.3

Title: "Asthma"

Organization: ACSM

Year published: 2009

Purpose: To improve fitness, to improve exercise tolerance without necessarily improving physiological fitness, and to improve musculoskeletal conditioning

Location: *ACSM's Exercise Management for Persons With Chronic Diseases and Disabilities, Third Edition*

Population: Individuals with *moderate asthma* (FEV$_1$ of 40%-60% of predicted value)

GUIDELINES

Component	Frequency	Intensity	Duration	Type
Aerobic exercise	3-5 d/wk; 1-2 sessions per day	RPE of 11-13 out of 20	30 min each session (shorter intermittent sessions may be necessary at first)	Large-muscle activities
Resistance training	2-3 times per week	Low resistance with high repetitions	Not specified	Isokinetic or isotonic exercises and free weights
Flexibility exercise	3 sessions per week	Not specified	Not specified	Not specified
Neuromuscular training	Daily	Not specified	Not specified	Walking, balance exercises, and breathing exercises

Special considerations include the following:

- Several minutes of warm-up and cool-down may reduce the likelihood of EIA.
- Many people with chronic asthma have worse symptoms in the early morning, so avoiding exercise at this time may prevent flare-ups.
- Extremes in temperature or humidity may cause flare-ups.
- RPE and subjective dyspnea are the preferred methods of monitoring intensity; many individuals cannot achieve a high target heart rate but will still exhibit physiological improvement.

Adapted, by permission, from C.J. Clark and L.M. Cochrane, 2009, Asthma. In *ACSM's exercise management for persons with chronic diseases and disabilities*, 3rd ed., edited by J.L. Durstine et al. (Champaign, IL: Human Kinetics), 147.

GUIDELINE 14.4

Title: "Asthma"

Organization: ACSM

Year published: 2009

Purpose: To perform exercise at a higher intensity, to correct physical deconditioning, to reduce lactic acidosis, and to reduce ventilatory requirement during exercise

Location: *ACSM's Exercise Management for Persons with Chronic Diseases and Disabilities, Third Edition*

Population: Individuals with *severe asthma* (FEV$_1$ of <40% of predicted value)

GUIDELINES

Component	Frequency	Intensity	Duration	Type
Aerobic exercise	3-5 d/wk; 1-2 sessions per day	RPE of 11-13 out of 20	30 min each session (shorter intermittent sessions may be necessary at first)	Large-muscle activities, including walking, cycling, swimming
Resistance training	2-3 times per week	Low resistance with high repetitions	Not specified	Isokinetic or isotonic exercises and free weights
Flexibility exercise	3 d/wk	Not specified	Not specified	Stretching or tai chi
Neuromuscular training	Not specified	Not specified	Not specified	Walk drills, balance drills, and breathing exercises

Special considerations include the following:

- Several minutes of warm-up and cool-down may reduce the likelihood of EIA.
- Many people with chronic asthma have worse symptoms in the early morning, so avoiding exercise at this time may prevent flare-ups.
- Extremes in temperature or humidity may cause flare-ups.
- Many individuals take sympathomimetic bronchodilators, which may cause tachycardia (better to use RPE than age-predicted HRmax for monitoring intensity).

Adapted, by permission, from C.B. Cooper, 2009, Chronic obstructive pulmonary disease. In *ACSM's exercise management for persons with chronic diseases and disabilities*, 3rd ed., edited by J.L. Durstine et al. (Champaign, IL: Human Kinetics), 134.

GUIDELINE 14.5

Title: *Exercise and Asthma*

Organization: ACE

Year published: 2001

Purpose: To reduce stress, promote better sleep, improve energy, and reduce the risk of developing many other diseases

Location: www.acefitness.org/fitfacts/fitfacts_display.aspx?itemid=2594

Population: Individuals with asthma

GUIDELINES

- Obtain physician clearance before beginning a program.
- Utilize a warm-up.
- Exercise toward the lower end of the target heart rate, and slowly increase intensity over time.
- Engage in strength training if desired; strength training exercises are acceptable.
- Avoid exercising in polluted environments or cold or dry air.
- Extend the cool-down.
- Consider a warm bath or shower after exercise.

Adapted, by permission from American Council on Exercise Fit Facts, San Diego 2001. www.acefitness.org/fitfacts Accessed 10/10/2009.

Types of Exercise and Likelihood for an Asthma Attack

Most likely
- Outdoor running

Least likely
- Pool swimming
- Walking
- Cycling
- Treadmill running

Adapted, by permission from American Council on Exercise Fit Facts, San Diego 2001. www.acefitness.org/fitfacts

target heart rate zone are less likely to precipitate an asthma attack. The ACE lists types of exercise with respect to their likelihood for inducing an asthma attack (see the highlight box on this page). Although outdoor running is the type of activity most likely to induce an asthma attack, it is possible to still participate in this higher-risk activity (and other higher-risk activities) if medications and other modifications are used to control asthma. Jackie Joyner-Kersee is a great example of an athlete who clearly excelled despite a diagnosis of asthma, winning Olympic gold!

Guidelines for Children With Asthma

Children with asthma may successfully participate in physical activity as long as they are closely monitored. The AAP first issued a statement on children with asthma and participation in sports and physical education in 1984. While prophylactic treatment with beta-adrenergic agonists may be needed, full participation in sports is otherwise recommended. The National Asthma Education and Prevention Program (NAEPP) of the NHLBI recommends a preventive strategy in treating asthma in order to ensure safe and productive activities. The NAEPP School Asthma Education Subcommittee has published guidelines for supporting schoolchildren with asthma while they are in school, including the time spent in physical activity. The foundation of successful management of asthma at school involves having, being familiar with, and using an asthma management plan (see the sidebar).

In order to be able to participate fully in physical education classes and other physical activity, children should have convenient access to their medication. Some will require regular use of medication before exercising, while others may need it only in times of suboptimal control of their asthma symptoms. Regardless of medication

Components of an Asthma Management Plan

- Brief history of the student's asthma
- Knowledge of asthma symptoms
- Information on how to contact the student's health care provider and parent or guardian
- Physician and parent or guardian signature
- List of factors that exacerbate the student's asthma
- The student's personal best peak flow reading if the student uses peak flow monitoring
- List of the student's asthma medications
- A description of the student's treatment plan, based on symptoms or peak flow readings, including recommended actions for school personnel to help handle asthma episodes

GUIDELINE 14.6

Title: "Managing Asthma Long Term in Children 0-4 Years of Age and 5-11 Years of Age" and "Managing Asthma Long Term in Youths ≥12 Years of Age and Adults," *Expert Panel Report 3: Guidelines for the Diagnosis and Management of Asthma*

Organization: NHLBI NAEPP

Year published: 2007

Purpose: To reduce impairment; maintain normal or near-normal lung function and activity levels; reduce the risk of exacerbations, emergency care, or hospitalization; and prevent loss of lung function or reduced lung growth

Location: www.nhlbi.nih.gov/guidelines/asthma/

Population: Individuals with asthma

GUIDELINES

- Full participation in physical activity at play or in organized sports should be encouraged.
- Pretreatment before exercise includes short-acting beta-agonists, leukotriene receptor antagonists, cromolyn, or nedocromil for children with EIB.
- Warm up or wear a mask or scarf over the mouth to protect against cold-induced EIB.

GUIDELINE 14.7

Title: *Asthma and Physical Activity in the School*

Organization: NAEPP School Asthma Education Subcommittee

Year published: 2004

Purpose: To help classroom teachers, physical education teachers, and coaches help their students participate fully and safely in sports and physical activities

Location: www.nhlbi.nih.gov/health/public/lung/asthma/phy_asth.pdf

Population: School-aged children with asthma

GUIDELINES

- Follow asthma management plan, including knowing triggers, best peak flow readings, and personal asthma medications.
- Allow convenient use of inhaled medication 5 to 10 min before exercise.
- Modify physical activities to match current asthma status based on symptoms or peak flow (may need to temporarily modify type, length, or frequency to reduce the risk of further symptoms).
- Include adequate warm-up and cool-down.
- Monitor the environment for potential allergens and irritants and consider temporary change in location if necessary.
- Keep the student involved in some way if temporary but major modification in exercise is necessary. Have student dress for class and participate at a lower level.

GUIDELINE 14.8

Title: *Breathing Difficulties Related to Physical Activity for Students With Asthma: Exercise-Induced Asthma*

Organization: NHLBI

Year published: 2005

Purpose: To help students with asthma participate in physical activity

Location: www.nhlbi.nih.gov/health/prof/lung/asthma/exercise_induced.pdf

Population: Students with EIA

GUIDELINES

- If prescribed preexercise treatment, take appropriate medication 5 to 10 min before exercise.
- Warm up briefly before exertion.
- When mild symptoms or triggers are present, consider modifying the intensity, location, or duration of physical activity. Intermittent, very light, or nonaerobic exercise is safer than intense continuous activity.
- If ozone alerts, high pollen counts, freshly cut fields, or sprayed fields are present, avoid outdoor physical activity if necessary.

usage, being aware of symptoms of decreased control facilitates defensive and proactive steps to improve breathing, allows for some level of activity, and avoids adverse effects to the child. While *not* participating in school physical education and activity might seem the safer alternative, over the long term, sitting out causes greater damage. One Dutch study found that children (aged 7-10 y) with diagnosed or undiagnosed asthma had similar levels of daily physical activity when compared with healthy controls (van Gent and others 2007). This finding suggests that children with asthma are able to participate in physical activity.

Both the American Academy of Allergy Asthma and Immunology (AAAAI) and the NHLBI have issued recommendations for individuals with EIA. Suggestions include premedicating, warming up and cooling down, and restricting or modifying activity that is more likely to induce bronchoconstriction. Also, the AAAAI suggests an anatomical alteration (pursed-lip breathing) to alter physiology in order to decrease airflow obstruction. In addition, the AAAAI recommends obtaining clearance from an allergist or immunologist before attempting scuba diving. The AAAAI's full guidelines are available online at www.aaaai.org/patients/publicedmat/tips/exerciseinducedasthma.stm.

SUMMARY

Physical activity and exercise are both integral parts of the management plan for children and adults with asthma. Awareness—and avoidance—of triggers during exercise can prevent flare-ups. Gradually warming up and premedicating with inhaled beta-agonists may reduce bronchoconstriction. EIB is a common and related entity that also can be lessened by warming up, inhaling warmed or humidified air, and using beta-agonists. Individuals with mild or well-controlled asthma can exercise at a level similar to that of people without asthma, while individuals with severe asthma need to limit duration, intensity, or environmental extremes.

Guidelines for Exercise Testing and Beyond

The final part of this text covers exercise testing and nutrition. The first chapter introduces standards for exercise testing in adults, including tests for aerobic fitness, strength, flexibility, and body composition. Chapter 15 also dedicates a section to exercise testing in children, which is important in the evaluation of cardiac, pulmonary, neuromuscular, and obesity-related issues. Chapter 16 addresses cardiac testing, including invasive and noninvasive treadmill stress tests and other newer modalities. Chapter 17 examines dietary guidelines as they pertain to—or are presented alongside—physical activity recommendations. There are perhaps just as many guidelines for nutrition and weight control as there are guidelines for physical activity, and there is not nearly enough room in this text to include all of them. Thus, chapter 17 focuses on the dietary recommendations that are associated with physical activity guidelines for specific populations or disease states. The final chapter deals with the particulars and logistics of exercise – finding and using appropriate facilities and equipment.

Exercise Testing

The ACSM provides guidelines for exercise testing, including treadmill and bicycle stress tests and assessments of body composition, flexibility, and strength. These parameters can help an individual develop a personalized exercise program. In addition to test results, professionals who are specialized or licensed in the exercise, health, and fitness fields can be helpful in implementing a program. There are several certifiable designations and each can play a role in helping individuals execute an appropriate plan. People are recommended to consult with their physician before initiating a regular exercise program, particularly if they have not been physically active in the past.

Exercise testing has many purposes, including clinical evaluation, fitness assessment, performance evaluation, and assessment of response to physical activity interventions. Tests are accompanied by databases of normative data so that comparison with appropriate standards is available. Because there is a strong relationship between health-related fitness and future health conditions, there is much interest in evaluating the different components of health-related fitness. Some widely used tests cover several of these components and provide an overall summary or score that can be used in a predictive fashion. Other tests utilize the measures of different health-related parameters to evaluate fitness and consequently suggest ways to improve these

parameters and ultimately improve overall health. Popular test series include the California Physical Fitness Test, the Fitnessgram, and the President's Challenge Adult Fitness Test.

Standards for Exercise Testing in Adults

Exercise testing is performed at the onset of a physical activity program not only to help an individual develop appropriate ranges for intensity but also to evaluate the safety and suitability of the program for the individual. Many people are familiar with a treadmill stress test, which is used to evaluate aerobic fitness. There are also other forms of aerobic fitness testing, such as bicycle testing or field tests. Additionally, there are strength tests that are used to assess maximum strength; these can also be useful in developing a proper program.

Aerobic Testing

Aerobic testing is used to measure aerobic capacity or power. When a person participates in exercise of increasing workload, oxygen consumption increases along with workload and then plateaus as the body starts to use sources of energy other than oxygen to make ATP. Aerobic capacity can be used to monitor and evaluate aerobic training and can also be used to assess the fitness and health status of individuals. Many studies have found a link between improved mortality rates and improved aerobic capacity. Aerobic exercise tests include treadmill and cycle ergometer protocols that can be performed in an exercise laboratory, doctor's office, or hospital. Other tests may be performed in the field, meaning outdoors or indoors. These tests commonly measure how much activity can be performed in a fixed amount of time and then compare the results with normative data in order to quantify percentile ranks and an estimated $\dot{V}O_2$max. For example, in the Cooper test, the participant is asked to cover as much distance as possible in 12 min. The formula used to estimate $\dot{V}O_2$max is 0.0268(distance covered) – 11.3, where the distance covered is in meters and the $\dot{V}O_2$max is in ml · kg^{-1} · min^{-1} (Hoffman 2006). Other, newer tests have been devised for easier estimation of cardiorespiratory fitness. These actually incorporate the resting heart rate and other variables including age, sex, BMI, and self-reported physical activity to come up with a fairly accurate assessment of fitness (Jurca 2005).

Strength Testing

Muscular strength and endurance can be developed by recruiting muscle mass in a way that requires more force or energy than usual; the result is a compensatory increase in strength, power, or endurance. As with the tests for aerobic fitness, muscular strength and endurance testing can be used to evaluate improvement, compare the participant with norms, and predict future sport performance. Tests are generally divided into those measuring strength and those measuring endurance. The ACSM considers both muscular strength and endurance to be part of the overall category of muscular fitness. While strength is the ability of the muscle to exert force, endurance is the ability of the muscle to continue to perform for successive exertions or many repetitions (Corbin and others 2000).

The standard of strength assessment has been the 1-repetition maximum (1RM). The 1RM is the greatest resistance that can be moved through the full range of motion in a controlled manner with good posture for any given muscle group.

The following is the generally accepted protocol for testing the 1RM:

1. Perform a warm-up with several submaximal repetitions.
2. Perform 1 repetition with a resistance that is between 70% and 90% of the expected 1RM.
3. Progressively increase the resistance until the selected repetition cannot be completed.
4. Rest for 3 to 5 min between the warm-up and the first repetition and between each subsequent repetition or attempt.

The final amount of weight successfully lifted is the 1RM. Another method used to measure muscular strength is a multiple repetition maximum, such as a 6RM, which is the amount of weight that can be lifted for 6 repetitions.

Muscular endurance measures also utilize the 1RM in that they consist of testing the number of repetitions an individual can perform for a submaximal load, such as a fixed percentage of the 1RM. For example, an individual may chart the number of repetitions performed at 70% 1RM to evaluate for progress. This measures muscular endurance because endurance is the ability of a muscle group to execute repeated contractions over time. Tests evaluating a maximum number of crunches (curl-ups), pull-ups, or push-ups are commonly used methods for evaluating muscular endurance of the abdominals and upper-body muscles. The curl-up and push-up tests are two that make up the muscular endurance portion of the Fitnessgram (see table 15.1).

Flexibility Testing

Most flexibility tests consist of measuring static flexibility, while dynamic flexibility assessment is limited to the research setting. Measures can be based on linear or angular tests on single joints or a group of joints (compound joint test). Angular measurements using goniometers or inclinometers are usually used in the medical professions, while linear measurements are used in the field. Examples of the latter include the toe touch or the sit and reach. Flexibility testing is designed to identify individuals with an increased risk of

TABLE 15.1 Fitnessgram Tests

Test	Measurement
PACER (progressive aerobic cardiovascular endurance run)	Aerobic capacity
Skinfold test	Body composition
Curl-up	Muscular strength and endurance
Trunk lift	Muscular strength
Push-up	Muscular strength and endurance
Back-saver sit and reach	Flexibility

These are the recommended tests; alternatives are available.

From FITNESSGRAM/Activitygram Reference Guide, www.fitnessgram.net/overview Accessed 10/10/2009.

TABLE 15.2 Weight Classification by BMI, Waist Circumference, and Associated Metabolic Disease Risk

Category	BMI (kg/m^2)	Obesity class	Men ≤102 cm (40 in.) Women ≤88 cm (35 in.)	Men >102 cm (40 in.) Women >88 cm (35 in.)
Underweight	<18.5			
Normal weight	18.5-24.9			
Overweight	25-29.9		Increased	High
Obese	30.0-34.9	I	High	Very high
	35.0-39.9	II	Very high	Very high
Extremely obese	≥40.0	III	Extremely high	Extremely high

From the National Institute of Health (NIH) and the National Heart, Lung, and Blood Institute, 1998. *Clinical Guidelines on the Identification, Evaluation, and Treatment of Overweight and Obesity in Adults (Executive Summary)*, NIH Publication 98-4083. Available at: www.nhlbi.nih.gov/guidelines/obesity/ob_gdlns.pdf.

injury due to a decreased range of motion for specific joints or muscle groups.

Body Composition

Given that the prevention of obesity is a major goal of a regular physical activity program, it is important to be able to quantify body composition. In the health and medical industry, BMI is one of the most widespread tools used to assess fatness. While BMI is a relatively good surrogate marker for central adiposity, it is by no means perfect in the evaluation of body composition. BMI is used more as an epidemiological tool to mark trends in obesity and to place individuals in certain weight classifications. The uses, goals, and shortcomings of BMI are discussed more in chapter 17. Risk for metabolic disease correlates with both BMI and waist circumference (see table 15.2).

Body composition, on the other hand, can more accurately reflect lean body mass. There are several techniques for measuring and estimating body composition, including anthropometric, densitometric, electrical, and radiographic methods. The gold standard for determining body composition is densitometry, which uses a measure of the whole-body density (body mass / body volume). Densitometry can be performed either underwater (hydrostatic densitometry or hydrodensitometry) or in a closed chamber (plethysmography). Anthropometric methods using skinfolds (at 3 or 7 different sites) correlate well with hydrodensitometry, particularly if the technician is experienced and the individual is not extremely lean or obese. Bioelectrical impedance analysis (BIA) has been promoted in many health clubs and for home use, but the accuracy of BIA depends on following a stringent protocol, as the results may vary significantly depending on hydration status and other variables. Other methods include dual-energy X-ray absorptiometry (DXA), total-body electrical conductivity (TOBEC), and magnetic resonance imaging

(MRI). While more accurate, these last techniques are expensive and are unlikely to be available for widespread use. Norms for body composition have been published in many places and vary according to age, sex, ethnicity, and sport. Data from The Cooper Institute were used to create percentile norms for specific populations and are available in the eighth edition of *ACSM's Guidelines for Exercise Testing and Prescription*. There is currently no consensus on a body fat value associated with optimal health, but an estimate of 20% to 32% for women and 10% to 22% for men has been considered satisfactory (ACSM 2010). Table 15.3 presents ideal and acceptable values, stratified by age and sex, that are used by the Cooper Aerobics Center and represent stringent standards for health. These represent data collected over time by the Cooper Institute for Aerobics Research. DXA measurement is starting to become more widely used and accepted for estimating body composition. However, as opposed to skinfold or hydrostatic methods, DXA also includes essential fat in its estimate and so has a different—and somewhat higher—normal range. One set of DXA norms is presented in table 15.4. For males, 5% is considered essential fat, while 8% is considered essential in females.

Benefits of Exercise Testing for Children

Although exercise testing was initially developed to evaluate CAD in adults, today there are also many uses for exercise testing in the evaluation of disease in children and adolescents. The utility of exercise testing for four chronic disorders was specifically addressed in a symposium in 1995 with the purpose of educating medical and exercise professionals on the importance of exercise testing in children with chronic disease (Tomassoni 1996). Chronic disease evaluated by exercise testing includes cardiovascular, pulmonary, and neuromuscular disease and obesity. Objective information from children's responses to exercise can help in the development of an exercise prescription that in turn can provide numerous benefits to children and their disease processes.

TABLE 15.3 Body Composition Standards

	MEN		WOMEN	
Age (y)	Ideal (% body fat)	Acceptable (% body fat)	Ideal (% body fat)	Acceptable (% body fat)
20-29	9	13	16	18
30-39	12.5	16.5	18	20
40-49	15	19	18.5	23.5
50-59	16.5	20.5	21.5	26.5
60+	16.5	20.5	22.5	27.5

Data from Cooper Aerobics Center.

TABLE 15.4 Body Composition Norms (DXA Method)

	MEN		WOMEN	
Age (y)	Athletic (% body fat)	Healthy (% body fat)	Athletic (% body fat)	Healthy (% body fat)
20-29	14	18	24	26
30-39	17	22	26	28
40-49	20	24	26	32
50-59	21	26	29	35
60+	21	26	30	36

Data from Cooper Aerobics Center.

There are a number of pediatric disorders in which exercise testing is clinically relevant. The basis for testing physiological parameters in the pediatric population has been discussed by many authors (Chang and others 2006; Stephens and Paridon 2004; Tomassoni 1996), and selected purposes are presented here:

- Evaluate physiological function or deficiency
- Establish a baseline fitness level, particularly for children with chronic disease

- Assess effectiveness of the exercise program
- Assess for exercise-induced pathophysiological changes (respiratory, cardiac)
- Evaluate dysrhythmias
- Assess effectiveness of surgical correction
- Differentiate symptoms and assess need for further invasive testing
- Motivate parent and child to participate in exercise program

---

SUMMARY

Exercise testing is used to evaluate aerobic fitness, strength, flexibility, and body composition. Different methods are used to assess and quantify these parameters and may be appropriate in different situations and individuals. Some methods are suitable for evaluating large groups of people, while others require expensive or detailed equipment but provide more in-depth or comprehensive information.

Cardiac Exercise Testing and Prescription

Cardiac exercise testing is available in many different forms as a method of evaluating cardiorespiratory function. The first exercise testing dates back to 1846. Cardiac exercise testing, also known as *graded exercise testing* or *stress testing,* is often used as an assessment tool before making recommendations for an exercise program. For many individuals with chronic disease who are at higher cardiovascular risk, exercise testing is performed to evaluate the safety of an exercise program before recommending specific parameters of the program. This is often referred to as *exercise clearance.* In addition, there are many other reasons to obtain an exercise test, including evaluating chest pain, assessing for CAD, assessing the severity of CAD in individuals already known to have the disease, evaluating cardiac function after myocardial infarction, and determining functional capacity. Each of these is discussed briefly in this chapter.

Candidates for Testing

There are several indications for performing exercise testing; the American College of Cardiology (ACC) has divided indications into class I and class II, depending on the strength of the justification and the appropriateness of the testing (see the sidebar). Although older individuals generally manifest more risk factors for CAD, exercise testing is not routinely recommended for asymptomatic men over 40 y, even those with two or more risk factors for CAD. On the other hand, the ACSM has recommended that individuals ≥75 y undergo exercise testing before beginning a regular exercise program. One study from the Mayo Clinic looked at the prognostic value of treadmill exercise testing in elderly persons and concluded that there is additional information, above and beyond that gained from clinical data, garnered from the stress test. Workload, as measured by exercise capacity in METs, is strongly associated with health outcomes, including decreased risk of death and cardiac events. For individuals 65 y and older, the association translates into an 18% reduction in cardiac events for each MET increase in exercise capacity. Other studies have found that additional measures, independent of myocardial ischemia, have good predictive value for future cardiovascular mortality (Gibbons and others 2000; Aktas and others 2004; Wei and others 1999; Mora 2003).

Despite this, currently neither the U.S. Preventive Services Task Force or the AHA recommends exercise testing as a routine screening modality in asymptomatic adults. According to the U.S. Preventive Services Task Force statement for exercise testing in asymptomatic individuals, testing should be considered for people at a 10% or higher 10 y risk of having a cardiac event; for people at a risk of less than 10% over the next 10 y, the number of false positives ultimately leading to invasive testing simply doesn't justify the possible benefits of uncovering previously unsuspected ischemia (USPSTF 2004). The ACSM guidelines on cardiac exercise testing for older adults—many of whom do indeed have a higher cardiovascular risk—have been challenged recently, because it is felt that large-scale testing may cause an unnecessary strain on health care resources and discourage many older individuals from becoming more physically active. Suggestions for safely initiating an

Appropriate Reasons for Exercise Testing

Class I: Universally Justified

- To aid in the diagnosis of CAD in patients with atypical ischemic symptoms
- To assess post–myocardial infarction (uncomplicated) prognosis and severity
- To assess revascularization effect and prognosis
- To evaluate exercise-induced cardiac arrhythmias
- To evaluate functional capacity of selected patients, to aid in developing an exercise prescription
- To evaluate patients with rate-responsive pacemakers
- To evaluate efficacy of drug therapy in special populations such as people with hypertension, ischemia, or dysrhythmia

Class II: Frequently Performed But Under Debate Regarding Value and Appropriateness

- To evaluate asymptomatic male patients over the age of 40 with special occupations (e.g., pilots, bus drivers, and others with critical processes)
- To evaluate asymptomatic males over the age of 40 with two or more CAD risk factors
- To evaluate sedentary male patients over the age of 40 who plan to start vigorous exercise
- To assist in the diagnosis of CAD in women with a history of typical or atypical chest pain
- To assist in the diagnosis of CAD in patients who are taking digitalis or who have complete right bundle branch block
- To evaluate the functional capacity and response to cardiovascular drug therapy in patients with CAD or heart failure
- To evaluate patients with variant angina
- To annually follow patients with CAD
- To evaluate individuals who have a class I indication but also have another factor limiting the value of the test (baseline ECG changes or coexisting medical problems)
- To evaluate post–myocardial infarction (complicated) in patients who have subsequently stabilized
- To annually follow asymptomatic patients after revascularization
- To evaluate functional capacity of selected patients with valvular heart disease
- To evaluate the exercise blood pressure response of patients with hypertension
- To evaluate children or adolescents with valvular or congenital heart disease

Based on ACC Foundation *Journal American College of Cardiology*, 1986, 8:725-738.

exercise program in this population are presented in chapter 7.

Contraindications to performing an exercise test do exist. The ACSM divides these into absolute and relative contraindications (see following lists). Whereas performing a test on an individual with an absolute contraindication is wholly unsafe, it may be reasonable to perform an exercise test on an individual with a relative contraindication so long as the benefits outweigh the risks.

Absolute Contraindications

- Acute myocardial infarction (within 2 d) or other acute cardiac event
- Significant change on the ECG suggesting ischemia, myocardial infarction, or other acute cardiac event
- Unstable angina not stabilized by medical therapy
- Uncontrolled cardiac dysrhythmias causing symptoms or hemodynamic compromise
- Symptomatic severe aortic stenosis
- Uncontrolled symptomatic heart failure
- Acute pulmonary embolus or pulmonary infarction
- Acute myocarditis or pericarditis
- Suspected or known dissecting ventricular or aortic aneurysm
- Acute infections (influenza, rhinovirus)

Relative Contraindications

- Left main coronary stenosis
- Moderate stenotic valvular heart disease
- Severe arterial hypertension of >200/110 mmHg (either or and)
- Tachycardic or bradycardic dysrhythmias
- Hypertrophic cardiomyopathy or other forms of outflow tract obstruction
- Mental or physical impairment that leads to inability to exercise adequately or is exacerbated with exercise
- High-degree atrioventricular block
- Ventricular aneurysm
- Chronic infectious disease (AIDS, mononucleosis, hepatitis)

- Electrolyte abnormalities (hypokalemia, hypomagnesemia)
- Uncontrolled metabolic disease (diabetes, myxedema, thyrotoxicosis)

Adapted, with permission, from Ehrman Clinical Exercise Physiology, Second Edition (2009), figure 6.4, page 83. Modified from Gibbons 2002 and Visich 2004

The sensitivity of an exercise test may not be high enough in individuals with a high risk for CAD or with resting ECG abnormalities such as the following:

- Complete left bundle branch block
- Preexcitation syndrome (Wolff-Parkinson-White syndrome)
- ST elevation or depression of more than 1 mm (often due to digoxin or left ventricular hypertrophy)

In these patients, a test using echocardiography or nuclear images may improve the ability to detect exertional myocardial ischemia. Of course, if the purpose of the exercise test is to evaluate physiological response to exercise or exercise-induced dysrhythmia, it may be acceptable to perform an exercise test without additional imaging beyond the ECG tracings. Because exertional myocardial ischemia commonly is missed in women (due to atypical symptoms) and women are more likely to have a false-positive stress test (based on ECG tracings), it is suggested that women with a high pretest probability for CAD undergo either nuclear stress testing or stress echocardiography as a frontline test.

Utility of Information Acquired From Exercise Testing

As noted, exercise testing has other applications beyond its diagnostic value for CAD. These include evaluating functional capacity, disease severity in individuals with known or highly suspected CAD, and prognosis in the early period after myocardial infarction. Also, exercise testing can evaluate the physiological responses to exertion, including blood pressure throughout exercise and heart rate during exercise and recovery. In fact, heart rate recovery in the first minute after exercise can be a very valuable tool in risk stratification or prognosis for heart disease.

• **Diagnosis of CAD:** Because the specificity of exercise testing is estimated to be about 77% (Gibbons 2002), performing tests on a high number of low-risk, asymptomatic individuals leads to a significant amount of false positives with concomitant adverse consequences (including the emotional stress caused to the person and the financial costs of subsequent testing). A concept known as *Bayes' theorem* states that the diagnostic efficacy of exercise testing is related to the probability that an individual has disease *even before undergoing testing.* That is, the pretest probability of disease dictates the posttest probability of disease based on the stress test results. In other words, an asymptomatic individual with a very low risk for CAD who has a positive stress test still has a somewhat low probability of having CAD, despite the positive stress test. Conversely, someone with a very high risk for CAD who has a normal stress test has only a marginally lower posttest probability of CAD—the disease may still be present despite the test results. For this reason, it is believed that a stress test should be performed for diagnostic purposes only if the individual's pretest probability is sufficiently high to aid in the diagnosis.

• **Functional capacity:** Results from a standardized exercise test can be used to determine functional, or exercise, capacity. The absolute value of METs achieved may be compared with age- and sex-predicted normative data to come up with a percentile ranking. Because lower exercise capacity is an independent risk factor for both cardiovascular and all-cause mortality, the result may be particularly useful in formulating an exercise prescription, encouraging physical activity, and estimating prognosis of future mortality. In one study of 527 men with CVD, those who achieved ≤4.4 METs on the exercise test had the highest all-cause and cardiovascular mortality, while those at ≥9.2 METs had no mortalities at all (Vanhees and others 1994). Another larger study evaluated exercise testing in more than 3,600 men with CAD. The relative risk of death was 4.1 times greater for those with an exercise capacity of ≤4.9 METs when compared with those with a capacity of ≥10.7 METs. In this study, each 1 MET increase conferred a 12% improvement in survival (Myers and others 2002).

• **Disease severity:** Among individuals who have already been diagnosed with CAD, an exercise test may be able to help determine the severity of the CAD. In general, the magnitude of ST segment depression is proportional to the degree of coronary ischemia or stenosis. Additionally, when a larger number of ECG leads show abnormalities, the severity of significant disease is greater. Disease severity is directly correlated to the length of time into the recovery that ST segment depression persists. Markers that are inversely proportional to ischemia include the HRmax, systolic blood pressure, METs achieved, ST segment slope, and double product at which the ST segment depression occurs.

• **Post–myocardial infarction testing:** There are several valuable pieces of information that can be collected from exercise testing during the time shortly after myocardial infarction. Both submaximal and maximal tests are used to assess prognosis, develop an exercise prescription, and evaluate for efficacy of medical therapy. Earlier tests are usually submaximal and can help determine readiness for a level of activity appropriate for performing ADLs. Tests at a higher level of exertion are generally performed 2 wk or more after myocardial infarction. Prognostic evaluation can be determined through several factors. The following are *poor* prognostic features: ischemic ST segment depression at a low level of exercise, reduced left ventricular systolic function, functional capacity of less than 5 METs, or a hypotensive response to exercise (Whaley 2005).

Protocols for Exercise Testing

There are several protocols for cardiac exercise testing, and which tests are used depends on the individual being tested. Each contributes information on physiological response to exercise, cardiac risk, and fitness level. The most common protocol is the Bruce protocol. Others include the Naughton, Balke-Ware, Ellestad, substandard Balke, standard Balke, and superstandard Balke protocols and the standard bicycle test. Standard tests are used in individuals who are willing and able to perform the test at the determined speed, while substandard tests are utilized for people who cannot necessarily walk comfortably at the standard test speed.

For each protocol, graded exercise tests may be executed to a maximal or submaximal level, depending on the characteristics of the test subject. For example, an individual who has just recovered from an acute myocardial infarction generally will perform a submaximal exercise test for evaluation of safety to participate in a cardiac rehabilitation program. People who are performing a stress test for exercise clearance, risk stratification, or fitness evaluation may perform to a maximal level of exertion (without being limited by abnormal signs or symptoms), which may be defined by voluntary or volitional fatigue, achieving an RPE greater than 17, achieving a rate–pressure product of 24,000, hitting a plateau in heart rate despite an increasing workload, or having attained 85% of predicted HRmax.

Standard protocols, such as the Bruce protocol, have several steplike increases in work rate. Each incremental increase is significantly harder than the last. Increases may be nonlinear; this type of test is effective in screening for ischemia. In contrast, continuously increasing, or ramping, protocols utilize a constant increase in work rate and may be more accurate in characterizing a submaximal exercise response. Nonlinear protocols may be more beneficial when clearing an individual for exercise or evaluating for ischemia, whereas ramp protocols are recommended when attempting to appropriately design a physical activity or exercise program.

While treadmills are the most commonly used mode of equipment for graded exercise tests, bicycle ergometers may also be used effectively. Most individuals attain a lower functional capacity when measured on a bicycle ergometer compared with treadmill testing. Like treadmill testing, bicycle testing has several different protocols that are selected depending on the goals of the test and the background of the individual being tested. A constant cadence is maintained on the bicycle and the work rate is incrementally increased by increasing the resistance. Blood pressure, ECG tracings, and RPE are recorded, just as they are during a treadmill test.

Multiple physiological parameters are monitored during exercise testing and contribute to the understanding of fitness and cardiovascular risk. These include blood pressure, heart rate, and ECG changes (of rhythm, ST segment, and conduction pattern). These allow calculation and interpretation of exercise capacity, chronotropic response, and heart rate recovery. Also, some exercise tests measure gas exchange or oxygen saturation to determine ventilatory response and allow for calculation of maximal oxygen uptake. Additionally, subjective symptoms such as chest pain, dyspnea, dizziness, or claudication are helpful in determining whether to terminate the test or whether there is a correlation between ECG change and symptoms. Lastly, RPE not only provides information on a person's subjective exercise experience but also indicates how much longer a person may be able to exercise. When used with the other data from the test, the RPE may help in devising the exercise prescription. See chapter 3 for details regarding using the RPE.

Exercise Testing Procedures

Resting ECG, blood pressure, and heart rate are obtained in supine and standing positions. If there are no contraindications to proceeding with the test, the patient may begin exercising. Blood pressure, heart rate, RPE, and any symptoms are recorded at the end of each stage of the test. Blood pressure may be auscultated manually or with an automated cuff. The test may be terminated for many reasons. Whether the test is to be maximal or submaximal, there are certain indications for stopping a test before the intended end point. These include hemodynamic compromise, severe angina, dizziness, and perfusion and ECG criteria. Relative indications for premature test termination include some arrhythmias, ST segment changes, or extensive hypertensive responses.

Absolute Indications

- Decrease in systolic blood pressure of >10 mmHg from baseline, despite an increase in workload, in the *absence* of evidence of ischemia
- Marked shift in QRS axis

- ST depression >2 mm horizontal or down-sloping
- Arrhythmias including multifocal premature ventricular contractions (PVCs), PVC triplets, supraventricular tachycardia, heart block, or bradyarrhythmias
- Symptoms including fatigue, dyspnea, wheezing, leg cramps, or claudication
- Bundle branch block that cannot be distinguished from ventricular tachycardia
- Increasing chest pain
- Hypertensive response exceeding a systolic blood pressure of 250 mmHg or a diastolic blood pressure of 115 mmHg

Relative Indications

- Decrease in systolic blood pressure of >10 mmHg from baseline, despite an increase in workload, when accompanied by evidence of ischemia
- Angina that is moderate to moderately severe, or very uncomfortable (rating of 2-3+)
- Worsening central nervous system symptoms such as ataxia, dizziness, or presyncope
- Cyanosis or pallor
- Technical difficulties, including monitoring the ECG or blood pressure
- Subject's desire to stop
- Sustained ventricular tachycardia
- ST elevation of >1 mm in leads without diagnostic Q waves (other than V1 or aVR)

At the termination of the exercise, the treadmill is reduced in speed and grade for a cooldown. ECG tracings and physiological data are recorded. It is important to continue to monitor the patient during this recovery as some abnormalities may not appear until this stage of the test.

Interpretation of the ECG tracings, hemodynamic response, and patient symptoms determines whether a test is considered to be normal, abnormal, equivocal, or inconclusive. A normal hemodynamic response to exercise includes (1) a gradual rise in heart rate until maximal effort is reached; (2) a decrease in heart rate during the recovery, with a decrease by at least 12 beats in the first minute of recovery; (3) a rise in systolic blood pressure with exercise; (4) no change or a slight decline in the diastolic blood pressure; and (5) a progressive decline in systolic blood pressure with recovery. In the absence of obstructive CAD, there are many reasons for abnormal ECG changes. These include resting repolarization abnormalities, ventricular hypertrophy, medication effects, electrolyte disturbances, mitral valve prolapse, anemia, and female sex. The major sign or symptom suggestive of ischemia is angina, which may or may not be accompanied by ECG abnormalities. For further details on interpreting data, predictive value, and prognosis of cardiac exercise testing, the *ACSM's Guidelines for Exercise Testing and Prescription, Eighth Edition* is an excellent resource.

Other Methods of Cardiac Testing

Additional cardiac imaging modalities may be indicated in specific situations. These include (1) a very high pretest probability of disease, where a negative exercise test would not be convincing enough to assure that CAD is not present; (2) a need to follow up on a nondiagnostic standard exercise test; or (3) a need to rule out disease after a positive exercise test occurs in an individual with low probability of disease. Both nuclear imaging and echocardiography can be performed in conjunction with standard exercise testing to yield information in addition to that obtained with simultaneous ECG tracings.

There are several types of protocols for nuclear imaging, including those using technetium or thallium, in which resting images are compared with stress images to evaluate for perfusion abnormalities. One main advantage of nuclear imaging is a high sensitivity and specificity, particularly when single photon emission computed tomography (SPECT) is used. Drawbacks include a large amount of time required for testing and exposure to ionizing radiation.

During exercise echocardiography, resting echo images are compared with those taken immediately after exercise; myocardial contractility as measured by wall motion generally increases with exercise whereas ischemia may cause hypokinesis, dyskinesis, or even akinesis. Echocardiography can be performed in a shorter amount of time than nuclear imaging requires and does not require ionizing radiation, but the test may be limited in individuals with poor echocardiographic windows or a low exercise heart rate reducing the amount of time available to capture abnormalities in wall motion before they normalize in recovery.

For individuals who cannot exercise or cannot achieve an appropriate workload due to medication or other medical issues, pharmacological stress testing is an option. In dobutamine stress echocardiography, dobutamine is used to increase heart rate and myocardial oxygen demand, potentially uncovering ischemic wall motion abnormalities. Like an exercise stress echocardiogram, physiological parameters as well as ECG tracings are recorded throughout the test in order to evaluate for arrhythmias, ST depression, blood pressure response, and heart rate response. Nuclear imaging can also be obtained in people who cannot exercise. Vasodilators such as adenosine or dipyridamole are used to maximally vasodilate coronary arteries; the normal arteries dilate whereas stenotic segments do not, creating a coronary steal phenomenon that can be observed on perfusion imaging.

Other forms of testing are available to evaluate for the presence of CAD. Electron beam computed tomography (EBCT) became available in the mid-1980s and has been used to screen asymptomatic or high-risk individuals for CAD. The presence and degree of coronary artery calcification may be compared with normative data for age and sex and used as a tool in risk stratification to determine how aggressive risk factor modification should be. The AHA released a statement in 2000 recommending that EBCT not be used as a routine CAD screening test for asymptomatic individuals, in light of the review of studies determining that the overall predictive accuracy of EBCT is similar to that of standard exercise testing—approximately 70%. More recently, the use of multidetector computed tomography (MDCT) with intravenous contrast to create computed tomography angiography (CTA) has made it possible to view the luminal condition of coronary arteries at a very high degree of resolution. The predictive accuracy of CTA is quite good, approaching that of standard cardiac catheterization with angiography, and may be used as a noninvasive surrogate for the more invasive angiography procedure. In the near future it is likely that CTA in conjunction with standard exercise testing will replace nuclear imaging or stress echocardiography as an effective method to diagnose CAD.

SUMMARY

Cardiac testing is used for exercise clearance, CAD risk stratification, evaluation for dysrhythmias, and measurement of physiological parameters during exercise. The traditional cardiac stress test is a treadmill test, although there are other methods of cardiac testing, both invasive and noninvasive. Depending on the characteristics and risk factors of the individual being tested, special considerations may be warranted not only to ensure safety but also to ensure that the information obtained from the test is as accurate as possible. Cardiac testing remains a very effective method for assessing cardiopulmonary function.

Diet and Weight Management

17

Dietary guidelines have been accompanied by physical activity recommendations for the past several years because of the indisputable relationships among dietary intake, physical activity, body weight, and health. Many studies have found that although *either* diet or exercise may create a caloric or energy deficit, weight loss or weight maintenance is difficult without *both.* This chapter describes the major U.S. dietary guidelines that have notable physical activity components presented earlier in this book (in chapter 2 and other chapters where appropriate). Also included are dietary guidelines for exercising individuals; it would be inappropriate to describe the benefits of physical activity and exercise while overlooking the accompanying dietary guidelines that might optimize those benefits. These guidelines include recommendations on basic nutrient requirements, supplements, and hydration both at rest and during exercise. Next the chapter introduces the food pyramid and its recent modifications. Lastly, the chapter includes short sections detailing specific dietary guidelines as they pertain to the populations addressed in earlier chapters. For some of these populations, such as pregnant women, there are established and specific guidelines, while for others there are less-detailed suggestions and for others, such as people with OA, there are none at all. Because diet plays such a strong role in the control of diseases such as hypertension, CAD, and diabetes, only general recommendations included as part of physical activity guidelines are presented here; the detailed dietary guidelines for each of these diseases would be enough to fill a whole new text!

As discussed in many of the preceding chapters, physical activity guidelines must be considered alongside dietary recommendations in the context of health and weight maintenance or improvement. Taken separately, the guidelines may enhance health, but considered together they afford the largest opportunity to improve health outcomes. For example, weight loss and maintenance are rarely achieved through diet or exercise alone, and thus the recommendations for physical activity and diet often come in parallel.

Basic Facts About Body Weight

Active individuals require more daily calories than sedentary people require. Energy balance is achieved by matching energy consumption and energy expenditure. Body weight is maintained when the sum of all energy gained from food, supplements, and fluids is equal to the sum of all energy expended to support bodily processes and movement. If caloric intake is insufficient to meet caloric expenditure, then muscle, bone, or fat is lost with a resultant reduction in body weight. On the other hand, when energy intake exceeds energy expenditure, weight gain occurs. An individual's energy requirement is determined by numerous variables, including age, sex, body composition, activity level, and activity intensity. Active individuals require more energy to repair and produce muscle, and these individuals burn more calories not only through their activity (which requires calories to fuel the work) but also through a higher basal metabolic rate (due to having a body composition with more muscle and less fat) that requires more fuel for ADLs.

There are several ways to classify weight, including BMI, body composition (with several methods to estimate this), waist circumference, and waist-to-hip ratio. A healthy weight is one that can be maintained, allows for physical activity, minimizes the risk of injury or illness, is consistent with long-term health outcomes, and reduces the risk factors for chronic disease. Once a person's ideal body weight is determined, an overall plan encompassing both physical activity and nutrition can be developed for that person. When losing or maintaining weight, it is important to maintain lean body mass. This is accomplished by a slow and steady reduction in the overall caloric balance. A reasonable goal for weight loss is 0.5 to 1 lb (0.2-0.5 kg) a week.

BMI can be determined from weight and height, and BMI charts are widely available. BMI is often reported in terms of kilograms per body area in meters squared, but it can be calculated using nonmetric units by dividing weight in pounds by height in inches squared and multiplying that number by 703. A BMI of 18.5 to 24.9 kg/m^2 is considered normal or healthy, while a BMI of 25 to 29.9 kg/m^2 is considered overweight and a BMI of 30 kg/m^2 or greater is obese (see table 17.1). Despite the considerable wealth of data linking overweight and obesity to an assortment of health problems, fascinatingly there is not a correlation between normal BMI and reduced all-cause mortality. Instead, there seems to be an increased risk of mortality at the extremes of BMI (<18.5 and >30 kg/m^2). In fact, people who are overweight have an improved lifespan over people who have a healthy BMI (Flegal and others 2005). The possible explanation for this is discussed in subsequent journal articles and may reflect the fact that some individuals with chronic disease are included in the normal BMI group, and this inclusion alters the results.

BMI certainly does not tell the whole story with respect to health status, but it can be a good measure of progress in a person's attempt to improve health. Although BMI does not take into account muscle mass versus fat mass, it seems to have a linear relationship with body fat. BMI also does not account for age, sex, or ethnic differences. For example, a professional football player might have a much higher BMI than an elderly woman who has very little muscle mass and might be categorized as overweight or even obese despite the fact that he likely has a significantly lower percentage of fat mass. Despite the shortcomings of BMI, it is widely used because the two variables used to calculate it are readily available and many individuals are aware of their current and target BMI. Measuring lean body mass using skin calipers, underwater weighing, DXA, or visceral adiposity using computed tomography (CT) or MRI may be more accurate in terms of predicting weight-related health risks but is impractical for widespread everyday use. For more information on methods used to estimate body composition, see chapter 15.

Establishing the proper weight for infants, children, and adolescents traditionally has relied

TABLE 17.1 Classification of Weight Status by BMI

BMI in kg/m^2	Classification
<18.5	Underweight
18.5-24.9	Healthy or normal
25.0-29.9	Overweight
30.0-34.9	Obese I
35.0-39.9	Obese II
≥40.0	Extremely obese

From Food and Nutrition Board, 2005, *Dietary reference intakes for energy, carbohydrate, fiber, fat, fatty acids, cholesterol, protein, and amino acids* (Macronutrients). Available: www.nap.edu/catalog.php?record_id=10490

on growth curves; the concept behind these curves is that percentiles for height should correspond to percentiles for weight. For example, a child who falls into the 50th percentile for weight but only the 25th percentile for height might be considered overweight, while another child who is in the same weight percentile but at the 95th percentile for height may be regarded as underweight. Recently, there has been more of an emphasis on BMI-for-age percentiles in these younger populations. For example, a child's BMI is calculated and then compared with normative data, and a percentile rank is produced for that child. Figure 17.1 presents one example of these percentiles. The BMI-for-age assessment is felt to take into account a child's relative risk of developing obesity and the myriad of health consequences that ensue. Guidelines categorize children in the 85th percentile rank as being *at risk for overweight* and children in the 95th percentile rank as *being overweight.* Online calculators are also available at http://apps.nccd.cdc.gov/dnpabmi/calculator.aspx, and other growth charts are available at www.cdc.gov/growthcharts.

Dietary Guidelines From National Organizations

There are many dietary guidelines available from national organizations, government organizations, and private organizations. The IOM is a private entity that has issued DRIs on a regular basis, beginning in 1989. The U.S. government has also published guidelines and references for dietary intakes—most recently the *Dietary Guidelines for Americans.* Given the prevalence of heart disease and stroke in the United States, along with the significant effect diet and weight have on both, the AHA has also issued dietary guidelines. This section of the chapter reviews all of these guidelines as well as includes selected international guidelines for comparison.

Dietary Reference Intakes

The IOM report on DRIs that was published in 2002 was the sixth in a series that was initially published in 1989 as the *Recommended Dietary Allowances.* The most recent report takes advantage of the information that has become available since the time of the initial publication. This information links healthier intakes with the prevention of not only diseases of deficiency but also diseases of excess such as diabetes, obesity, heart disease, hypertension, and so on. The report was a comprehensive effort by the IOM Food and Nutrition Board in collaboration with Health Canada. The study undertaken to prepare the text was coordinated through the Office of Disease Prevention and Health Promotion of the USDHHS in collaboration with Health Canada. The DRIs are designed to meet the nutritional needs of healthy individuals; people with specific diseases may need to follow guidelines for their particular condition, which are available through either a disease-specific guideline or a health care provider.

The 2002 IOM report presents guidelines for total energy, major nutrients, fiber, water, vitamins, minerals, and, of course, physical activity. Starting with birth, there are age- and sex-specific recommendations for each component listed above. A new concept introduced in this report is that there are different reference points for each component, including the Estimated Average Requirement (EAR), Adequate Intake (AI), Recommended Dietary Allowance (RDA), and Tolerable Upper Limit (TUL). For total caloric intake, there is an Estimated Energy Requirement (EER) that takes into account healthy physical activity and any needs for deposition of tissue or milk secretion to maintain a balance of energy.

Highlights of key nutrients, selected vitamins and minerals, and EER are included in the following guideline. The full report is available online.

It is also possible to determine the acceptable range for nutrients by looking at the percentage of overall energy the component provides. Since today's food labels often list the percentage of calories from fat, individuals are more familiar with this concept and can use this system to help gauge their dietary goals. The ranges for adults are presented in the following guideline; the ranges for children are included later in the chapter. Overall, younger children need a higher proportion of energy from fat, although this need declines into adulthood. Both children and adults should get adequate amounts of polyunsaturated

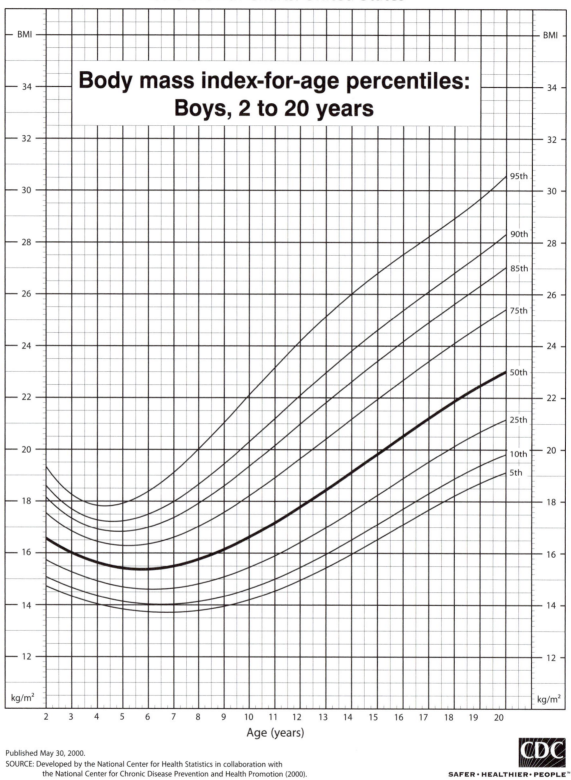

CDC Growth Charts: United States

Body mass index-for-age percentiles: Boys, 2 to 20 years

Published May 30, 2000.
SOURCE: Developed by the National Center for Health Statistics in collaboration with the National Center for Chronic Disease Prevention and Health Promotion (2000).

SAFER · HEALTHIER · PEOPLE™

Figure 17.1 BMI-for-age percentiles.

From www.cdc.gov/healthyweight/assessing/bmi/childrens_bmi/about_childrens_bmi.html.

GUIDELINE 17.1

Title: *Dietary Reference Intakes for Energy, Carbohydrate, Fiber, Fat, Fatty Acids, Cholesterol, Protein, and Amino Acids*

Organization: IOM Food and Nutrition Board

Year published: 2002

Purpose: To promote health and vigor and to balance food energy intake with total energy expenditure

Location: www.nap.edu/books/0309085373/html

Population: Healthy adults 18 y and older

GUIDELINES

Component	Male	Female	Female who is pregnant/lactating
Estimated energy requirement (kcal/d)	3,067 (subtract 10 kcal/d for each year of age above 19 y)	2,403 (subtract 7 kcal/d for each year of age above 19 y)	2,368-2,855 (depending on age and trimester)/2,698-2,803 (depending on age and time since birth)
Carbohydrate RDA (g/d)	130	130	175/210
Protein RDA (g/d)	56	46	71
Fiber AI (g/d)	30 (19-50 y); 38 (51 y and older)	21 (19-50 y); 25 (51 y and older)	28/29
Linoleic acid AI (g/d)	17 (19-50 y); 14 (51 y and older)	12 (19-50 y); 11 (51 y and older)	13
Alpha-linolenic acid AI (g/d)	1.6	1.1	1.4/1.3
Calcium AI(mg/d)	1,000 (19-50 y); 1,200 (51 y and older)	1,000 (19-50 y); 1,200 (51 y and older)	1,000 (1,300 if pregnant or lactating and younger than 19 y)
Vitamin D AI µg/d	5 (19-50 y); 10 (51-70 y); 15 (71 y and older)	5 (19-50 y); 10 (51-70 y); 15 (71 y and older)	5
Folate µg/d	400	400	600/500

fat while limiting saturated and trans fat. The range for carbohydrate is 45% to 65% of overall energy intake, while protein should make up 10% to 35% of overall intake. There seems to be some inconsistency in the IOM guidelines, given that the absolute protein RDA is only 46 to 56 g/d. Since 50 g of protein provides 200 kcal, in a 2,500 kcal diet 50 g of protein contributes only 8% to overall energy intake. On the other hand, 35% of a 2,500 kcal diet is 875 kcal, which is well over 200 g of protein! Other points made in the IOM report are that added sugar should not total to more than 25% of total daily calories.

Dietary Guidelines for Americans

Since 1980, dietary guidelines have been published every 5 y by the USDHHS and the USDA. The most recent (sixth) edition guidelines are referred to as the *Dietary Guidelines for Americans 2005.* The guidelines emphasize health promotion and chronic disease prevention through diet. In the 2005 edition, these guidelines for the first time contain not only nutrient information but also specific recommendations on physical activity as part of the weight maintenance strategy. Fruit,

GUIDELINE 17.2

Title: *Dietary Reference Intakes for Energy, Carbohydrate, Fiber, Fat, Fatty Acids, Cholesterol, Protein, and Amino Acids*

Organization: IOM Food and Nutrition Board

Year published: 2002

Purpose: To promote health and vigor and to balance food energy intake with total energy expenditure

Location: www.nap.edu/books/0309085373/html

Population: Healthy adults

GUIDELINES

Component	Percentage of calories from specific component
Fat	20%-35% (5%-10% should be omega-6 polyunsaturated fatty acids and 0.6%-1.2% should be omega-3 polyunsaturated fatty acids)
Protein	10%-35%
Carbohydrate	45%-65%

(Fruit, vegetables, whole grains, nuts, beans, and low-fat or nonfat dairy products are encouraged, while simple carbohydrate, alcohol, and saturated fat are recommended only in moderation. Dietary guidelines can be categorized in different ways. One way focuses on utilizing the percentage of total energy coming from specific macronutrients such as carbohydrate, protein, and fat. Another way is to focus on food groups, types of foods, or specific foods themselves.

Just as the 2002 IOM physical activity suggestion of *at least* 60 min of daily activity surprised and puzzled many people because it seemed to be inconsistent with the 1996 Office of the Surgeon General's report, the 2002 IOM nutritional guidelines are also different from the USDHHS dietary guidelines. The 2005 *Dietary Guidelines for Americans* differs from its previous versions in that this most recent report is a detailed scientific analysis. As a result, although the information is available to the general public, much of it is oriented toward policy makers, nutrition educators, nutritionists, and health care providers. The 2005 report contains several key recommendations, which are grouped into nine interrelated focus areas. Each contains generalized recommendations for all Americans, followed by recommendations for specific populations.

For most Americans, the 2005 *Dietary Guidelines for Americans* encourages eating fewer calories, being more active, and getting required nutrients through food consumption.

Adequate Nutrients Within Caloric Needs

Obtaining adequate nutrients within caloric needs is important as many Americans eat *too many* calories. Finding foods that fulfill nutrient needs may be a challenge for some individuals. Many foods, such as soft drinks or refined sugar and flour, contain empty calories that do not contain significant amounts of protein or fiber but certainly contribute to the daily caloric total. Inherent in meeting this guideline is limiting the intake of saturated and trans fat, cholesterol, added sugar, salt, and alcohol. A balanced diet, such as that advocated in the DASH (dietary approaches to stop hypertension) eating plan, is encouraged in order to meet this goal. Briefly, the DASH eating plan encompasses the following:

- Grains, which are emphasized as a major source of energy and fiber

- Fruit and vegetables, which are rich sources of potassium, magnesium, and fiber

Title: *Dietary Guidelines for Americans*

Organization: USDHHS (jointly with the USDA)

Year published: 2005

Purpose: To promote health and reduce risk for chronic disease through diet and physical activity

Location: www.health.gov/dietaryguidelines/dga2005/document/pdf/DGA2005.pdf

Population: Everyone

GUIDELINES

Follow the key recommendations of the focus areas, which are listed here and outlined in the text.

Focus Areas

- Adequate nutrients within caloric needs
- Weight management
- Physical activity
- Food groups
- Fat
- Carbohydrate
- Sodium and potassium
- Alcoholic beverages
- Food safety

- Fat-free and low-fat milk and milk products, which are recommended as a source of both calcium and protein
- Lean meats, poultry, and fish, which are a good source of protein
- Nuts, seeds, and legumes, which in moderation provide energy, magnesium, protein, and fiber
- Fat and oil, which in the DASH study provided 27% of daily calories
- Sweets and added sugar, which should be minimized and low in fat

Older adults may need to consume additional vitamin B_{12} and vitamin D through fortification or supplements. Women of childbearing age who may become pregnant should consume foods high in heme iron or iron-rich plant foods along with an iron absorption enhancer such as vitamin C as well as folic acid from foods or supplements.

Weight Management

The second focus point in the 2005 *Dietary Guidelines for Americans* is maintaining a healthy weight through a balance of caloric expenditure and intake. People aiming to lose weight are encouraged to do so in a slow but steady fashion, com-

bining increased physical activity with adequate nutrient intake. Children who are overweight are encouraged to maintain growth and development and *reduce the rate* of weight gain.

Physical Activity

The *Dietary Guidelines for Americans* physical activity recommendations for various populations have been presented in the corresponding sections in this text. They are summarized here as part of the full 2005 report recommendations. The report encourages 30 min of moderate-intensity physical activity, above usual physical activity, on most days of the week. People aiming to manage body weight and prevent unhealthy weight gain should get 60 min of moderate- to vigorous-intensity activity on most days of the week. For sustained weight loss, 60 to 90 min of daily moderate-intensity physical activity is the goal. Additional health benefits may be obtained by increasing the duration or intensity of the exercise. Adding stretching and resistance exercises as well as cardiorespiratory conditioning may improve flexibility, muscle strength and endurance, and physical fitness. Children are encouraged to be physically active for 60 min/d, while pregnant women are encouraged to incorporate 30 min of moderate-intensity activities with a low risk for abdominal trauma or falls. Exercise poses no risk for breast milk production, and thus women who are breast-feeding are encouraged to follow the general adult guidelines.

Food Groups

Fruit and vegetables are emphasized, with a target of consuming 2 1/2 cup of vegetables and 2 cup of fruit daily. Vegetables from all five of the following subgroups should be consumed when possible:

- Dark green vegetables
- Orange vegetables
- Legumes
- Starchy vegetables
- Other vegetables such as eggplant, onion, or mushrooms

Whole-grain products are emphasized, with a target of 3 oz (90 g) or more daily. Fat-free, low-fat, or other milk products are recommended at a target of 3 cup (710 ml) daily, as they provide both calcium and a lean source of protein. Smaller or greater amounts of these may be necessary depending on a person's daily caloric requirement; the recommendations covered here are based on a goal of 2,000 kcal/d.

Fat

Total fat intake should make up 20% to 35% of daily calories, with most fat coming from polyunsaturated and monounsaturated sources, such as fish, nuts, and vegetable oils. When selecting and preparing meat, poultry, dry beans, and milk or milk products, lean, low-fat, or fat-free choices should be made whenever possible. Fat sources that are high in saturated or trans fat, as well as products that contain them, should be limited. Less than 10% of daily calories should come from saturated fat and less than 300 mg/d should come from cholesterol; trans fat should be avoided. Children and adolescents need a slightly higher fat intake: children aged 2 to 3 y should aim to get 30% to 35% of their calories from fat, while children and adolescents aged 4 to 18 y should get between 25% and 35% of their calories from fat. In both age groups the primary source of fat should contain polyunsaturated and monounsaturated fatty acids.

Carbohydrate

Key recommendations regarding carbohydrate include choosing fiber-rich fruit, vegetables, and whole grains whenever possible. Individuals should avoid added sugar or caloric sweeteners in prepared foods and beverages. Both the USDA food guide pyramid and DASH eating plan suggest specific amounts for these items. By limiting the intake of simple sugar and starch, individuals can reduce the incidence of dental caries—this is one intended outcome of the recommendations.

Sodium and Potassium

One tsp of salt or 2,300 mg of sodium daily is the maximum recommended intake. Both salt in preprepared foods and added salt must be counted toward this daily goal—both are significant sources of sodium in the American diet. Fruit and vegetables are rich sources of potassium and should be consumed. Because of the correlation between sodium intake and hypertension, specific groups of the population should aim to consume no more than 1,500 mg of sodium daily. These include people with hypertension, African Americans, and middle-aged or older adults. These individuals should also aim to meet the daily potassium recommendation (4,700 mg) with food.

Alcoholic Beverages

Moderation is the key recommendation with respect to alcohol intake: up to 1 drink per day for women and up to 2 drinks per day for men. The following are examples of what constitutes a drink:

- Beer: 12 oz (360 ml)
- Wine: 5 oz (150 ml)
- Hard liquor: 1.5 oz (45 ml)

Several population groups are recommended to avoid alcohol consumption. These include people who cannot restrict their alcohol intake, women of childbearing age who may become pregnant, pregnant and lactating women, children and adolescents, individuals taking medications that can interact with alcohol, and people with specific medical conditions. Also, individuals engaging in activities requiring attention, skill, or coordination (such as driving or operating machinery) should avoid alcohol consumption within the time frame that it would affect these activities.

Food Safety

There are several ways to avoid the risk of a microbial food-borne illness. Meat and poultry should not be washed or rinsed. Food contact surfaces, fruit and vegetables, and hands should all be washed before food is prepared. When a person is purchasing, preparing, or storing foods, raw, cooked, and ready-to-eat foods should be separated. To kill microorganisms, individuals should cook foods to a safe temperature. Perishable foods should be refrigerated promptly and foods should be defrosted properly. Unpasteurized milk (or milk products), raw or partially cooked eggs (or foods containing raw eggs), raw or undercooked meat and poultry, unpasteurized juices, and raw sprouts should be avoided. This last recommendation is highlighted for individuals at a higher risk for infections, and raw or undercooked fish or shellfish is also added to their list of foods to avoid. Higher-risk individuals include infants and young children, pregnant women, older adults, and people who are immunocompromised. Lastly, pregnant women, older adults, and people who are immunocompromised should eat only certain deli meats and frankfurters that have been reheated to steaming hot.

American Heart Association

The AHA has issued diet and lifestyle recommendations as part of the strategy to prevent CVD. The most recent update, released in 2006, includes specific dietary guidelines in addition to physical activity guidelines as part of an overall plan to achieve and maintain a healthy weight throughout life. The 2006 update not only presents dietary guidelines reflecting the most recent scientific data but also provides practical guidance on how to achieve these dietary guidelines, particularly when eating away from home, as has become more and more common. The AHA recommendations are similar to those reflected in the USDA food pyramid but are presented differently. Individuals are encouraged to choose fruit, vegetables, and whole-grain, high-fiber foods.

Saturated fat intake is restricted to <7% of total energy, trans fat intake to <1% of energy, and cholesterol to <300 mg daily. In order to achieve these goals, individuals should select lower-fat (or nonfat) dairy products, choose lean meats and vegetable alternatives to the meats and to the fattier meats, and consume oily fish at least twice a week. Moderation is recommended for alcohol, salt, and beverages and foods with added sugar. Although separate guidelines have been published for children, the 2006 dietary guidelines are appropriate for children aged 2 y and older, and children who follow them will maintain appropriate growth while lowering their risk for future CVD. Other populations addressed in the report include older adults, people with the metabolic syndrome, and people with chronic kidney disease (CKD). Salient points for these groups are summarized in the following guideline, as are the full AHA recommendations.

Food Guide Pyramid

The food guide pyramid was first introduced in 1992, at which point in time it replaced the earlier concept of the four food groups. The initial food guide pyramid had levels of varying size corresponding to intake recommendations for different types of food. This initial pyramid caused some confusion because each level gave a fairly wide range of servings to consume for a given type of food. For example, the base of the old food guide pyramid was formed by bread, grains, cereal, rice, and pasta, the recommended consumption of which was very broad—6 to 11 servings daily. Also, for the other components of the pyramid (including fruit and vegetables; milk products; meat and beans; and fat, oils, and sweets), the actual amount of food in a single serving was difficult for many people to remember, much less know. At that point in time, the average serving sizes at restaurants—particularly fast-food joints—seemed to be growing and growing, adding even more to the confusion. As early as 2001 the USDA and NIH admitted that the food pyramid was a total failure and based on uncertain scientific evidence. This, coupled with the increasing rates of obesity and cardiometabolic disease among Americans, led to the development of a new concept.

GUIDELINE 17.4

Title: "Diet and Lifestyle Recommendations Revision 2006"

Organization: AHA Nutrition Committee

Year published: 2006

Purpose: To prevent CVD

Location: *Circulation* 114: 82-96

Population: Adults and children aged 2 y and older

GUIDELINES

- Balance caloric intake and physical activity to achieve or maintain a healthy body weight.
- Consume a diet rich in vegetables and fruit.
- Choose whole-grain, high-fiber foods.
- Consume fish, especially oily fish, at least twice a week.
- Limit intake of saturated fat to <7% of daily energy, limit trans fat to <1% of daily energy, and keep cholesterol intake below 300 mg/d by
    - choosing alternatives such as vegetables and lean cuts of meat;
    - selecting fat-free (skim), 1% fat, and low-fat dairy products; and
    - minimizing intake of partially hydrogenated fat.
- Minimize intake of beverages and foods with added sugars.
- Choose and prepare foods with little or no salt.
- Consume alcohol in moderation, if at all.
- When eating food prepared outside of the home, follow the AHA diet and lifestyle recommendations.

Adapted, by permission, from A.H. Lichtenstein et al., 2006, "Diet and lifestyle recommendations revision 2006: A scientific statement from the American Heart Association Nutrition Committee," *Circulation* 114(1): 82-96.

GUIDELINE 17.5

Title: "Diet and Lifestyle Recommendations Revision 2006"

Organization: AHA Nutrition Committee

Year published: 2006

Purpose: To prevent CVD

Location: *Circulation* 114: 82-96

Population: People with metabolic syndrome, people with CKD, and people who are older

GUIDELINES

Population	Recommendations
Older adults	Select nutrient-dense choices within each food group.
Persons with metabolic syndrome	Avoid diets really low in fat if triglyceride levels are elevated or HDL cholesterol levels are depressed.
Persons with CKD	Reduce salt intake and replace meat with dairy and vegetable alternatives. Reduce intake of protein, phosphorus, and potassium if CKD is advanced.

The new food guide pyramid was introduced in 2005 with the 2005 *Dietary Guidelines for Americans.* Several changes were made to the original model in order to clarify and update the nutritional recommendations. There is still a pyramid, but the different food components now run in stripes from bottom to top, and the relative thickness of the stripes represents the corresponding quantity of the food to be consumed. The new pyramid includes a Web-based tool that helps individuals to personalize their own requirements and obtain precise measurements on recommended food intake. Another interesting component of the new food guide pyramid is that a specific amount of physical activity is incorporated into the personalized pyramid to reflect the energy needs of the individual. The older pyramid merely made reference to the need for physical activity.

International Dietary Guidelines

Canadian guidelines are similar to those in the United States and reflect Canada's increasing obesity rates. *Canada's Food Guide,* which is endorsed by Health Canada, makes recommendations for nutrient intake and physical activity to balance caloric intake. Canadian guidelines include foods to incorporate into the diet and foods to avoid or at least limit. The latter category includes foods and beverages high in calories, fat, sugar, or salt.

The four food groups described in *Canada's Food Guide* are vegetables and fruit, grains, milk products, and meat and meat alternatives. The food guide is presented alongside the physical activity recommendations for an overall strategy targeting improved health. The number

---

GUIDELINE 17.6

Title: *Canada's Food Guide*

Organization: Health Canada

Year published: 2007 (updated)

Purpose: To contribute to overall health and vitality; meet needs for vitamins, minerals, and other nutrients; and reduce risk of obesity, diabetes, heart disease, cancer, and osteoporosis

Location: www.hc-sc.gc.ca/fn-an/food-guide-aliment/index-eng.php

Population: Everyone

GUIDELINES

Food group	Recommendations
Vegetables and fruit	Eat at least 1 dark green and 1 orange vegetable each day. Enjoy vegetables and fruit prepared with little or no added fat, sugar, or salt. Have vegetables and fruit more often than juice.
Grain products	Make at least half of your grain products whole grain each day. Choose grain products that are low in fat, sugar, and salt.
Milk and alternatives	Drink skim, 1%, or 2% milk each day. Select lower-fat milk alternatives.
Meat and alternatives	Have meat alternatives such as beans, lentils, and tofu often. Eat at least 2 servings of fish each week, limiting mercury exposure. Select lean meat and meat alternatives prepared with little or no added fat and salt.

Other considerations include the following:

- Enjoy a variety of foods from the four food groups.
- Satisfy thirst with water; drink more water during hot weather or when very active.

of recommended servings for each food group varies depending on age and sex. Details are available on the Health Canada Web site at www. healthcanada.gc.ca/foodguide. Pregnant and breast-feeding women are recommended to eat 2 to 3 additional servings from any of the food groups each day as well as to get adequate iron and 400 µg of folic acid daily. All adults aged 50 y and older are recommended to take a vitamin D supplement of 10 µg (400 IU).

In 2004, the WHO issued the *Global Strategy on Diet, Physical Activity and Health*. Along with providing physical activity guidelines (which are detailed in chapter 2), the report made dietary recommendations for both populations and individuals. These guidelines were based on the fact (as described in the *World Health Report 2002)* that six major risk factors contribute to the bulk of the burden of noncommunicable disease worldwide. Five of these risk factors are very closely related to physical activity and diet. Goals of the global strategy include reducing these risk factors through public policy, increased awareness, and scientific evaluation and intervention. The dietary guidelines recommend achieving a healthy weight; limiting fat, free sugar, and salt consumption; and increasing

consumption of fruit, vegetables, legumes, whole grains, and nuts.

Hydration, Energy, and Supplementation During Activity

Physical activity and exercise require special attention to nutrition status. Hydration is particularly important, as loss of fluid increases during activity. Energy requirements increase as well, with carbohydrate and fat providing different proportions of energy at different exercise intensities. Lastly, supplementation with various vitamins or minerals is considered important, depending on requirements increased through metabolism or body losses.

Hydration

Position statements on hydration before, during, and after exercise have been issued by the ACSM (ACSM 1996; Sawka and others 2007) and National Athletic Trainers' Association (Casa 2002); these statements provide guidelines as well as detail the physiology behind those guidelines.

GUIDELINE 17.7

Title: *Global Strategy on Diet, Physical Activity and Health*

Organization: WHO

Year published: 2004

Purpose: To reduce mortality, morbidity, and disability attributable to noncommunicable disease

Location: www.who.int/dietphysicalactivity/strategy/eb11344/strategy_english_web.pdf

Population: Adults

GUIDELINES

- Achieve energy balance and a healthy weight.
- Limit energy intake from fat and shift fat consumption away from saturated fat to unsaturated fat; work to eliminate trans fat.
- Increase consumption of fruit and vegetables and legumes, whole grains, and nuts.
- Limit the intake of free sugar.
- Limit salt (sodium) consumption from all sources and make sure that salt is iodized.

Adapted, by permission, from WHO, *Global Strategy on Diet, Physical Activity and Health,* 2004, *Global strategy on diet, physical activity and health.* Available: www.who.int/dietphysicalactivity/strategy/eb11344/strategy-english_web.pdf

---

In addition, the ACE has published a *Fit Facts* information sheet titled "Healthy Hydration." The ACSM recommends the following: In the 24 h before exercise, consume generous amounts of fluid; consume an additional 400 to 600 ml (14-20 oz) of fluid 2 h before exercise. During exercise, drink 150 to 350 ml (6-12 oz) of fluid every 15 to 20 min. When exercising in hot or humid environments or for durations lasting longer than 1 h, use sports drinks containing electrolytes and carbohydrate. Table 17.2 compares the ACSM recommendations with the ACE recommendations.

Fluid and electrolyte replacement during physical activity is a common concern. Dehydration of as little as 1% of body weight can significantly alter muscle and metabolic function, impairing performance and increasing the risk for injury or adverse effects during exercise. Electrolyte loss can also cause an imbalance that compromises performance. Thus, replacing fluids lost while exercising is important. For exercise lasting 30 min or more,

TABLE 17.2 Hydration for Exercise

Time	ACE recommendations	ACSM recommendations
24 h before exercise		Generous amounts of fluid
2 h before exercise	17-20 oz (500-600 ml) of water	400-600 ml (14-20 oz) of fluid
During exercise	7-10 oz (300 ml) of fluid every 10-20 min	150-350 ml (6-12 oz) of fluid every 15-20 min
After exercise (within 30 min)	8 oz (240 ml) of fluid	
Other recommendations	Drink 16-24 oz (480-710 ml) of fluid for each pound of body weight lost after exercise; use sports drinks during high-intensity exercise exceeding 45-60 min.	Use sports drinks with carbohydrate and electrolytes in hot or humid environments.

Based on American Council on Exercise Fit Facts, San Diego 2001. www.acefitness.org/fitfacts Accessed 10/10/2009. and M.N. Sawka, L. M. Burke, E.R. Eichner, et. al. 2007, "Exercise and Fluid Replacement," *Medicine & Science in Sports & Exercise.* 39(2):384-386.

GUIDELINE 17.8

Title: "Exercise and Fluid Replacement"

Organization: ACSM

Year published: 2007

Purpose: To sustain appropriate hydration of individuals performing physical activity

Location: *Med Sci Sports Exerc* 39(2): 377-390

Population: Individuals performing physical activity

GUIDELINES

Timing	Recommendations
Before exercise	• Exercisers should drink 5 to 7 ml/kg (about 0.1 oz/lb) of body weight at least 4 h before exercise. • If urine is dark or concentrated, exercisers should slowly drink another 3 to 5 ml/kg (about 0.1 oz/lb) about 2 h before exercise. • Individuals should consume beverages with sodium (20-50 mEq/L) or small amounts of salted snacks at meals to stimulate thirst and retain the consumed fluids. • Enhancing palatability (temperature, flavor, sodium content) promotes fluid consumption; preferred water temperature for palatability is between 59 and 70 °F (15 and 21 °C).

>continued

GUIDELINE 17.8 >*continued*

During exercise	• Individuals should develop customized fluid replacement programs that prevent excessive (<2% body weight reduction) dehydration. • The routine measurement of pre- and postexercise body weight is useful for determining sweat rates and customizing fluid replacement programs. • Consumption of beverages containing electrolytes and carbohydrate (20-30 mEq/L sodium chloride, 2-5 mEq/L potassium, and 5%-10% carbohydrate) can help sustain fluid and electrolyte balance and exercise performance. • Individuals should ingest 0.5 to 1.0 L (16.9-33.8 oz) of a sports drink consisting of 6% to 8% carbohydrate (providing 30-80 g carbohydrate) each hour. • To maximize carbohydrate delivery, the beverage should contain a mixture of sugars (glucose, sucrose, fructose, and maltodextrin) and should not exceed 8% carbohydrate.
After exercise	• Individuals should consume normal meals and beverages after exercise. • Individuals with excessive dehydration can drink 1.5 L (50 oz) of fluid for each kilogram (or each 2.2 lb) of body weight lost. • Consuming beverages and snacks with sodium expedites rapid and complete recovery. • Intravenous fluid replacement is not advantageous unless medically indicated (due to nausea, vomiting, or diarrhea).

fluid losses should be replaced with 8 oz (240 ml) of water. Then, for every additional 15 min spent exercising, an additional 8 oz (240 ml) should be ingested. If exercise lasts more than 60 min, the ACSM recommends that the fluid contain some carbohydrate and electrolytes to replace what is metabolized and lost in sweat. As a general rule, sports drinks should contain 4% to 10% solution of carbohydrate, or roughly 13 to 19 g of carbohydrate per 8 oz (240 ml) fluid. For every hour spent in exercise, a person should drink 12 to 32 oz (360-1,000 ml) of this concentration of sports drink, depending on the rate of fluid loss from sweat. Research indicates that drinks containing more than one type of carbohydrate may improve the amount of carbohydrate that eventually gets utilized by the muscle. These recommendations on fluid and electrolyte replacement can be found in the ACSM position stand "Exercise and Fluid Replacement."

Fuel Sources

Energy bars, gels, and other sources of carbohydrate are often needed before and during exercise. A lot has changed since 1986, when the first PowerBar was introduced. Today there are thousands of varieties of bars, gels, and carbohydrate and energy replacement systems specific to athletes, athletic events, sex, and an array of other categories. Guidelines have been issued by the ACSM for proper use of these products.

Gels usually come in single-serving disposable packages and contain sugars and maltodextrin, which is a complex carbohydrate made up of several glucose units. Some gels also contain caffeine, electrolytes, and other ingredients such as ginseng or amino acids. A standard serving of gel contains 25 g of carbohydrate and has 100 kcal. For longer-duration exercise, a single serving of gel can be ingested every 20 to 60 min during activity. Because a gel pack is a concentrated source of sugar, it should be taken with 4 to 8 oz (120-240 ml) of water. Caffeine content varies depending on the brand and flavor of gel; some individuals prefer to ingest caffeine during endurance activities, while other people develop palpitations, nervousness, or gastrointestinal upset. Multiple studies have shown some performance improvement when caffeine is utilized as an ergogenic aid.

Energy bars may also be used during exercise, although they are more commonly used about 1 h before exercise in order to fuel the upcoming workout. Most bars contain a combination of fat, carbohydrate, and protein. High-carbohydrate bars are good fuel sources for before and during a workout; protein and fat typically slow digestion and therefore are not the ideal source of fuel for muscles during exercise. Most carbohydrate in energy bars comes from whole grains and sugars.

The ACSM recommends choosing a bar with 25 to 40 g of carbohydrate and less than 15 g of protein. One bar should be eaten an hour before exercise and another bar should be eaten during each hour of exercise. As with the energy gels, ample water should be consumed with the bar to prevent dehydration.

One other source of carbohydrate commonly used during exercise is fruit. Fresh fruit provides readily available sugars along with a natural source of fluid and electrolytes. One serving size (1 banana or orange) contains about 15 g of carbohydrate; an equivalent amount of dried fruit is usually 1/4 cup. The ACSM recommends eating 1 or 2 servings of fruit before a workout and 2 or 3 servings during each hour of exercise. Although fruit may contain some fluid, it should still be consumed with additional water.

Supplements

Estimates of supplement use among athletes range from 50% to 100% (Sobal 1994). The Dietary Supplement Health and Education Act of 1994 allows supplement manufacturers to make unsubstantiated claims regarding the effects of their supplements on the structure and function of the body, and thus manufacturers often make a broad range of health statements that may or may not be accurate. Many individuals—athletes in particular—utilize supplements in a hopeful attempt to optimize performance. There are not, however, any established guidelines governing the products, as this would require scientific proof to substantiate claims. This lack of guidelines only helps to fuel the supplement craze. The PCPFS Research Digest publication "Nutrition and Physical Activity: Fueling the Active Individual" suggests using a multivitamin and multimineral supplement that contains micronutrients in amounts close to recommended amounts. Sticking to the recommended amounts allows the athlete to avoid potential toxicity of specific nutrients or interactions with other nutrients or medications. One practical suggestion is to select supplements from a reputable, well-established company that carries the U.S. Pharmacopeia (USP) or National Formulary (NF) designation.

GUIDELINE 17.9

Title: *Selecting and Effectively Using Sports Drinks, Carbohydrate Gels and Energy Bars*

Organization: ACSM

Year published: 2005

Purpose: To limit loss of body fluids, drop in blood sugar, and depletion of muscle carbohydrate stores

Location: www.acsm.org/AM/Template.cfm?Section=brochures2&Template=/CM/ContentDisplay.cfm&ContentID=12036

Population: Active individuals

GUIDELINES

Type	Serving size	Sugar content	Timing and frequency of consumption
Sports drinks	8 oz (240 ml)	13-19 g	Drink 8 oz (240 ml) before exercise and 8 oz (240 ml) every 15-45 min during exercise.
Carbohydrate gels	1 packet	25 g	Consume 1 to 3 packets per hour during exercise.
Energy bars	1 bar	25-40 g	Eat 1 bar an hour before a long workout and 1 bar per hour of exercise; consume with ample water.
Fruit	1 piece of whole fruit or 1/4 cup of dried fruit	15 g	Eat 1 to 2 servings before a workout and 2 or 3 servings every hour during a workout; consume with water.

These indicate compliance with a strict set of standards regarding product purity, strength, packaging, labeling, and weight variation.

Amino acids, protein supplements, and specific vitamins and minerals are all commonly used supplements. While there are 22 amino acids used by the human body, several of which are essential, there is no evidence that healthy individuals can benefit from large doses of a single amino acid. Many athletes take creatine supplements, an amino acid that does not occur in proteins but is found in the muscle tissue of vertebrates, to augment muscle mass; the potential risks and benefits of this are beyond the scope of this text but have been debated in the literature. Taking additional vitamins beyond what is found in a multivitamin is also common; for a few vitamins this is unlikely to be harmful and may actually be helpful. Epidemiological data have suggested that individuals who ingest higher amounts of vitamins C and E and beta-carotene (along with a healthy diet containing fruit, vegetables, and other dietary sources of vitamins) may have a lower level of harmful oxidants as well as a lower risk of heart disease and cancer. Randomized controlled trials with antioxidant supplementation have failed to find a protective benefit for vitamin A or E, but vitamin C seems to be acceptable in terms of potentially having a protective benefit without inducing harm. Vitamin D is a hormone, supplement, and vitamin that has begun to take center stage in the context of preventive medicine and public health. Recent reports suggest that vitamin D deficiency—which ranges in prevalence from 40% to 100% depending on the population demographics and time of year (Holick 2007)—is linked to an increased rate of heart disease, diabetes, autoimmune disease, osteoporosis, multiple sclerosis, and cancers of the prostate, lung, colon, and breast! In fact, a recent meta-analysis reported that all-cause mortality was decreased for individuals taking vitamin D supplements (Autier and Gandini 2007). In light of the high prevalence of vitamin D deficiency in the United States, it is unlikely that the various components of the American diet, lifestyle, and supplement behaviors are adequate to provide for optimal vitamin D status. The current RDA and reference daily intake (RDI) may be insufficient for many segments of the population. After having their serum vitamin D level checked, people who are deficient—which is likely to be a large number of individuals—benefit from supplementation. Current guidelines on vitamin D supplementation are usually presented with calcium guidelines and are part of the recommendations for the prevention or treatment of osteoporosis. In light of the myriad effects of vitamin D on bone and heart health as well as its link to a lower risk for chronic disease, including cancer, it seems appropriate to include vitamin D in this text about physical activity and overall health.

At one time or another, specific minerals have been in vogue for various groups of individuals. Chromium, selenium, magnesium, and zinc have all been promoted for their various metabolic benefits, although there is no indication that a considerable deficiency in any of these is prevalent among the healthy adult population.

One mineral that deserves mention is calcium. A significant proportion of American adults do not meet the daily RDA for calcium, which varies depending on age, hormonal status, and sex. Calcium is essential for heart and skeletal muscle function, bone health, and modulation of blood vessel tone. Dietary sources of calcium include dairy products such as milk, yogurt, or cheese; canned fish; soy products; and some green leafy vegetables. Also, products such as orange juice, energy bars, and cereal increasingly are becoming supplemented with calcium in an attempt to boost consumption. Several convenient calcium supplements, such as chewable calcium carbonate or calcium citrate, are available to enhance compliance with the recommended intake values. Like vitamin D, calcium influences a number of metabolic functions and thus deserves inclusion in this text. Specific calcium recommendations are presented in the section on osteoporosis, although calcium is mentioned in the framework of several other topics.

Dietary Guidelines for Special Populations

This section focuses on the guidelines that exist for specific populations, such as athletes, children, pregnant women, and older individuals. Some guidelines stand alone and are issued by organizations specific to that population (such as the

ACSM guidelines for athletes), while others are just a specific segment of the national guidelines for all Americans. International guidelines, issued by the WHO, focus on special populations such as older individuals.

Athletes

The dietary guidelines on proper nutrition for athletes relate to the DGA the way the ACSM guidelines for *physical fitness* relate to the CDC and ACSM guidelines for *health promotion.* As opposed to a goal of merely preventing disease or promoting health, the goal of these athletic guidelines is to optimize performance improvement. Athletes typically have body weights within the normal range for their height, but they often seek improvements in body composition in order to meet the demands of their sport. Guidelines were published by the ACSM along with the American Dietetic Association and Dietitians of Canada in 2000 and updated in 2009; the PCPFS issued a report in 2004. Each set of guidelines addresses the recommended amount of macronutrients and makes suggestions for hydration. The exact amounts differ somewhat but the goals are similar: to provide adequate and appropriate sources of energy for performance and muscle growth and maintenance. A balance of adequate carbohydrate for glycogen replacement, protein for maintenance and repair of lean tissue, and essential fat for other functions (detailed shortly) is recommended to meet caloric needs. Both "Nutrition and Athletic Performance" and "Nutrition and Physical Activity: Fueling the Active Individual" take into account the athlete's level of activity; the amount of macronutrients recommended differs according to moderate-intensity versus high-intensity endurance activities.

Carbohydrate

The intake of carbohydrate, or *CHO,* varies depending on the athlete's sex, type of sport, and environmental conditions. During exercise, carbohydrate helps accomplish both blood glucose maintenance and glycogen repletion. While the ACSM position statement suggests that the daily carbohydrate demand varies from 6 to 10 g for every kilogram of body weight, the PCPFS guidelines suggest that athletes participating in

moderate-intensity exercise should consume 5 to 7 $g \cdot kg^{-1} \cdot d^{-1}$. This value increases to 7 to 12 $g \cdot kg^{-1} \cdot d^{-1}$ for people doing high-intensity endurance activities to account for the increased caloric needs of athletes exercising at high intensities. Thus, the absolute amount of carbohydrate needed to maintain muscle glycogen stores is adequate despite variation in the actual percentage of calories coming from carbohydrate sources.

Protein

Protein recommendations are also based on body weight. It is true that active individuals need to consume diets higher in protein, reflecting the need to cover the building and repair of muscle tissue, support gains in muscle mass that occur with exercise, and provide an additional energy source during exercise (Lemon 1998). Demand for protein depends on the type of exercise performed (endurance versus resistance), the intensity of activity, the athlete's body composition, and the athlete's desire to gain or lose weight. Both the ACSM and PCPFS suggest that individuals participating in endurance sports eat 1.2 to 1.4 $g \cdot kg^{-1} \cdot d^{-1}$. Those who participate in resistance training require a higher amount of protein to allow for both the accumulation and maintenance of lean tissue. These individuals should strive for 1.6 to 1.7 $g \cdot kg^{-1} \cdot d^{-1}$ (according to the ACSM; the range is 1.6-1.8 $g \cdot kg^{-1} \cdot d^{-1}$ in the PCPFS guidelines). Like most nonathlete Americans, many athletes do not have any difficulty getting an adequate amount of protein; only active individuals who restrict energy intake for weight loss or follow strict vegetarian diets may be at risk for low protein intake. The ACSM believes that as long as energy intake is sufficient, there is no need for amino acid or protein supplementation.

Fat

The advice regarding type and amount of fat intake for good health has changed many times over the past few decades. This is just as true for athletes as it is for the rest of the population. One world-class rower (personal communication) recalls, while in training for the Olympic trials only 50 years ago, being encouraged to eat a T-bone steak; a loaded baked potato stuffed with butter, cheese, and sour cream; and full-cream cottage cheese as a snack! Although often perceived as something to avoid, fat is a necessary part of a normal diet, especially for athletic individuals. Actually, low amounts of fat—less than 15% to 17% of energy intake—may decrease energy and nutrient intake as well as exercise performance and thus are not recommended for active individuals (Horvath and others 2000a, 2000b). Recommendations for dietary fat for active individuals are similar to those for all individuals, regardless of activity level. This includes keeping saturated and trans fat to a minimum while providing adequate amounts of the essential fatty acids (EFAs) linoleic acid and alpha-linolenic acid. The ACSM does not specify types and amounts of fat intake, while the PCPFS goes into more detail with respect to the EFAs.

The EFAs are necessary in order to regulate immune response, heart rate, blood pressure, and blood clotting. Most people do not have trouble meeting the recommendation for 14 to 17 g/d (men) or 11 to 12 g/d (women) of linoleic acid, as it is found in vegetable and nut oils, including salad dressing, margarine, and mayonnaise. The alpha-linolenic acid is found primarily in fish, canola and soy oil, walnuts, and leafy green vegetables. While only 1.6 g/d for men and 1.1 g/d for women are recommended, this goal may be more challenging to fulfill, particularly when individuals consume very low-fat diets.

Supplements and Hydration

Maintenance of fluid balance is essential to achieving peak exercise performance. As little as 1% loss of body fluid can have a substantial deleterious effect on performance and can seriously hamper thermoregulation, leading to an increased risk for heat injury, muscle fatigue, and loss of coordination. The PCPFS suggestions for water intake—even before taking into account losses incurred during exercise—were recommended in the 2004 Dietary Reference Intakes for Water, Potassium, Sodium, Chloride, and Sulfate. Adults should strive to meet total water intake through beverages and food sources (80% through beverages and 20% through food). Men should consume 125 oz/d (3,700 ml/d) and women 91 oz/d (2,700 ml/d). Good sources include water, juice, tea, fruit, vegetables, soups, and smoothies. The ACSM's recommendations are the same (see table

17.2); these are reflected in the 2007 guidelines specific to exercise and fluid replacement that are presented in the following section.

In general supplements are not a requirement. In fact, the ACSM does not recommend supplements for athletic performance unless the athlete has an issue—unrelated to exercise—that necessitates additional supplementation. Examples include a specific micronutrient deficiency, pregnancy, or illness. The PCPFS differs somewhat in that it does support the use of supplements, so long as they are appropriate and from a trustworthy source. The suggestion is to use a supplement containing amounts close to the RDA and to avoid supraphysiological or megavitamin doses. One specific mineral addressed by the PCPFS is calcium, which is necessary for muscular, vascular, and myocardial function. Many individuals, including athletes, do not ingest enough of this mineral and so the PCPFS recommends that a supplement be taken separately if the dietary intake is insufficient.

Children

The USDA has created a colorful, fun food pyramid (MyPyramid) for youths. In addition to the information on the food groups, it includes a section at the bottom stating that physical activity is also a part of the pyramid. Keeping in line with the IOM recommendations, the MyPyramid points out that children should aim for at least 60 min of activity on most days. The pyramid accompanies its recommendations for specific intake amounts of food groups with suggestions to consume specific items within these groups and ideas on how to go about achieving the goals. For example, beneath the grains category is the catchphrase "Make half your grains whole" and then a suggestion to start breakfast with grains such as whole-grain cereals. Also, there is a hint about reading labels to determine whether the source is a whole grain. For the milk group, the pyramid states that milk provides calcium, which builds strong bones, and includes the hint to eat low-fat or fat-free dairy products to meet this need. For each of the major food groups (grains, vegetables, fruit, milk, and meat or beans), the pyramid lists a recommended total daily amount in specific measurements as opposed to number of servings. Finally, fat, sugars, and oils are listed separately, with information on healthier choices and suggested upper limits.

The 2002 *Dietary Reference Intakes for Energy, Carbohydrate, Fiber, Fat, Fatty Acids, Cholesterol, Protein, and Amino Acids* also has comprehensive nutrient information for children. One notable recommendation is the carbohydrate RDA. Whereas the recommendation for children aged 1 y to adults over 70 y is the same, at 130 g/d, many adults consume 2 or 3 times that much! Acceptable macronutrient distribution ranges follow those guidelines for specific components.

GUIDELINE 17.10

Title: "Nutrition and Physical Activity: Fueling the Active Individual"

Organization: PCPFS

Year published: 2004

Purpose: To optimize health and exercise performance

Location: PCPFS Research Digest or www.fitness.gov/publications/digests/digest-march2004.pdf

Population: Active individuals

GUIDELINES

Protein	Carbohydrate	Fat	Hydration	Supplements
• 1.2-1.4 g · kg^{-1} · d^{-1} (endurance exercise) • 1.6-1.8 g · kg^{-1} · d^{-1} (resistance or speed exercise) • Typically represents 15% of total calories	• 5-7 g · kg^{-1} · d^{-1} (moderate intensity) • 7-12 g · kg^{-1} · d^{-1} (high-intensity endurance activity)	• 1.6 g/d of alpha-linolenic acid and 14-17 g/d of linoleic acid (men) • 1.1 g/d of alpha-linolenic acid and 11-12 g/d of linoleic acid (women)	• 400-600 ml (13.5-20.3 oz) of fluid 2 h before exercise • 150-350 ml (5.1-11.8 oz) of fluid every 15-20 min of exercise	• Use a multi-vitamin or multimineral containing amounts close to RDAs. • Use calcium supplement separately if not getting adequate dietary calcium. • Use supplements with USP or NF notation.

GUIDELINE 17.11

Title: *Nutrition and Athletic Performance*

Organization: American Dietetic Association and Dietitians of Canada and ACSM

Year published: 2009

Purpose: To enhance physical activity, athletic performance, and recovery from exercise

Location: www.acsm-msse.org

Population: Active individuals

>continued

GUIDELINES

Protein	Carbohydrate	Fat	Hydration	Supplements
• 1.2-1.4 g · kg^{-1} · d^{-1} (endurance athletes) • 1.6-1.7 g · kg^{-1} · d^{-1} (resistance and strength training athletes) • Should not need to take amino acid or protein supplements	6-10 g · kg^{-1} · d^{-1}	• Should not be restricted • 20%-25%, and no less than 15%, of energy should come from fat	• 400-600 ml (13.5-20.3 oz) of fluid 2-3 h before exercise • 150-350 ml (5.1-11.8 oz) of fluid every 15-20 min of exercise • After exercise, 450-675 ml (15.2-22.8 oz) of fluid per pound of body weight lost during exercise	• No vitamin or mineral supplements are required unless athlete is dieting, vegetarian, or treating a specific medical diagnosis. • Ergogenic aids should be used with caution only if legal, safe, potent, and efficacious.

Special considerations include the following:

- A preexercise meal should consist of 200 to 300 g of carbohydrate 2 to 4 h before exercise.
- During exercise lasting more than 60 min, 0.7 g · kg^{-1} · h^{-1} of carbohydrate will extend endurance performance (start ingesting shortly after onset of activity).
- Postexercise carbohydrate intake should start immediately after exercise: Ingest 1.5 g/kg of carbohydrate and repeat at 2 h intervals.

GUIDELINE 17.12

Title: *MyPyramid for Kids*

Organization: USDA

Year published: 2005

Purpose: To promote healthy growth

Location: www.mypyramid.gov/kids/index.html

Population: Children

GUIDELINES

Food group	Suggested servings (based on 1,800 kcal/d diet)
Grains	6 oz (170 g), half of which should be whole grains
Vegetables	2 1/2 cup
Fruit	1 1/2 cup
Milk	3 cup (2 cup for children aged 2-8 y)
Meat and beans	5 oz (142 g)
Oils	Not specified, but fish and nuts are recommended sources, along with corn oil, soybean oil, and canola oil
Fat and sugar	Limit solid fat; choose food and beverages low in added sugar and other caloric sweeteners

GUIDELINE 17.13

Title: *Dietary Reference Intakes for Energy, Carbohydrate, Fiber, Fat, Fatty Acids, Cholesterol, Protein, and Amino Acids*

Organization: IOM Food and Nutrition Board

Year published: 2002

Purpose: To promote health and vigor and to balance food energy intake with total energy expenditure

Location: www.nap.edu/books/0309085373/html

Population: Healthy children from birth to age 18 y

GUIDELINES

Component	Children aged 0-12 mo	Children aged 1-3 y	Children aged 4-8 y	Children aged 9-13 y	Children aged 14-18 y
Carbohydrate RDA (g/d)	60 (0-6 mo); 95 (7-12 mo)	130	130	130	130
Protein RDA (g/d)	9.1 (0-6 mo); 11+ (7-12 mo)	13	19	34	52 (boys); 46 (girls)
Fat RDA (g/d)	31 (0-6 mo); 30 (7-12 mo)	Not determined	Not determined	Not determined	Not determined
Fiber AI (g/d)	Not determined	19	25	31 (boys); 26 (girls)	38 (boys); 26 (girls)
Linoleic acid AI (g/d)	4.4 (0-6 mo); 4.6 (7-12 mo)	7	10	12 (boys); 10 (girls)	16 (boys); 11 (girls)
Alpha-linolenic acid AI (g/d)	0.5	0.7	0.9	1.2 (boys); 1.0 (girls)	1.6 (boys); 1.1 (girls)
Calcium AI (mg/d)	210 (0-6 mo); 270 (7-12 mo)	500	800	1,300	1,300
Vitamin D AI (µg/d)	5	5	5	5	5
Folate AI (µ/d)	65 (0-6 mo); 80 (7-12 mo)	150	200	300	400

GUIDELINE 17.14

Title: *Dietary Reference Intakes for Energy, Carbohydrate, Fiber, Fat, Fatty Acids, Cholesterol, Protein, and Amino Acids*

Organization: IOM Food and Nutrition Board

Year published: 2002

Purpose: To promote health and vigor and to balance food energy intake with total energy expenditure

Location: www.nap.edu/books/0309085373/html

Population: Healthy children from birth to age 18 y

>continued

GUIDELINE 17.14 >continued

GUIDELINES

Component	Children aged 1-3 y	Children aged 4-18 y
Fat	30%-40% (5%-10% should be omega-6 polyunsaturated fatty acids and 0.6%-1.2% should be omega-3 polyunsaturated fatty acids)	25%-35% (5%-10% should be omega-6 polyunsaturated fatty acids and 0.6%-1.2% should be omega-3 polyunsaturated fatty acids)
Protein	5%-20%	10%-30%
Carbohydrate	45%-65%	45%-65%

GUIDELINE 17.15

Title: *Keep Fit for Life. Meeting the Nutritional Needs of Older Persons*

Organization: WHO

Year published: 2002

Purpose: To prevent both nutrient deficiency and chronic disease

Location: www.who.int/nutrition/publications/olderpersons/en/

Population: Older adults aged 60 y and older

GUIDELINES

Type	Nutrient intake
Total energy	1.4-1.8 multiples of the basal metabolic rate (corresponding to different levels of physical activity)
Fat	30% of energy for sedentary individuals and 35% for active persons; saturated fat should be limited to 8% of energy
Protein	0.9-1.1 g· kg^{-1} · d^{-1}

Special recommendations include the following:

- Emphasize healthy traditional vegetable- and legume-based dishes.
- Limit dishes that are heavily preserved in salt.
- Select nutrient-dense foods such as fish, lean meat, liver, eggs, soy products, low-fat dairy products, yeast-based products, fruit and vegetables, herbs and spices, whole-grain cereals, and nuts and seeds.
- Consume fat from whole foods such as nuts, seeds, beans, olives, and fatty fish. Select from a variety of liquid oils, including those high in omega-3 and omega-9 polyunsaturated fatty acids. Avoid fatty spreads.
- Avoid the regular use of celebratory foods.
- Eat several (5-6) small, nonfatty meals.
- Avoid dehydration, especially in warm climates, by regularly consuming fluids and foods with a high water content (consume at least 50.7 oz or 1,500 ml of fluid daily).

Older Adults

The WHO has issued dietary guidelines for older adults. These were jointly organized by the Food and Agriculture Organization of the United Nations (FAO) and the WHO in order to increase the relevance and effectiveness of nutritional recommendations in everyday life.

Pregnant and Postpartum Women

During pregnancy, women who are a normal weight require, on average, an extra 300 kcal/d to meet metabolic needs. The additional demand is lower during the first trimester and peaks at 390 kcal/d between week 20 and week 30 before declining to 250 kcal/d during the last 10 wk of gestation. Active women may require additional calories depending on the type and intensity of their exercise. Because pregnant women utilize carbohydrate at a greater rate (both at rest and with exercise) than nonpregnant women, adequate carbohydrate intake is important. Complex carbohydrate provides glucose and amino acids to the growing fetus as well as free fatty acids, ketones, and glycerol to the mother as sources of fuel. In addition to the general increased need for calories in the form of protein, carbohydrate, and fat, pregnant women require higher amounts of some nutrients such as folic acid, vitamin D, iron, and calcium.

In the postpartum period, many women focus on returning to their prepregnancy weight. Exercise and diet both play a role in this weight loss effort. Breast-feeding women require between 300 and 500 kcal/d additional energy to provide adequate protein and nutrients for producing breast milk. Restriction of calories or fluid may significantly reduce production and may adversely affect the growth of the infant. Dietary guidelines for nursing mothers address extra caloric intake as well as recommend intake of specific nutrients such as vitamin B_6, calcium, vitamin D, iron, and folic acid. The ACSM's consensus statement on pregnancy and postpartum physical activity recommends that women wait until lactation is established before restricting caloric intake. At least 2 mg/d of vitamin B_6 should be ingested either through dietary intake or supplements; this amount is essential for infant growth and maternal health (Mottola 2002).

Dietary Guidelines for Various Diseases

Just as there are physical activity guidelines for individuals with various diseases, there are also dietary guidelines for these individuals. Together, physical activity and diet are felt to account for the majority of all preventable disease. Existing guidelines address people who wish to prevent some of the more common diseases, such as cancer, diabetes, and heart disease, as well as people who are already diagnosed with these conditions.

Cancer

Cancer is the second leading cause of death in the United States, and diet plays an integral role both in the prevention and in the survival of cancer. Guidelines for both have been issued by the ACS and the World Cancer Research Fund.

Cancer Prevention

Chapter 8 presents the guidelines on physical activity needed as part of a healthy lifestyle to prevent cancer. Both nutrition and physical activity play a role in the maintenance of health and the prevention of cancer. Every 5 y, the ACS publishes guidelines for cancer prevention that represent the most current evidence relating dietary and activity patterns to cancer risk. Because approximately 35% of cancer deaths in the United States may be avoidable through dietary modification (Willett 1994; Steinmetz and Potter 1991; Ames and others 1995), it is compelling to minimize cancer risk through prudent dietary choices. Both the ACS and the World Cancer Research Fund, along with the American Institute for Cancer Research, have issued specific dietary guidelines along with those for physical activity. Like the physical activity guidelines, the dietary guidelines provide both individual recommendations and recommendations for community action. The guidelines suggest eating a variety of healthy foods and emphasizing plant sources when possible.

Cancer Survival

Chapter 9 provided details about physical activity guidelines for cancer survivors. Because both activity and nutrition play a role in maintaining weight and because weight loss reduces both

GUIDELINE 17.16

Title: *American Cancer Society Guidelines on Nutrition and Physical Activity for Cancer Prevention: Reducing the Risk of Cancer With Healthy Food Choices and Physical Activity*

Organization: ACS

Year published: 2002

Purpose: To decrease cancer incidence and mortality and to improve the quality of life of cancer survivors

Location: http://caonline.amcancersoc.org/cgi/reprint/56/5/254

Population: Children, adolescents, and adults

GUIDELINES

Recommendations for Individual Action

- Eat a variety of healthful foods and emphasize plant sources.
    - Eat 5 or more servings of a variety of vegetables and fruit each day.
    - Choose whole grains in preference to processed (refined) grains and sugars.
    - Limit consumption of red meats, especially those high in fat and processed.
    - Choose foods that help maintain a healthful weight.
- Maintain a healthful weight throughout life.
    - Balance caloric intake with physical activity.
    - Lose weight if currently overweight or obese.

Recommendations for Community Action

- Public, private, and community organizations should work to create social and physical environments that support the adoption and maintenance of healthful nutrition and physical activity.
- Public, private, and community organizations should increase access to healthful foods in schools, workplaces, and communities.

GUIDELINE 17.17

Title: *Food, Nutrition, Physical Activity, and the Prevention of Cancer: A Global Perspective*

Organization: World Cancer Research Fund and the American Institute for Cancer Research

Year published: 2007 (1997 first report)

Purpose: To reduce the incidence of cancer

Location: www.dietandcancerreport.org/downloads/chapters/chapter_12.pdf

Population: Children, adolescents, and adults around the world

GUIDELINES

Personal Recommendations

- Limit consumption of energy-dense foods.
    - Consume energy-dense foods sparingly.
    - Avoid sugary drinks.
    - Consume fast foods sparingly, if at all.

Personal Recommendations *(continued)*

- Eat mostly foods of plant origin.
  - Eat at least 5 servings (at least 400 g or 14 oz) of a variety of nonstarchy vegetables and fruit every day.
  - Eat relatively unprocessed cereals (grains) or pulses (legumes) with every meal.
  - If starchy roots or tubers are consumed as staples, ensure sufficient intake of nonstarchy vegetables, fruit, and pulses.
- Limit intake of red meat and avoid processed meat: People who eat red meat should consume less than 500 g (18 oz) a week, very little of which, if any, is processed meat.
- Limit alcoholic drinks: If alcoholic drinks are consumed, limit consumption to no more than 2 drinks a day for men and 1 drink a day for women.
- Limit consumption of salt.
  - Avoid salt-preserved, salted, or salty foods; preserve foods without using salt.
  - Limit consumption of processed foods with added salt to ensure an intake of less than 6 g (2.4 g of sodium) a day.
  - Do not eat moldy cereals or pulses.
- Aim to meet nutritional needs through diet alone: Dietary supplements are not recommended for cancer prevention.

Public Health Recommendations

- Limit consumption of energy-dense foods and avoid sugary drinks.
  - The average energy density of diets should be lowered toward 125 kcal for every 100 g of food consumed.
  - The average consumption of sugary drinks among the world population should be halved every 10 y.
- Eat mostly foods of plant origin.
  - The average consumption of nonstarchy vegetables and fruit among the world population should be at least 600 g (21 oz) daily.
  - Relatively unprocessed cereals or pulses and other foods that are a natural source of dietary fiber should contribute to an average of at least 25 g nonstarchy polysaccharide daily among the world population.
- Limit intake of red meat and avoid processed meat: The average consumption of red meat among the world population should be no more than 300 g (11 oz) a week, very little, if any, of which should be processed.
- Limit alcoholic drinks: The proportion of the world population drinking more than the recommended limits should be reduced by one-third every 10 y.
- Limit consumption of salt.
  - The average consumption of salt from all sources among the world population should be less than 5 g (2 g of sodium) a day.
  - The proportion of the world population consuming more than 6 g of salt (2.4 g of sodium) a day should be halved every 10 y.
  - Minimize exposure to aflatoxins from moldy cereals or pulses.
- Aim to meet nutritional needs through diet alone: Maximize the proportion of the world population achieving nutritional adequacy without dietary supplements.

cancer recurrence and incidence of new cancers, the ACS has also advocated dietary strategies to improve weight and health. Goals of nutritional care for cancer survivors are to prevent or reverse nutrient deficiencies, to preserve lean body mass, to minimize nutrition-related side effects of cancer (such as decreased appetite, nausea, taste changes, or bowel changes), and to maximize quality of life. While a multivitamin and mineral supplement containing 100% of the RDAs may help cancer survivors meet nutrient needs, an excess of some vitamins such as antioxidants and folic acid may actually be harmful. These supplements can interfere with chemotherapy and promote precancerous or cancerous growths. While moderate use of alcohol may attenuate the risk for CVD, the benefits of alcohol consumption for CVD should be weighed against the increased risk for new cancers, including cancers of the breast, liver, pharynx, larynx, esophagus, pancreas, mouth, colon, and rectum.

Nutrition recommendations include using omega-3 polyunsaturated fatty acids as sources for dietary fat as much as possible and avoiding saturated and trans fat. Carbohydrate recommendations are to consume foods rich in essential nutrients, phytochemicals, and fiber, such as whole grains, fruit, vegetables, and legumes. Individuals should eat at least 5 servings of fruit and vegetables per day. Protein sources should be low in fat whenever possible and should include nuts, seeds, and legumes in addition to leaner meats and low-fat dairy products. Alcohol intake should be limited to no more than 1 drink per day for women or 2 drinks per day for men. There is an emphasis on maintaining an appropriate weight. Cancer guidelines also address proper food safety, including both food preparation and types of food ingested. There are also nutritional guidelines for specific cancers that focus on reducing the risk of recurrence, moderating side effects, and decreasing the later risk for CVD. The various sources from which these guidelines can be found are all listed in appendix A.

Hypertension and Coronary Artery Disease

One of the themes of this text is how essential healthy diet and physical activity are in treat-

ing and preventing chronic disease and yet how prevalent poor diet and physical inactivity are throughout the world. Hypertension and CAD are so common, and diet and exercise play such a central role in determining an individual's risk and lifetime course for these processes. The DASH eating plan presented earlier in this chapter is advocated by the NHLBI and supported by many other national and international entities—not just for control and prevention of hypertension but also for cancer prevention and treatment of multiple types of heart disease, including CAD and heart failure. The other important guidelines with respect to CAD are those issued by the Third Report of the Expert Panel on Detection, Evaluation, and Treatment of High Blood Cholesterol in Adults (Adult Treatment Panel III or ATP III) in conjunction with the guidelines for cholesterol treatment. The therapeutic lifestyle changes (TLC) recommended include a heart-healthy diet, and these are endorsed by the AHA:

- Diet
  - Saturated fat intake of <7% of calories per day, cholesterol intake of <200 mg/d
  - Consider increased viscous (soluble) fiber (10-25 g/d) and plant stanols or sterols (2g/d) as therapeutic options for lowering LDL levels
- Weight management
- Increased physical activity

Osteoporosis

Diet plays a role in osteoporosis. A poor diet that is lacking in specific nutrients or overabundant in others is a risk factor for the development of osteoporosis. In addition, individuals who have osteoporosis need to optimize their diet in order to provide an ideal environment for bone formation and regrowth. Calcium and vitamin D are important nutrients in fighting osteoporosis. Calcium intake is recommended at 1,000 mg/d for all adults or 1,200 mg/d for adults 51 y and older. Women who are pregnant or lactating should consume 1,000 mg/d of calcium or 1,300 mg/d if they are teenagers. Currently vitamin D intake is recommended at 200 IU each day; however, many individuals who consume this amount are deficient in vitamin D as measured by their

25-hydroxyvitamin D levels. For this reason, it has been suggested that the RDA for vitamin D should be increased. In the 2000 *NIH Consensus Statement* on osteoporosis, the NIH recommended an intake of 400 to 600 IU of vitamin D daily. Vitamin D is essential for many functions, one of which is to enhance the absorption of calcium. Other vitamins and minerals important in bone formation are vitamins A and K, boron, magnesium, and zinc. Drinking alcohol may cause loss of calcium, magnesium, and zinc in the urine and thus may harm bone density. Additionally, a diet high in sodium can increase calcium excretion. For the elderly, who are already at increased risk for bone loss, dietary sodium seems to be an important factor in osteoporosis. One study found that reduced sodium intake reduces urinary calcium losses (Blackwood 2001), and it has been speculated that these changes may have a substantial effect on loss of bone mass.

Diabetes

Diet is a cornerstone in the management and treatment of diabetes. While complete dietary recommendations for people with diabetes are beyond the scope of this book, a few key recommendations are emphasized here. Because being overweight is such a risk factor for developing type 2 diabetes and because weight control is a convincing treatment, the role of diet in achieving an ideal weight should be emphasized. Also, given that type 2 diabetes is a significant risk factor for CAD, a heart-healthy diet—one that optimizes lipid profile—is prudent in order to reduce heart disease. Just as there is a food pyramid for adults, a diabetes food pyramid is also available and can be viewed at www.diabetes.org/food-and-fitness/food/planning-meals/diabetes-food-pyramid.html.

The diabetes food pyramid differs from the USDA food pyramid in that the six groups of

---

GUIDELINE 17.18

Title: *Prevent Osteoporosis Now*

Organization: ACE

Year published: 2001

Purpose: To prevent bone loss and encourage bone growth

Location: www.acefitness.org/fitfacts

Population: All individuals

GUIDELINES

Age group	Calcium (mg/d)	Vitamin D$_3$ (daily IU)
Infants 0-6 mo	210	200
Infants 6-12 mo	270	200
Children 1-3 y	500	200
Children 4-8 y	800	200
Children 9-18 y	1,300	200
Adults 18-50 y	1,000	200
Adults 51-70 y	1,200	400
Adults 70+ y	1,200	600

Special considerations include the following:

- A high-calcium, high-fiber, low-fat diet may bolster bone strength.
- Smoking, drugs, alcohol, caffeine, and high-phosphate soft drinks may cause osteoporosis.

Adapted, by permission from American Council on Exercise Fit Facts: Prevent osteoporosis now, San Diego 2001. www.acefitness.org/fitfacts Accessed 6/28/2008.

food are based on carbohydrate and protein content instead of food classification. At the base of the pyramid are the breads, grains, and starches. These foods remain at the bottom of the pyramid, although the pyramid stresses that the starches should be complex carbohydrate such as corn, oats, and cereal as well as beans such as black-eyed peas and pinto beans. The pyramid recommends 6 to 11 servings a day; most individuals should aim toward the minimum unless they are expending a significant amount of energy via physical activity. Vegetables make up the next level in the diabetes food pyramid. Vegetables are an important source of fiber, minerals, and vitamins; 3 to 5 servings are recommended daily. Many people with diabetes incorrectly assume that they should avoid the various types of fruit because of their natural sugar content. On the contrary, fruit are a good source of complex carbohydrate, vitamins, minerals, and fiber, and they make up the next level of the pyramid. People with diabetes should aim for 2 to 4 servings of fruit a day. Milk products are a source of protein as well as calcium; consuming 2 or 3 servings a day helps fulfill protein and calcium requirements. Choosing nonfat or low-fat varieties helps decrease the amount of saturated fat in the diet. Meat and meat substitutes, including beef, chicken, turkey, fish, eggs, tofu, dried beans, cheese, cottage cheese, and peanut butter,

are great sources of protein. Only 4 to 6 *oz* (125-175 g) daily are recommended. Again, choosing lean products is emphasized. The daily intake should be divided between meals. Lastly, at the top of the pyramid are fat, sweets, and alcohol. Because these foods contain significant amounts of fat or sugar, they should be consumed only in moderation.

Exercising with diabetes provides numerous benefits but also requires caution (as detailed in chapter 12). Dietary guidelines regarding energy intake *during* exercise reflect the glucose-lowering effect of exercise. In the long term, exercise improves glycemic control for people with type 2 diabetes; exercise will also transiently lower blood sugar levels for several hours afterward. Thus, most physical activity guidelines for people with diabetes also contain information on eating before, during, and after exercise in order to prevent adverse effects. Depending on its duration and intensity, exercise usually requires either an increase in carbohydrate ingestion or a reduction in preexercise insulin bolus. Table 17.3 shows the estimated number of minutes of activity a youth must complete in order to match an intake of 15 g of carbohydrate in terms of glucose equilibrium. The different physiology of the body's response to (and production of) insulin in type 1 versus type 2 diabetes necessitates slightly different

TABLE 17.3 Number of Minutes of Activity Required to Maintain Glucose Level After Ingestion of 15 g Carbohydrate

Activity	BODY WEIGHT		
	20 kg (44 lb)	40 kg (66 lb)	60 kg (132 lb)
Basketball (game)	30	15	10
Cross-country skiing	40	20	15
Cycling at 6 mph (9.7 kph)	65	40	25
Cycling at 9 mph (14.5 kph)	45	25	15
Running 1 mi (1.6 km) in 12 min (5 mph or 8.0 kph)	25	15	10
Soccer	30	15	10
Swimming	55	25	15
Tennis	45	25	15
Walking 2.5 mph (4.0 kph)	60	40	30
Walking 4 mph (6.4 kph)	40	30	25

Reprinted, by permission from K. Robertson, P. Adolfsson, M.C. Riddell, G. Scheiner, and R. Hanas, 2008, "Exercise in children and adolescents with diabetes," *Pediatric Diabetes* 9(1):65-77.

recommendations for exercise energy intake in these different populations. Consensus guidelines for people with type 1 diabetes are followed by recommendations for people with type 2 diabetes.

GUIDELINE 17.19

Title: "Exercise in Children and Adolescents with Diabetes: Clinical Practice Consensus Guidelines 2006-2007"

Organization: ISPAD

Year published: 2008

Purpose: To address the issue of blood glucose regulation during various forms of sport and exercise

Location: *Pediatric Diabetes* 9: 65-77

Population: Children and adolescents with type 1 diabetes

GUIDELINES

- Consume up to 1.5 g/kg of carbohydrate for each hour of strenuous or longer-duration exercise when circulating insulin levels are high.
- Avoid alcohol, which increases the risk for hypoglycemia.
- Consume sugar-free fluids (up to 1.3 L/h or 44 oz/h for adolescents in hot and humid environments) in order to avoid dehydration.
- If endurance or prolonged activity is being performed, ingest glucose-sweetened water or carbohydrate before, during, and after exercise.
- In case of unforeseen physical activity, increase glucose consumption immediately before, during, and after the activity.
- Always carry sugar in case of hypoglycemia; adults or other accompanying people should be aware of signs and treatment of hypoglycemia.

GUIDELINE 17.20

Title: "Exercise Prescription for Other Clinical Populations": Diabetes Mellitus

Organization: ACSM

Year published: 2010

Purpose: To prevent hypoglycemia associated with exercise

Location: *ACSM's Guidelines for Exercise Testing and Prescription, Eighth Edition*

Population: Individuals with type 2 diabetes mellitus

GUIDELINES

- If pre- or postexercise blood glucose is <100 mg/dL (<5.55 mmol/L), the individual should ingest 20 to 30 g of additional carbohydrate.
- Adjust carbohydrate intake and/or medication before and after exercise based on blood glucose levels and exercise intensity.

SUMMARY

Dietary guidelines are provided alongside physical activity guidelines due to the tightly interwoven role of the two in preventing and treating chronic disease. Guidelines for special populations, including children, pregnant women, older individuals, athletes, and people with chronic disease, focus on meeting nutritional needs and balancing energy requirements while providing the highest quality of food.

Exercise Equipment and Facilities

Purchasing equipment for a home or a facility requires attention to cost, durability, safety, space constraints, and user needs. Both aerobic and resistance training equipment can be bought new or used. Stationary bicycles, treadmills, and weights are the most commonly used pieces of equipment; all are suitable for home or facility use. Heart rate monitors and pedometers are relatively inexpensive pieces of equipment that can make a significant difference in physical activity not only by indicating intensity or duration of exercise but also by providing feedback and improving compliance. Due to the advent of the Internet, people today can comparison shop and greatly increase their options while reducing their costs. The ACE, IDEA, and ACSM provide information on purchasing and using a variety of types and qualities of fitness equipment. Once the desired type of equipment is known, *Consumer Reports* is a good source for product reviews. After obtaining suitable equipment, the user must become familiar with its proper use in order to get the most out of the program.

This chapter begins with general guidelines for purchasing equipment, followed by suggestions for purchasing ancillary accessories such as CDs, DVDs, and videos. In addition to using exercise equipment and other fitness tools, many individuals elect to use the services of a personal trainer. Personal trainers can help their clients to optimize the time spent exercising and to focus on any specific concerns regarding age, risk factors, or disease processes. Personal trainers are certified by any number of professional entities, and guidelines for choosing a suitable personal trainer are also included in this chapter.

Aerobic Exercise Machines

As many ways as there are to exercise outdoors, there are ways to exercise indoors. These indoor options are made possible by various types of equipment. Exercise equipment can be used in a home or fitness center setting and comes in a wide range of cost, weight, size, and features. This section includes information on aerobic exercise equipment and discusses the issues to consider in their selection, such as cost, frequency of use, safety, durability, power, and desired performance.

Treadmills

Treadmills are one of the most popular pieces of home exercise equipment. Despite their popularity, at least 20% of home treadmills are not used at all! When used properly and consistently, however, they can be a very effective and convenient way of meeting physical activity recommendations. In fact, a 1996 study (Zeni and others 1996) found that treadmills are the most efficient way to burn calories when compared with other indoor exercise machines such as the cross-country skiing simulator, cycle ergometer, rowing machine, and stair stepper. Although treadmills can be either motorized or human powered, the ACSM recommends considering a motorized treadmill for the following reasons: (1) The smaller running belts of a manual treadmill make it more difficult for jogging or running and (2) the running belt often doesn't move as smoothly on a manual treadmill, which can cause an inconsistent pace and the need to hold onto the handrails in order to generate more power. Holding onto the handrails can make it more difficult to elevate heart rate and may also increase the risk for injury due to muscle strain. Inexpensive motorized treadmills may cost as little as $300 U.S., although high-end machines with multiple programming features may reach well over $4,000 U.S.!

Primary considerations for selecting a treadmill include safety, maintenance and durability, power and performance, and operation.

- **Safety:** Using a treadmill requires some basic safety practices. The back of the treadmill should be positioned at least 6 to 8 ft (1.8-2.4 m) away from a ledge, wall, or window. There should be adequate space around the treadmill as well as enough space above the treadmill to accommodate the height of the individual who intends to use it. Power supply and wiring should be hidden or secured. Keeping the belt and deck underneath the belt maintained according to the manufacturer's directions prevents excessive wear and tear. The platform should be stable, feel solid, and allow for a full range of motion. When walking, the individual's posture should be upright with the shoulders back. Staying in the center of the belt and paying enough attention to prevent drifting off the end of the treadmill are important. Handrails or safety bars should be accessible but should not impinge on the normal arm motion. The handrails can be used for support but may decrease the heart rate or caloric expenditure achieved. An automatic shutoff feature—whether a key, clip, or tether—is important in case of an emergency. If the treadmill is stored in a different location or position while not in use, it should be balanced properly to avoid injury.

- **Maintenance and durability:** Length of warranty, ease of assembly and maintenance, and the reliability and reputation of the manufacturer are important. If the treadmill is heavy it should have adequate support from the ground beneath it.

- **Power and performance:** Power supply or circuit alterations may be needed to meet the power requirements of the treadmill. Some machines require 220 V. According to the ACE, the ideal belt size is at least 17 in. (43 cm) wide and 49 in. (124 cm) long (see table 18.1 for other recommendations from the ACE and ACSM). Manual treadmills often have shorter belts that increase the chance of tripping or falling off of the belt. Ideally, motorized treadmills should be 2.5 to 3.0 hp, although 1.5 hp is the minimum recommended by the ACSM. Motorized treadmills should also provide a wide range of speeds. The recommended slowest starting speed is 0.1 mph (0.2 kph) to allow for a gradual and safe increase in speed. For home treadmills, a maximum speed of 8.0 mph (13 kph) is usually adequate, although experienced runners may want a higher capacity. Incline should range from 0° to 10°, although, once again, experienced individuals may desire the capability to exercise at a higher incline. Some treadmills in fitness facilities can go as high as 50°! Lastly, the ability to deliver a smooth stop is recommended, as this can prevent injury.

- **Operation:** As with other home equipment, the treadmill noise level should be evaluated before purchasing to be sure it is tolerable for the home. Many treadmills have automated programs, but it is recommended that manual use be an option as well. Some consoles provide information such as time, distance, incline, calories expended, power output, and pace; if specific elements used to track exercise are desired, they should be available. If the treadmill is to be used frequently, the belt should be heavy duty to prevent stretching.

Stationary Bicycles

There are two main types of stationary bicycles: upright and semirecumbent. Also, the method of resistance differs among models, much like resis-

TABLE 18.1   ACSM and ACE Recommended Treadmill Features

Feature	ACSM recommendations	ACE recommendations
Minimum belt width	18-20 in. (46-51 cm)	17 in. (43 cm)
Minimum belt length	48 in. (122 cm)	49 in. (124 cm)
Motor power	1.5 hp	2.0 hp, continuous duty
Frame type	Not specified	Steel or aluminum
Other key features	Smooth-running, relatively quiet motor Storage potential Weight of treadmill	Low-impact deck for shock absorption At least 10% grade range

From ACE FitFacts, ACSM template

tance varies among rowing or weight machines. The appropriate choice depends on the individual's needs and budget. For people with low back pain or mobility or balance concerns, the semirecumbent bicycle is a better option because it does not put as much pressure on the lower spine. As with other exercise equipment, major considerations in purchasing a stationary bicycle include cost, safety, maintenance and durability, and performance options. Some bicycles cause a significant amount of noise (depending on the mode of resistance), while others may require a significant amount of power or space. Some exercise bicycles actually *generate* electricity that can be used to power televisions, computers, or lights and can be a gentle motivator to encourage more activity!

Proper position is important to avoid injury during stationary biking. An appropriate seat height ensures an even distribution of body weight. When the user is in the proper position, there should be a slight bend in the knee when the pedal is at the bottom of its motion. Handlebars should be adjusted so that the user is leaning slightly forward. Many bicycles have different notches for adjustment, so that multiple users can adjust the bicycle to their preferred settings. Many individuals use a stationary bicycle to cross-train and avoid injury, while others use it as their primary source of physical activity. Others replicate a spinning class by watching exercise videos and varying the intensity of the workout to improve fitness!

Elliptical Trainers

Elliptical trainers are relatively new in the landscape of aerobic exercise machines. They combine the motion of stair-climbing with that of cross-country skiing and engage the legs in an elliptical motion—hence the name. Some machines also have upper-extremity poles that can increase the caloric expenditure by recruiting upper-body muscles. The newer models currently available have an adaptable mechanism that allows them to provide a stepping motion, an elliptical motion, or some hybrid of both.

As with other exercise equipment, safety is a primary consideration in choosing an elliptical trainer. Safety issues include sturdiness, proper fit (both the machine itself and the area around the machine that the head, arms, and legs occupy), and the availability of a shutoff switch. Purchasing a trainer from a reputable manufacturer that provides a warranty is also important. Because elliptical trainers are large, they require adequate space for either use or storage. If the trainer requires power, it must be placed close enough to an appropriate outlet. Also, maintenance issues should be considered; if the trainer is not easily maintained, local technicians who are familiar with the machine should be available.

Trying out a trainer at least once before committing to a purchase is highly recommended. Some machines have wider or narrower pedal tracks that may not fit all individuals. Good pedals should be textured to avoid slippage, particularly if sweat drips onto them during activity. The maximum possible stride length also varies somewhat from model to model, so testing before purchasing will allow the user to determine if it is adequate. A good fit between trainer and user allows for upright posture with hand placement where the shoulders can be pulled back. Arms and legs should not bump into each other or any part of the console. Some trainers are loud, which is another reason to try them out before purchasing, especially if the model is intended for home use where other individuals will be present. Lastly, there are multiple options for the control panel, including time, distance, resistance level, elevation, caloric expenditure, and power output, and those features that are desired by the user should be offered by the trainer.

Rowing Machines

Because rowing is an effective low-impact exercise that provides excellent high-intensity exercise for the upper and lower body, rowing machines are a good option for many individuals. Many machines have a monitor displaying pace, distance, power output, heart rate, and caloric expenditure. Some also have preprogrammed workouts with intervals or other variables. The four types of resistance available are air, water, magnetic, and piston. Table 18.2 lists the pros and cons of each type.

In choosing a rowing machine it is important to determine the level of use the machine will get. Periodic home use probably doesn't require the industry standard, expensive air resistance. There should be adequate space for the machine to allow for a full arc of rowing motion, and the machine should be placed on a solid, level surface. Since many individuals store rowing machines at home in an upright position when not in use, the machine must be stable in this position so as not to cause accidental injury. Rowing handles should be covered with a nonslip rubber surface and should fit comfortably in the user's hand. After use, the handle should be placed against the flywheel to avoid unnecessary pulling against the resistance cord or chain. The cord or chain should not be twisted. Regular cleaning extends the life of the machine and reduces transmission of bacteria. Lastly, the machine should have a warranty and a manual with clear instructions on maintenance and troubleshooting.

Rowing machines can provide an excellent full-body workout. To avoid stressing the lower back, strokes should be smooth without sudden maximal effort. As with any form of exercise, a gradual warm-up can reduce the risk of injury. The two main variables in rowing are resistance level and stroke rate. Injury risk correlates with resistance level in that lower levels of resistance are associated with less injury. Conversely, a higher resistance and slower stroke rate place greater stress on the lower back. The seat positioning should allow for a full extension and flexion of the knees and a full extension of the shoulders and arms. Proper rowing technique includes moving the handle back before sliding back on the seat and passing the handle over the knees before flexing the knees. Not doing either of these may lead to injury due to strain from uneven movement.

Stair Steppers

Stair steppers utilize repetitive stepping to strengthen the lower body and improve aerobic fitness. The most important feature of a stepper is safety. A good stair stepper should be ergonomically sound and allow an upright workout posture with the knees behind the toes; other positions can lead to injuries from back, knee, or even wrist stress. Ensuring that the machine fits in the desired location, has suitable floor support, and can access the appropriate electrical voltage

TABLE 18.2 Rowing Machines

Type	Benefits	Drawbacks
Air	Industry standard Allows for fast change in resistance May be folded for easy storage	Most expensive May cause injury if flywheel doesn't have a mesh cover
Water	Provides closest replication to the feeling of rowing on water	Very heavy
Magnetic	Quietest type of machine Allows for fast change in resistance	
Piston	Least expensive May be folded for easy storage	

is also important. Stepping machines should be placed in a location that is level and has a high enough ceiling to allow the user to assume proper posture. Steppers should have side rails or a front rail for balance, and the frame should be solid. Handrails should be used for balance only and not to support the body weight, as doing so reduces caloric expenditure and increases the likelihood of poor posture or an upper-extremity injury.

A basic console should display the distance covered, time elapsed, calories burned, and intensity level attained. Some machines have only a manual program, while others have a selection of preprogrammed workouts designed to meet various goals. These include interval workouts, workouts that gradually increase in intensity, and workouts with built-in warm-ups and cool-downs. Durability concerns include the time and depth of the warranty, the costs for replacement parts or labor, and the expected use. A machine expected to be used multiple times daily should be covered for significantly more wear and tear than one that may be used only a few times a week in the home. It is also important to consider the reputation the manufacturer has for reliability.

Step height may vary depending on the exercise speed, but in general it should always feel comfortable and should not strain the knees or ankles. Pedals should be self-leveling so that the foot can stay flat throughout the range of motion. The stepping action should be smooth and independent such that pushing down on one pedal does not cause the other pedal to rise. Cheaper models are less likely to have an entirely smooth action. The initial stepping rate should be set such that the intensity may be progressively increased after the user gets accustomed to the machine. The stepper should allow for an increase (or decrease) in difficulty, depending on the goals of the workout. When desiring to end the workout, the user can ride the pedals to the floor and then slowly step off of one pedal at a time to release the pedals to the upper position.

Weights

As with aerobic exercise equipment, there are many kinds of resistance training equipment. Individuals may choose from free weights, a wide assortment of weight machines, resistance bands, and other arrangements, depending on preference.

Free Weights

Because resistance exercise is an essential part of a well-rounded physical activity program, it is important to consider owning and using weights. Free weights have long been associated with resistance training, although individuals now use many other types of resistance equipment, including rubber tubing, body weight and calisthenics, kettlebells, and medicine balls.

The two major types of free weights are barbells, which are generally 4 to 6 ft (1.2-1.8 m) long, and dumbbells, which are smaller weights that can be held in a single hand. In general, barbells have an area to add various weight plates to modify the weight lifted. Dumbbells come in weight increments ranging from 2.5 to 150 lb (1.1-68 kg), while a standard barbell before any extra weight is added is usually 25 to 45 lb (11.3-20.4 kg). The total barbell weight may vary significantly, depending on the amount of weight added with the weight plates. According to the ACSM, the most important considerations for using and choosing free weights are technique, safety precautions, and the exercises that can be performed with them.

- **Technique:** When starting to use free weights, it is helpful to have an experienced individual guide and observe the lifter's technique. Balancing exercises in the front and back of the body as well as on both sides can help prevent injury. Proper breathing technique for weightlifting is to inhale during the easy portion of the exercise and exhale through the hardest part. Moving the weight smoothly through the full range of motion can prevent injury and maximize benefit. Proper lifting technique when picking weight up from the floor entails using the legs instead of the back. Also, maintaining a straight spine instead of hyperextending can prevent injury. Lastly, when just starting a weight program or lifting relatively heavier weights, it is good practice to use a spotter as a margin of safety.

- **Safety:** The risk of injury is higher with free weights than it is with machines, primarily because free weights require more muscular coordination and do not restrict movement. Maintaining a good grip and stable position can help prevent injury. The grip of the free weight is an important consideration; the handle should feel comfortable, not cause undue muscle fatigue, and provide friction for a good grip. Weight plates should be secured on the barbell so that they do not slide off. The greatest risk for injury occurs when an individual tries to lift too much weight too soon. Thus, a gradual increase in weight is recommended. Additionally, when utilizing heavy weights or *heavier weights than accustomed to,* it is important to use a spotter, as mentioned earlier. Other safety considerations include making sure the weights are stowed in an area where they cannot inadvertently slip or be moved, causing harm to animals, children, or household items.

• **Exercises:** Using a combination of dumbbells and barbells affords the ability to perform a variety of exercises. As detailed in chapter 3, a good resistance training program exercises the major muscle groups, including the upper body, torso, and legs. A lower number of repetitions may enhance muscular strength, while higher numbers improve muscular endurance.

• **Other considerations:** Durability varies, depending on the materials from which the dumbbells or barbells are made. They come in chrome, metal, plastic, concrete, and with a thin foam covering. Chrome weights may require more maintenance. Some weight sets require assembly in order to provide a variety of different loads. Depending on the amount of equipment purchased, there may be a requirement for a significant amount of storage space. As noted previously, space should be available for safe storage in order to avoid unintended injury.

Weight Machines

A home weight machine is one large machine that has multiple stations that can be used to work a variety of muscle groups. There are several different types of machines; table 18.3 compares the different styles. Features that are important in choosing an appropriate machine for home use are the space available for the unit, the maximum amount of weight that can be lifted, the ease of use and manipulation needed to modify the machine when desired, and the need for maintenance and warranty.

Tools for Monitoring Physical Activity

Individuals often use various pieces of gear to monitor and analyze physical activity. Pedometers are an inexpensive way to gain an objective measure of steps walked (or run!); information on step counts is often included in physical activity recommendations. Heart rate monitors range from simple to complex, with extremely detailed data available, and are used to quantify intensity over the course of an activity session.

TABLE 18.3 Pros and Cons of Home Weight Machines

Style	Advantages	Disadvantages
Weight stacks	Easy to chart progress Resistance feels natural through range of motion	Heavy and bulky Requires significant assembly
Hydraulic pistons	Light Easy to move	Difficult to chart progress Resistance varies with pace and effort
Flexible rods and bands	Light Easy to move Resistance is lower at the beginning of movement	Resistance varies throughout range of motion (harder at end) May feel unnatural due to variable resistance
Body weight	Easy to assemble, move, and store Changing body position can easily vary muscle groups worked	Maximum resistance is limited to 50% of body weight May be awkward to get into starting position

Pedometers

Pedometers are a good tool for tallying the accumulation of steps on a daily basis. Many individuals respond well to the instant feedback a pedometer provides and put in extra effort to incorporate additional activity such as taking the stairs, walking during breaks, and so on. According to the ACSM pamphlet on pedometer use, approximately 2,000 steps equals 1 mi (1.6 km). The media have widely publicized a target of 10,000 steps a day for adults. The ACSM recommends 12,000 to 16,000 daily steps for children. The President's Challenge suggests 11,000 steps daily for girls and 13,000 steps daily for boys.

Pedometers have either spring-suspended or accelerometer-type mechanisms that count steps. Fancier models may be able to calculate total distance traveled and calories burned, depending on information input by the owner. Some very new models utilize GPS technology to give a very accurate measurement of distance covered. These models can also calculate pace time per mile. This new technology has significantly greater accuracy than the older pedometers have. Older models may incorporate stride length into the calculation of distance covered, and small discrepancies in the stride length that is input by the user can multiply over hundreds and thousands of steps, giving a highly inaccurate total distance. The same type of errors may result from attempting to calculate caloric expenditure by using body weight input. Lastly, inaccuracy may result when nonambulatory movements such as shaking up and down accidentally register as counted steps. On the other hand, certain active movements such as cycling, rowing, and upper-body exercise do not register because they are not ambulatory activity.

Accuracy is one of the key features to look for in a pedometer. According to the ACSM, accuracy in step count is the most important feature. The accuracy of pedometers is reported to be greatest at a speed of 2.5 mph (4.0 kph) or greater. When testing a pedometer for accuracy, the user should walk for 20 steps with the pedometer attached to the belt or waistband. If the pedometer is not attached to a firm or snug waistband in the upright position, the steps counted will likely be less than the true number of steps taken. Because walking is one of the most popular forms of exercise, pedometers are an effective tool in quantifying exercise and helping individuals gradually increase step counts and total volume of activity. More information on how pedometer step counts can be used to fulfill physical activity recommendations and earn the Presidential Active Lifestyle Award is presented in chapters 2 and 5.

Heart Rate Monitors

Heart rate monitors were introduced in the 1980s and have since become a popular fitness tool. A heart rate monitor can be worn during many different types of physical activity, including walking, running, biking, rowing, yard work, dancing—even swimming. The features available on heart rate monitors vary from basic models that provide beat-to-beat heart rate to models that cost several hundred dollars and have downloadable data such as time spent in various zones of intensity, caloric expenditure, recovery rate, interval splits, and so on. Most models consist of a band worn around the chest just below breast level that senses heartbeat and a watch monitor that displays the information. Many have programmable target training zones; individuals who know their actual HRmax or desired heart rate zone can override the preset program and input their precise figures. Depending on the user's goals, the monitor can be used to fine-tune an exercise program and give an individual a significant amount of feedback regarding intensity. Combining the heart rate during activity with time or distance can help an individual track improvements in fitness throughout the course of a program.

As discussed in chapter 3, heart rate is a very accurate method to track intensity, although there are several factors that may affect heart rate. Anything that requires oxygen or energy may increase the exercise heart rate. These factors include illness, injury, fatigue, food or drink, warm or humid ambient temperature, and various medications. Other medications such as beta-blockers may blunt the heart rate response to exercise.

Exercising at Home

Many individuals choose to exercise at home and thus set up various different types of equipment.

While there are lots of different types of equipment, it is important to consider what type the individual is most likely to use. Other considerations for those exercising at home are DVDs or videos, which may help guide the activity.

Designing a Home Gym

People who are purchasing various pieces of exercise equipment for the home have a few ideas to consider. Depending on the goal of exercise, the need for aerobic, strength training, and flexibility equipment may vary. Although all three components are inherent to a good exercise program, not all three need to be present in a home gym. For example, individuals who plan on getting most if not all of their aerobic activity by walking outside may not need an aerobic exercise machine at home. The ACE has some suggestions for building a good home gym. These include knowing the budget, making sure any piece of cardiorespiratory equipment can function continuously and smoothly for 20 min, being able to accommodate the sizes and abilities of all potential users, and making sure equipment is adjustable if needed. Planning for adequate space is important too. Table 18.4 lists the space requirements of individual pieces of home exercise equipment.

Using Exercise Videos, CDs, and DVDs

Sorting through the overabundance of exercise videos, CDs, and DVDs can be challenging. Products created by health professionals with exercise certifications may have a more scientific basis. There currently are no national guidelines in this area. Important features to evaluate before using an exercise video include the participant's fitness level, the type of exercise (cardiorespiratory versus kickboxing versus Pilates) desired, the variety provided, and the cost. Reading online reviews may be helpful, particularly if there are multiple reviews available. Renting before purchasing can also aid in the decision-making process.

The ACE has published recommendations on choosing an exercise video. After determining that home exercise with a video is the desired mode of exercise, an individual should pick a video at the appropriate level of intensity and expertise. The instructor on the video should be certified and experienced. The ACE cautions against buying a video with a celebrity instructor if the celebrity status is the primary selling point. The video should offer options for exercise modifications, if needed. These modifications allow for a gradual increase in the intensity or type of exercise while minimizing the risk for injury from participating in a program that is beyond the appropriate level. The ACE also recommends building a collection of videos that emphasize the different components of the exercise program—aerobics, strength training, and flexibility—in order to achieve overall balance and conditioning. One other issue is logistics; the user should consider the additional equipment (steps, weights, stretch rope, chair, and so on) that might be needed and the space requirements for performing the exercises.

TABLE 18.4 Space Requirements for Selected Pieces of Home Exercise Equipment

Type	Approximate space requirement
Bicycle	10 ft² (1 m²)
Cross-country ski machine	25 ft² (2.3 m²)
Free weights	20-50 ft² (1.9-4.6 m²)
Multistation gym	50-200 ft² (4.6-18.6 m²)
Rowing machine	20 ft² (1.9 m²)
Single-station gym	35 ft² (3.3 m²)
Stair stepper	10-20 ft² (1-1.9 m²)
Treadmill	30 ft² (2.8 m²)

Adapted with permission from ACE Fit Facts How to design your own home gym www.acefitness.org/fitfacts Accessed October 10, 2009.

Exercising in a Workout Facility

Some individuals choose to work out in a facility. Just as there are many options for home equipment and exercise, the range of services and equipment at workout facilities is wide as well. Personal training is another choice often available at a facility. Information on important qualities and qualifications is included below.

Choosing the Right Facility

Many people prefer to exercise in an environment with several choices of exercise equipment, in a group atmosphere, away from home, or in the presence of fitness personnel. According to the International Health, Racquet and Sportsclub Association (IHRSA), there are more than 33 million Americans who belong to one or more than 17,000 health clubs in the United States. For people who want to start or continue a physical activity program in this setting, joining a health club or fitness center is a good decision. The guidelines of both the ACE and ACSM on choosing a health or fitness facility stress selecting a facility that fits the selected goals of the individual. Various health or fitness clubs may offer personal training, personalized fitness assessments, age-appropriate programs, qualified exercise instructors, and specific classes such as aerobics, spinning, martial arts, yoga, Pilates, and strength training. Other considerations include location, hours of operation, costs (including monthly dues, initiation, parking, child care, and so on), qualified staff, state-of-the-art equipment, and accommodation of special needs. Safety criteria include environmental issues such as appropriate lighting, heating, cooling, ventilation, and maintenance; medical and first aid equipment such as automated external defibrillators (AEDs); and qualified staff. Lastly, preparticipation screening should be adequate to ensure that individuals may exercise safely. Both the ACSM and ACE criteria, along with those suggested by the NCPAD, are summarized in the following guidelines. The International Council on Active Aging (ICAA) has a comprehensive checklist that may be accessed online to help older adults choose an appropriate fitness facility. Desired features for the older population are accessibility and safety as well as availability of suitable programs, equipment, and staff. The ICAA age-friendly facility checklist is available at www.icaa.cc/consumer/age-friendlyguides.htm.

GUIDELINE 18.1

Title: *Selecting and Effectively Using a Health/Fitness Facility*

Organization: ACSM

Year published: 2005 (updated 2007)

Purpose: To make an informed decision when choosing a safe health or fitness facility

Location: www.acsm.org/AM/Template.cfm?Section=brochures2

Population: Not given

GUIDELINES

Factor	Suggestions
Safety	Emergency response and evacuation plan posted and able to be executed AEDs on site Clean and well maintained No physical or environmental hazards Appropriately lit, heated, cooled, and ventilated Availability of adequate parking, particularly at peak times

>continued

GUIDELINE 18.1 >continued

Factor	Suggestions
Preactivity screening	Screening, such as PAR-Q, available Availability of health screening method appropriate for the type of exercise Availability of fitness assessments
Personnel	Appropriate education, certification, and training of staff Sufficient numbers of on-site staff members Recognizable, available, and friendly staff members Staff trained in first aid, CPR, and AED use Staff available to instruct in equipment use and exercise goals Staff members receive ongoing professional training
Youth services	Appropriate supervision Child care available, if needed
Programs	Availability of desired programs Programs developed by qualified exercise instructors Modification of programs available Specific needs such as weight loss, hypertension, diabetes, smoking cessation are addressed Age-appropriate activities or programs Fitness assessments or personalized exercise prescription available
Special needs	Any special equipment, facilities, or programs available Conforms to ADA and OSHA standards, codes, and regulations
Business practices	No high-pressure tactics for membership Trial membership available Different membership options Written set of rules and policies available for facility and members Procedures to inform members of changes in policy, charges, services Reasonable membership fee

GUIDELINE 18.2

Title: *How to Choose a Health Club*

Organization: ACE

Year published: 2009

Purpose: To pick a quality health club and to stick to an exercise program

Location: www.acefitness.org/fitfacts/fitfacts_display.aspx?itemid=2583

Population: Not given

GUIDELINES

Factor	Suggestions
Location	Close to either home or work
Classes	Desired selection Class times that fit desired schedule
Staff	Trainers and instructors certified through a nationally recognized certification organization

Factor	Suggestions
Hours	Desired hours of operation
Trial membership	Trial day for testing the club's services
Payments	Meet personal budget requirements Deposits for a new club kept in escrow account until club opening
Reputation	Better Business Bureau member with no registered complaints IHRSA member
Other details	Clean Desired sound level Number of working machines and other equipment Orientation and instruction available for new members

Adapted, by permission from American Council on Exercise Fit Facts, San Diego 2009. www.acefitness.org/fitfacts Accessed 10/10/2009.

GUIDELINE 18.3

Title: *Choosing a Fitness Center*

Organization: NCPAD

Year published: 2006 (updated)

Purpose: To find the right fitness center

Location: www.ncpad.org/exercise/fact_sheet.php?sheet=359

Population: Not given

GUIDELINES

Factor	Suggestions
Location	Choose a facility near home or work. Consider a facility that can be reached by an accessible means of public transportation.
Cost	Avoid signing up for membership that extends beyond 1 y. Read the fine print on the contract regarding fees.
Hours	Visit the facility during the time you are most likely to use it. Make sure the facility's hours match your needs.
Equipment	Look for a variety of well-maintained equipment. Look for well-located equipment that is accessible.
Type	Consider a multipurpose facility versus a gym. Ask what programs, trainers, amenities, courts, and pools are available.
Classes	Look for an adequate number of class sessions to allow for participation. If other programs beyond basic classes are desired, look for their availability.
Staff	Make sure that group or personal orientation to equipment is available at no charge upon joining. Make sure trainers are certified through the ACSM, ACE, or NSCA.

>continued

GUIDELINE 18.3 *>continued*

Factor	Suggestions
Accessibility	Check for parking that is accessible and near the entrance. Make sure elevator, lockers, showers, restrooms, and pool (with lift that can be operated independently if needed) are accessible. Check for cardiorespiratory equipment designed to work only the upper body if needed.

Other selected questions to consider include the following:

- How often is equipment replaced, cleaned, and maintained?
- Do staff members have continuing education or degree requirements?
- What is the busiest time for the facility?
- Do instructors know how to adapt programs for disabilities?
- Is one-on-one staff interaction available and how much would it cost?
- What is the emergency situation protocol? Does the facility have an AED?
- Is parking or child care available, and, if so, for what charge?
- Is medical clearance or health history required for some or all members?
- Are there showers or changing facilities?
- Are service animals allowed in the facility?

Choosing a Personal Trainer

Personal trainers can help a person to design, start, and maintain an exercise program. Although many trainers work for or in a specific fitness facility, others are independent and travel to homes or other locations. There are many different entities that certify personal trainers. The amount of experience, level of education, and other requirements of certification vary based on the organization, and there is no national standard a personal trainer must meet in order to hold this title. As a result, many individuals claiming to be a trainer are not fully qualified to be one. It is important to evaluate the experience and the type of certification a personal trainer has. Appropriate certifying organizations include the ACSM, ACE, National Academy of Sports Medicine (NASM), NSCA, and Cooper Institute. Selected qualities necessary for personal trainer certifications include the following:

- Four-year degree in health and fitness field or related area
- Ongoing continuing education
- Certification in first aid and CPR

After determining that a trainer is appropriately certified, the next goal is to find someone who is a good fit individually. A trainer who can communicate and motivate while still being sensitive to individual needs is important. Some people prefer trainers of a certain sex or a trainer who is very tough and creates demanding, rigorous workouts, while others prefer a trainer who makes conversation and casually guides them through their workouts. While finding a good personality fit plays an important role, it is equally—if not more—important to ensure that the trainer

- is able to screen the participant before exercise to evaluate for medical issues and injuries;
- can offer references and previous experience;
- has a current certification obtained from a nationally recognized organization;
- has contacts and a network with other health professionals such as nutritionists, physical therapists, and physicians;
- has a clearly written policy on contracts, billing, scheduling, and cancellations; and
- can accommodate any special needs such as training children or modifying exercise programs.

SUMMARY

Implementing a physical activity program requires more than just deciding to exercise! Choosing appropriate equipment, tools, location, and aids can make all the difference in having a successful and sustained program. Aerobic and resistance training equipment, exercise monitors, fitness facilities, and personal trainers are some of the variables that require careful consideration.

APPENDIX A

This appendix contains a list of the information sources that are used throughout this book so that the reader can refer to these sources for greater detail if desired. The sources are grouped by topic, and the topics are arranged in the order in which they are discussed in this book. The Web sites listed here not only provide more detailed information but also are the best place to find the most up-to-date physical activity guidelines. This is particularly appropriate because many guidelines are updated on a regular basis.

General Population

- American College of Sports Medicine: www.acsm.org
- Centers for Disease Control and Prevention: www.cdc.gov
- Exercise Is Medicine: www.exerciseismedicine.org
- Healthfinder.gov (links to health-related government agencies): www.healthfinder.gov
- U.S. Physical Activity Statistics: State Legislative Information (searchable database for information on proposed legislation related to nutrition and physical activity from all 50 states): http://apps.nccd.cdc.gov/DNPALeg

Adults

- American College of Sports Medicine Position and Consensus Statements (click on position stands): www.acsm-msse.org/
- American Heart Association: www.americanheart.org
- Australian Government Department of Health and Ageing: www.healthyactive.gov.au
- British Heart Foundation: www.bhf.org.uk/
- *Dietary Guidelines for Americans:* www.health.gov/dietaryguidelines/dga2005/document/html/chapter4.htm
- Health Canada, *Canada's Physical Activity Guide to Healthy Active Living:* www.paguide.com
- HealthierUS.gov: www.healthierus.gov
- National Heart, Lung, and Blood Institute: www.nhlbi.nih.gov
- *Physical Activity Guidelines for Americans:* www.health.gov/PAGuidelines
- President's Council on Physical Fitness and Sports: www.fitness.gov
- The Compendium of Physical Activities Tracking Guide (MET values): http://prevention.sph.sc.edu/tools/compendium.htm
- The Obesity Society: www.obesity.org
- The President's Challenge: www.presidentschallenge.org

- World Health Organization (on diet and physical activity information): www.who.int/dietphysicalactivity/en/

Personal Exercise Prescription

- America on the Move: www.americaonthemove.org
- American Council on Exercise *Fit Facts:* www.acefitness.org/fitfacts
- ExRx.net: www.exrx.net
- IDEA Health & Fitness Association: www.ideafit.com
- National Strength and Conditioning Association: www.nsca.com
- President's Council on Physical Fitness and Sports: www.fitness.gov
- Shape Up America! Healthy Weight for Life: www.shapeup.org

Children and Youth

- Action for Healthy Kids: www.actionforhealthykids.org
- American Academy of Pediatrics: http://aappolicy.aappublications.org
- American Council on Exercise *Fit Facts* (Youth Fitness section, especially "Strength Training for Kids: A Guide for Parents and Teachers"): www.acefitness.org/fitfacts
- American Public Health Association Tools for Parents, Teachers, Students, and Community Leaders to Eliminate Childhood Obesity: www.apha.org/programs/resources/obesity/obesityparenttools.htm
- Australian Government Department of Health and Ageing: www.healthyactive.gov.au
- California Center for Physical Activity's Guidelines for Physical Activity Across the Lifespan: www.caphysicalactivity.org/facts_recomm1.html
- *Canada's Physical Activity Guide to Healthy Active Living Family Guide:* www.phac-aspc.gc.ca/pau-uap/paguide/child_youth/pdf/yth_family_guide_e.pdf
- *CATCH for Improved Physical Activity and Diet in Elementary School Children:* www.cdc.gov/prc/prevention-strategies/adoptable-interventions/catch-improved-physical-activity-diet-elementary-school.htm
- Centers for Disease Control and Prevention Division of Nutrition, Physical Activity and Obesity: www.cdc.gov/nccdphp/dnpa/policy/nutrition.htm
- Centers for Disease Control and Prevention Growth Charts (BMI-for-age percentiles): www.cdc.gov/growthcharts
- Centers for Disease Control and Prevention Guidelines for School and Community Programs to Promote Lifelong Physical Activity Among Young People: www.cdc.gov/HealthyYouth/physicalactivity/guidelines
- Centers for Disease Control and Prevention National Center for Chronic Disease Prevention and Health Promotion Division of Nutrition, Physical Activity and Obesity: www.cdc.gov/nccdphp/dnpa/about.htm
- Children's Nutrition Research Center: www.kidsnutrition.org
- Eat Smart. Play Hard.: http://teamnutrition.usda.gov/Resources/eatsmartmaterials.html
- Institute of Medicine Preventing Childhood Obesity: Health in the Balance: www.nap.edu/catalog.php?record_id=11015#toc
- Institute of Medicine Prevention of Obesity in Children and Youth: www.iom.edu/Activities/Children/ObesPrevention.aspx
- Maternal and Child Health Library (list of resources): www.mchlibrary.info/KnowledgePaths/kp_phys_activity.html

- MyPyramid for Kids: http://mypyramid.gov/kids/index.html
- MyPyramid for Preschoolers: http://mypyramid.gov/preschoolers
- National Association for Sport and Physical Education: www.aahperd.org/naspe/
- National Coalition for Promoting Physical Activity (model program and policy efforts addressing physical activity among youths): www.ncppa.org
- *Physical Activity and Health: A Report of the Surgeon General:* www.cdc.gov/nccdphp/sgr/sgr.htm
- *Physical Activity Guidelines for Americans:* www.health.gov/PAGuidelines
- President's Council on Physical Fitness and Sports (links to resources and information about The President's Challenge and The President's Challenge Physical Activity and Fitness Awards Program): www.fitness.gov
- Office of the Surgeon General Childhood Overweight and Obesity Prevention Initiative: www.surgeongeneral.gov/obesityprevention/
- Sports, Play and Active Recreation for Kids (SPARK): www.sparkpe.org
- The President's Challenge (award for achieving physical activity goals): www.presidentschallenge.org
- VERB Youth Media Campaign: www.cdc.gov/youthcampaign/
- We Can! Ways to Enhance Children's Activity & Nutrition: www.nhlbi.nih.gov/health/public/heart/obesity/wecan

Pregnancy
- American College of Obstetricians and Gynecologists: www.acog.org
- American College of Sports Medicine Position and Consensus Statements (click on position stands): www.acsm-msse.org/
- Maternal and Child Health Library (list of resources): www.mchlibrary.info/KnowledgePaths/kp_phys_activity.html
- National Women's Health Information Center: www.4woman.gov
- *Physical Activity Guidelines for Americans:* www.health.gov/PAGuidelines
- Royal College of Obstetricians and Gynaecologists: www.rcog.org.uk
- Society of Obstetricians and Gynaecologists of Canada: www.sogc.org/

Older Adults
- American Academy of Family Physicians Exercise for the Elderly: www.aafp.org/afp/20020201/427ph.html
- American Association of Retired Persons (physical activity resources): www.aarp.org/health/fitness/
- American College of Sports Medicine Active Aging Partnership and the Strategic Health Initiative on Aging: www.agingblueprint.org/tips.cfm
- American College of Sports Medicine Position and Consensus Statements: www.acsm-msse.org/
- Health Canada, *Canada's Physical Activity Guide to Healthy Active Living:* www.paguide.com
- Healthy Aging: www.healthyaging.net
- International Council on Active Aging: ICAA.cc/
- Lifelong Fitness Alliance: www.50plus.org
- London Department of Health: www.bhfactive.org.uk/older-adults/publications.html#ALL
- National Blueprint: Increasing Physical Activity Among Adults Aged 50 and Older: www.agingblueprint.org/overview.cfm

- National Center on Physical Activity and Disability: www.ncpad.org
- National Council on Aging: www.healthyagingprograms.org/
- National Institute on Aging Exercise for Older Adults: www.nihseniorhealth.gov/exercise/toc.html
- Pep Up Your Life: www.fitness.gov/pepup.htm
- *Physical Activity Guidelines for Americans:* www.health.gov/PAGuidelines
- The State of Aging and Health in America 2007: www.cdc.gov/aging/saha.htm
- *What You Need to Know About Balance and Falls:* www.apta.org/AM/Images/APTAIMAGES/ContentImages/ptandbody/balance/BalanceFall.pdf
- World Health Organization Ageing and Life Course: www.who.int/hpr/ageing/publications.htm#Active%20Ageing

Diabetes
- American Diabetes Association: www.diabetes.org
- FamilyDoctor.org Diabetes and Exercise: http://familydoctor.org/online/famdocen/home/common/diabetes/living/351.htm
- National Diabetes Information Clearinghouse: www.diabetes.niddk.nih.gov
- National Institute of Diabetes and Digestive and Kidney Diseases: www.niddk.nih.gov

Cancer
- American Cancer Society: www.cancer.org
- American Cancer Society Guidelines on Nutrition and Physical Activity for Cancer Prevention: http://caonline.amcancersoc.org/cgi/reprint/56/5/254
- American Cancer Society Nutrition and Physical Activity During and After Cancer Treatment: http://caonline.amcancersoc.org/cgi/content/full/56/6/323
- American Institute for Cancer Research: www.aicr.org
- National Cancer Institute: www.cancer.gov
- World Cancer Research Fund: www.wcrf.org

Hypertension and Coronary Artery Disease
- American Heart Association: www.americanheart.org
- British Hypertension Society Guidelines for Management of Hypertension: www.bhsoc.org/pdfs/BHS_IV_Guidelines.pdf
- Canadian Hypertension Education Program: www.hypertension.ca/
- DASH Eating Plan: www.nhlbi.nih.gov/health/public/heart/hbp/dash/new_dash.pdf
- Your Guide to Lowering High Blood Pressure: www.nhlbi.nih.gov/hbp/index.html

Arthritis
- American Council on Exercise *Fit Facts:* www.acefitness.org/fitfacts
- Arthritis Foundation: www.arthritis.org
- Johns Hopkins Arthritis Center: www.hopkins-arthritis.org
- National Guideline Clearinghouse: www.guideline.gov
- National Institute of Arthritis and Musculoskeletal and Skin Diseases: www.niams.nih.gov
- National Osteoporosis Foundation: www.nof.org/osteoporosis

Osteoporosis
- *Bone Health and Osteoporosis: A Report of the Surgeon General:* www.surgeongeneral.gov/library/bonehealth/docs/exec_summ.pdf

- FRAX: www.shef.ac.uk/FRAX
- International Osteoporosis Foundation: www.osteofound.org
- National Osteoporosis Foundation: www.nof.org

Asthma

- American Academy of Allergy Asthma and Immunology: www.aaaai.org
- Guidelines for the Diagnosis and Management of Asthma: www.nhlbi.nih.gov/guidelines/asthma

Disability

- American Heart Association: www.americanheart.org
- Muscular Dystrophy Association: www.mda.org/disease/
- National Center on Physical Activity and Disability: www.ncpad.org
- National Institute of Neurological Disorders and Stroke: www.ninds.nih.gov/disorders
- National Parkinson Foundation: www.parkinson.org
- National Rehabilitation Information Center: www.naric.com
- United Cerebral Palsy: www.ucp.org

Exercise and Cardiac Testing

- American College of Sports Medicine: www.acsm.org
- American Heart Association: www.americanheart.org
- Body Mass Index Table: www.nhlbi.nih.gov/guidelines/obesity/bmi_tbl.pdf

Nutrition

- American Cancer Society Guidelines on Nutrition and Physical Activity for Cancer Prevention: http://caonline.amcancersoc.org/cgi/content/full/56/5/254
- American Cancer Society Nutrition and Physical Activity During and After Cancer Treatment: http://caonline.amcancersoc.org/cgi/content/full/56/6/323
- American College of Sports Medicine Position and Consensus Statements (click on position stands): www.acsm-msse.org/
- American Council on Exercise *Fit Facts* (Nutrition and Supplements section): www.acefitness.org/fitfacts
- Body Mass Index Table: www.nhlbi.nih.gov/guidelines/obesity/bmi_tbl.pdf
- *Canada's Food Guide:* www.hc-sc.gc.ca/fn-an/food-guide-aliment/index-eng.php
- Centers for Disease Control and Prevention Growth Charts (BMI-for-age percentiles): www.cdc.gov/growthcharts
- *Dietary Guidelines for Americans:* www.health.gov/dietaryguidelines/dga2005
- *Dietary Reference Intakes for Energy, Carbohydrate, Fiber, Fat, Fatty Acids, Cholesterol, Protein, and Amino Acids:* www.nap.edu/catalog/10490.html
- National Heart, Lung, and Blood Institute (DASH eating plan): www.nhlbi.nih.gov/health/public/heart/hbp/dash/
- Nutrition.gov: www.nutrition.gov
- U.S. Department of Agriculture Food Guide Pyramid: www.mypyramid.gov/

Physical Activity and Exercise Peripherals

- American Council on Exercise: www.acefitness.org
- American Council on Exercise *Fit Facts* (Getting Started section): www.acefitness.org/fitfacts

- American College of Sports Medicine (brochures on choosing workout equipment): www.acsm.org/AM/Template.cfm?Section=Brochures2
- *Consumer Reports:* www.consumerreports.org
- International Health, Racquet and Sportsclub Association: http://cms.ihrsa.org
- National Academy of Sports Medicine: www.nasm.org
- National Strength and Conditioning Association: www.nsca.com
- The Cooper Institute: www.cooperinst.org

APPENDIX B

This appendix contains selected tables of information central to the text. These tables provide examples of different measures of activity, such as PALs and METs, that are described in the preceding chapters and are fundamental to understanding the formation of many of the guidelines covered within the book. The PAR-Q is included because it is perhaps the most commonly used approach for clearing a person to participate in a physical activity program. The BMI table is included to provide a convenient at-a-glance guide that circumvents the need for converting units or calculating height and weight.

Estimated MET Levels for Selected Physical Activities

METs	Category	Activity
0.9	Inactivity	Sleeping
1.0	Inactivity	Sitting quietly and watching television
2.0	Transportation	Driving an automobile or light truck
3.0	Walking	Walking very slowly, strolling, household walking
4.0	Lawn and garden	Raking the lawn, gardening
5.0	Home repair	Cleaning gutters, painting outside of home
6.0	Occupation	Using heavy power tools (jackhammer)
7.0	Conditioning	Using stationary bicycle, ski machine, or rowing machine
8.0	Sports	Playing in a competitive basketball game, playing touch football
9.0	Walking	Climbing hills with a 42 lb (19 kg) backpack
10.0	Water	Freestyle lap swimming, using vigorous effort
11.0	Running	Running 1 mi (1.6 km) in 9 min
12.0	Bicycling	Road cycling at 14-16 mph (23-26 kph), fast or general racing
13.0-14.0	Running	Running 1 mi (1.6 km) in 7.0-7.5 min
15.0	Winter sports	Speed skating competitively

Reprinted from President's Council on Physical Fitness and Sports, 2003, *Research Digest*, Series 4, No. 2. Available: www.fitness.gov/publications/digests/digest-june2003.pdf

Physical Activity Level (PAL) Categories

PAL category	PAL value	Walking equivalent at 3-4 mph (5-6 kph) daily for a 154 lb (70 kg) person
Sedentary	1-<1.4	0 mi (0 km)
Low active	1.4-<1.6	2.2 mi (3.5 km)
Active	1.6-<1.9	7.3 mi (11.7 km)
Very active	1.9-2.5	16.7 mi (26.9 km)

PAL is the ratio of total energy expenditure (TEE) to basal energy expenditure (BEE).

Adapted, with permission, from *American Journal of Clinical Nutrition* Vol. 79, No. 5, 921S-930S, May 2004, "Chronicle of the Institute of Medicine physical activity recommendation: How a physical activity recommendation came to be among dietary recommendations," by George A Brooks, Nancy F Butte, William M Rand, Jean-Pierre Flatt and Benjamin Caballero.

Physical Activity Readiness
Questionnaire - PAR-Q
(revised 2002)

PAR-Q & YOU

(A Questionnaire for People Aged 15 to 69)

Regular physical activity is fun and healthy, and increasingly more people are starting to become more active every day. Being more active is very safe for most people. However, some people should check with their doctor before they start becoming much more physically active.

If you are planning to become much more physically active than you are now, start by answering the seven questions in the box below. If you are between the ages of 15 and 69, the PAR-Q will tell you if you should check with your doctor before you start. If you are over 69 years of age, and you are not used to being very active, check with your doctor.

Common sense is your best guide when you answer these questions. Please read the questions carefully and answer each one honestly: check YES or NO.

YES	NO		
☐	☐	1.	Has your doctor ever said that you have a heart condition __and__ that you should only do physical activity recommended by a doctor?
☐	☐	2.	Do you feel pain in your chest when you do physical activity?
☐	☐	3.	In the past month, have you had chest pain when you were not doing physical activity?
☐	☐	4.	Do you lose your balance because of dizziness or do you ever lose consciousness?
☐	☐	5.	Do you have a bone or joint problem (for example, back, knee or hip) that could be made worse by a change in your physical activity?
☐	☐	6.	Is your doctor currently prescribing drugs (for example, water pills) for your blood pressure or heart condition?
☐	☐	7.	Do you know of __any other reason__ why you should not do physical activity?

If you answered

YES to one or more questions

Talk with your doctor by phone or in person BEFORE you start becoming much more physically active or BEFORE you have a fitness appraisal. Tell your doctor about the PAR-Q and which questions you answered YES.

- You may be able to do any activity you want — as long as you start slowly and build up gradually. Or, you may need to restrict your activities to those which are safe for you. Talk with your doctor about the kinds of activities you wish to participate in and follow his/her advice.
- Find out which community programs are safe and helpful for you.

NO to all questions

If you answered NO honestly to all PAR-Q questions, you can be reasonably sure that you can:
- start becoming much more physically active — begin slowly and build up gradually. This is the safest and easiest way to go.
- take part in a fitness appraisal — this is an excellent way to determine your basic fitness so that you can plan the best way for you to live actively. It is also highly recommended that you have your blood pressure evaluated. If your reading is over 144/94, talk with your doctor before you start becoming much more physically active.

DELAY BECOMING MUCH MORE ACTIVE:
- if you are not feeling well because of a temporary illness such as a cold or a fever — wait until you feel better; or
- if you are or may be pregnant — talk to your doctor before you start becoming more active.

PLEASE NOTE: If your health changes so that you then answer YES to any of the above questions, tell your fitness or health professional. Ask whether you should change your physical activity plan.

Informed Use of the PAR-Q: The Canadian Society for Exercise Physiology, Health Canada, and their agents assume no liability for persons who undertake physical activity, and if in doubt after completing this questionnaire, consult your doctor prior to physical activity.

No changes permitted. You are encouraged to photocopy the PAR-Q but only if you use the entire form.

NOTE: If the PAR-Q is being given to a person before he or she participates in a physical activity program or a fitness appraisal, this section may be used for legal or administrative purposes.

"I have read, understood and completed this questionnaire. Any questions I had were answered to my full satisfaction."

NAME _____

SIGNATURE _____ DATE_____

SIGNATURE OF PARENT _____ WITNESS _____
or GUARDIAN (for participants under the age of majority)

> **Note: This physical activity clearance is valid for a maximum of 12 months from the date it is completed and becomes invalid if your condition changes so that you would answer YES to any of the seven questions.**

CSEP
SCPE © Canadian Society for Exercise Physiology   Supported by: [🍁] Health   Santé
Canada   Canada

continued on other side...

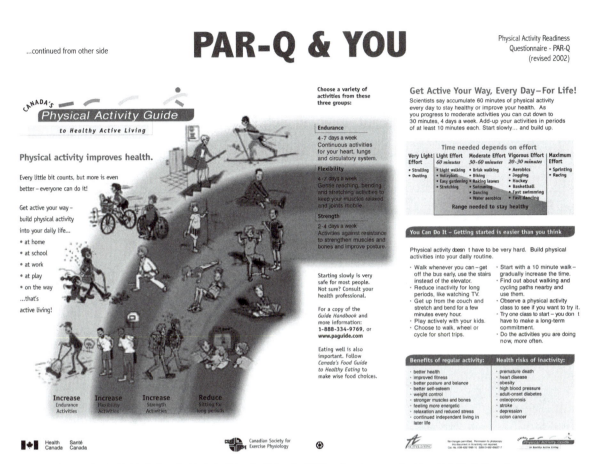

Source: Canada's Physical Activity Guide to Healthy Active Living, Health Canada, 1998 http://www.hc-sc.gc.ca/hppb/paguide/pdf/guideEng.pdf
© Reproduced with permission from the Minister of Public Works and Government Services Canada, 2002.

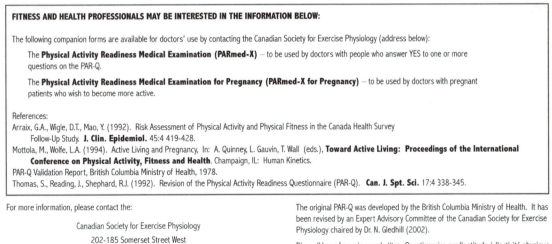

Body Mass Index

Weight (lb)	HEIGHT (IN.) 48	49	50	51	52	53	54	55	56	57	58	59	60	61	62	63	Weight (kg)
100	30.6	29.3	28.2	27.1	26.1	25.1	24.2	23.3	22.5	21.7	21.0	20.2	19.6	18.9	18.3	17.8	45.5
105	32.1	30.8	29.6	28.4	27.4	26.3	25.4	24.5	23.6	22.8	22.0	21.3	20.5	19.9	19.2	18.6	47.7
110	33.6	32.3	31.0	29.8	28.7	27.6	26.6	25.6	24.7	23.9	23.0	22.3	21.5	20.8	20.2	19.5	50.0
115	35.2	33.7	32.4	31.2	30.0	28.8	27.8	26.8	25.8	24.9	24.1	23.2	22.5	21.8	21.1	20.4	52.3
120	36.7	35.2	33.8	32.5	31.3	30.1	29.0	27.9	27.0	26.0	25.1	24.3	23.5	22.7	22.0	21.3	54.5
125	38.2	36.7	35.2	33.9	32.6	31.4	30.2	29.1	28.1	27.1	26.2	25.3	24.5	23.7	22.9	22.2	56.8
130	39.8	38.1	36.6	35.2	33.9	32.6	31.4	30.3	29.2	28.2	27.2	26.3	25.4	24.6	23.8	23.1	59.1
135	41.3	39.6	38.0	36.6	35.2	33.9	32.6	31.4	30.3	29.3	28.3	27.3	26.4	25.6	24.7	24.0	61.4
140	42.8	41.1	39.5	37.9	36.5	35.1	33.8	32.6	31.5	30.4	29.3	28.3	27.4	26.5	25.7	24.9	63.6
145	44.3	42.5	40.9	39.3	37.8	36.4	35.0	33.8	32.6	31.4	30.4	29.3	28.4	27.5	26.6	25.7	65.9
150	45.9	44.0	42.3	40.6	39.1	37.6	36.2	34.9	33.7	32.5	31.4	30.4	29.4	28.4	27.5	26.6	68.2
155	47.4	45.5	43.7	42.0	40.4	38.9	37.5	36.1	34.8	33.6	32.5	31.4	30.3	29.3	28.4	27.5	70.5
160	48.9	47.0	45.1	43.3	41.7	40.1	38.7	37.3	35.9	34.7	33.5	32.4	31.3	30.3	29.3	28.4	72.7
165	50.5	48.4	46.5	44.7	43.0	41.4	39.9	38.4	37.1	35.8	34.6	33.4	32.3	31.2	30.2	29.3	75.0
170	52.0	49.9	47.9	46.0	44.3	42.6	41.1	39.6	38.2	36.9	35.6	34.4	33.3	32.3	31.2	30.2	77.3
175	53.5	51.4	49.3	47.4	45.6	43.9	42.3	40.8	39.3	37.9	36.7	35.4	34.2	33.1	32.1	31.1	79.5
180	55.0	52.8	50.7	48.8	46.9	45.1	43.5	41.9	40.4	39.0	37.7	36.4	35.2	34.1	33.0	32.0	81.8
185	56.6	54.3	52.1	50.1	48.2	46.4	44.7	43.1	41.6	40.1	38.7	37.4	36.2	35.0	33.9	32.8	84.1
190	58.1	55.8	53.5	51.5	49.5	47.7	45.9	44.3	42.7	41.2	39.8	38.5	37.2	36.0	34.8	33.7	86.4
195	59.6	57.2	55.0	52.8	50.8	48.9	47.1	45.4	43.8	42.3	40.8	39.5	38.2	36.9	35.7	34.6	88.6
200	61.2	58.7	56.4	54.2	52.1	50.2	48.3	46.6	44.9	43.4	41.9	40.5	39.1	37.9	36.7	35.5	90.9
205	62.7	60.2	57.8	55.5	53.4	51.4	49.5	47.7	46.1	44.5	42.9	41.5	40.1	38.8	37.6	36.4	93.2
210	64.2	61.6	59.2	56.9	54.7	52.7	50.7	48.9	47.2	45.5	44.0	42.5	41.1	39.8	38.5	37.3	95.5
215	65.7	63.1	60.6	58.2	56.0	53.9	51.9	50.1	48.3	46.6	45.0	43.5	42.1	40.7	39.4	38.2	97.7
220	67.3	64.6	62.0	59.6	57.3	55.2	53.2	51.2	49.4	47.7	46.1	44.5	43.1	41.7	40.3	39.1	100.0
225	68.8	66.0	63.4	60.9	58.6	56.4	54.4	52.4	50.5	48.8	47.1	45.5	44.0	42.6	41.2	39.9	102.3
230	70.3	67.5	64.8	62.3	59.9	57.7	55.6	53.6	51.7	49.9	48.2	46.6	45.0	43.5	42.2	40.8	104.5
235	71.9	69.0	66.2	63.7	61.2	58.9	56.8	54.7	52.8	51.0	49.2	47.6	46.0	44.5	43.1	41.7	106.8
240	73.4	70.4	67.6	65.0	62.5	60.2	58.0	55.9	53.9	52.0	50.3	48.6	47.0	45.4	44.0	42.6	109.1
245	74.9	71.9	69.0	66.4	63.8	61.5	59.2	57.1	55.0	53.1	51.3	49.6	47.9	46.4	44.9	43.5	111.4
250	76.4	73.4	70.5	67.7	65.1	62.7	60.4	58.2	56.2	54.2	52.4	50.6	48.9	47.3	45.8	44.4	113.6
Height (m)	1.22	1.24	1.27	1.30	1.32	1.35	1.37	1.40	1.42	1.45	1.47	1.50	1.52	1.55	1.57	1.60	

HEIGHT (IN.)

Weight (lb)	64	65	66	67	68	69	70	71	72	73	74	75	76	77	78	Weight (kg)
100	17.2	16.7	16.2	15.7	15.2	14.8	14.4	14.0	13.6	13.2	12.9	12.5	12.2	11.9	11.6	45.5
105	18.1	17.5	17.0	16.5	16.0	15.5	15.1	14.7	14.3	13.9	13.5	13.2	12.8	12.5	12.2	47.7
110	18.9	18.3	17.8	17.3	16.8	16.3	15.8	15.4	14.9	14.5	14.2	13.8	13.4	13.1	12.7	50.0
115	19.8	19.2	18.6	18.0	17.5	17.0	16.5	16.1	15.6	15.2	14.8	14.4	14.0	13.7	13.3	52.3
120	20.6	20.0	19.4	18.8	18.3	17.8	17.3	16.8	16.3	15.9	15.4	15.0	14.6	14.3	13.9	54.5
125	21.5	20.8	20.2	19.6	19.0	18.5	18.0	17.5	17.0	16.5	16.1	15.7	15.2	14.9	14.5	56.8
130	22.4	21.7	21.0	20.4	19.8	19.2	18.7	18.2	17.7	17.2	16.7	16.3	15.9	15.4	15.1	59.1
135	23.2	22.5	21.8	21.2	20.6	20.0	19.4	18.9	18.3	17.8	17.4	16.9	16.5	16.0	15.6	61.4
140	24.1	23.3	22.6	22.0	21.3	20.7	20.1	19.6	19.0	18.5	18.0	17.5	17.1	16.6	16.2	63.6
145	24.9	24.2	23.5	22.8	22.1	21.5	20.8	20.3	19.7	19.2	18.7	18.2	17.7	17.2	16.8	65.9
150	25.8	25.0	24.3	23.5	22.9	22.2	21.6	21.0	20.4	19.8	19.3	18.8	18.3	17.8	17.4	68.2
155	26.7	25.8	25.1	24.3	23.6	22.9	22.3	21.7	21.1	20.5	19.9	19.4	18.9	18.4	17.9	70.5
160	27.5	26.7	25.9	25.1	24.4	23.7	23.0	22.4	21.7	21.2	20.6	20.0	19.5	19.0	18.5	72.7
165	28.4	27.5	26.7	25.9	25.1	24.4	23.7	23.1	22.4	21.8	21.2	20.7	20.1	19.6	19.1	75.0
170	29.2	28.3	27.5	26.7	25.9	25.2	24.4	23.8	23.1	22.5	21.9	21.3	20.7	20.2	19.7	77.3
175	30.1	29.2	28.3	27.5	26.7	25.9	25.2	24.5	23.8	23.1	22.5	21.9	21.3	20.8	20.3	79.5
180	31.0	30.0	29.1	28.3	27.4	26.6	25.9	25.2	24.5	23.8	23.2	22.5	22.0	21.4	20.8	81.8
185	31.8	30.8	29.9	29.0	28.2	27.4	26.6	25.9	25.1	24.5	23.8	23.2	22.6	22.0	21.4	84.1
190	32.7	31.7	30.7	29.8	28.9	28.1	27.3	26.6	25.8	25.1	24.4	23.8	23.2	22.6	22.0	86.4
195	33.5	32.5	31.5	30.6	29.7	28.9	28.0	27.3	26.5	25.8	25.1	24.4	23.8	23.2	22.6	88.6
200	34.4	33.4	32.3	31.4	30.5	29.6	28.8	28.0	27.2	26.4	25.7	25.1	24.4	23.8	23.2	90.9
205	35.3	34.2	33.2	32.2	31.2	30.3	29.5	28.7	27.9	27.1	26.4	25.7	25.0	24.4	23.7	93.2
210	36.1	35.0	34.0	33.0	32.0	31.1	30.2	29.4	28.5	27.8	27.0	26.3	25.6	25.0	24.3	95.5
215	37.0	35.9	34.8	33.7	32.8	31.8	30.9	30.0	29.2	28.4	27.7	26.9	26.2	25.5	24.9	97.7
220	37.8	36.7	35.6	34.5	33.5	32.6	31.6	30.7	29.9	29.1	28.3	27.6	26.8	26.1	25.5	100.0
225	38.7	37.5	36.4	35.3	34.3	33.3	32.4	31.4	30.6	29.7	28.9	28.2	27.4	26.7	26.1	102.3
230	39.6	38.4	37.2	36.1	35.0	34.0	33.1	32.1	31.3	30.4	29.6	28.8	28.1	27.3	26.6	104.5
235	40.4	39.2	38.0	36.9	35.8	34.8	33.8	32.8	31.9	31.1	30.2	29.4	28.7	27.9	27.2	106.8
240	41.3	40.0	38.8	37.7	36.6	35.5	34.5	33.5	32.6	31.7	30.9	30.1	29.3	28.5	27.8	109.1
245	42.1	40.9	39.6	38.5	37.3	36.3	35.2	34.2	33.3	32.4	31.5	30.7	29.9	29.1	28.4	111.4
250	43.0	41.7	40.4	39.2	38.1	37.0	35.9	34.9	34.0	33.1	32.2	31.3	30.5	29.7	29.0	113.6
Height (m)	1.63	1.65	1.68	1.7	1.73	1.75	1.78	1.80	1.83	1.85	1.88	1.91	1.93	1.96	1.98	

Reprinted, by permission, from J. Morrow, A. Jackson, J. Disch, and D. Mood, 2005, Measurement and evaluation in human performance, 3rd ed. (Champaign, IL: Human Kinetics), 242-243; Adapted from Gold-ing, Myers, and Sinning 1989.

REFERENCES

Chapter 1

Abbott RD, White LR, Ross GW, Masaki KH, Curb JD, Petrovitch H. Walking and dementia in physically capable elderly men. *JAMA* 292: 1447-1453, 2004.

A burden at work. *Businessweek* 44. 2008 March 17.

[ACSM] American College of Sports Medicine. The recommended quantity and quality of exercise for developing and maintaining fitness in healthy adults. *Med Sci Sports Exerc* 10: vii-x, 1978.

[ACSM] American College of Sports Medicine. Position stand: Exercise for patients with coronary artery disease. *Med Sci Sports Exerc* 26: i-v, 1994.

[ACSM] American College of Sports Medicine. Position stand: The recommended quantity and quality of exercise for developing and maintaining cardiorespiratory and muscular fitness in healthy adults. *Med Sci Sports Exerc* 22: 265-274, 1990.

[ACSM] American College of Sports Medicine. Position stand: The recommended quantity and quality of exercise for developing and maintaining cardiorespiratory and muscular fitness, and flexibility in healthy adults. *Med Sci Sports Exerc* 30(6): 975-991, 1998.

Ades PA, Pashkow FJ, Nestor JR. Cost-effectiveness of cardiac rehabilitation after myocardial infarction. *J Cardiopulm Rehabil* 17: 222-231, 1997.

[AHA] American Heart Association. Medical/scientific statement on exercise: Benefits and recommendations for physical activity for all Americans. *Circulation* 85(1): 2726-2730, 1992.

Ainsworth BE, Haskell WL, Whitt MC, Irwin ML, Swartz AM, Strath SJ, O'Brien WL, Bassett DR Jr, Schmitz KH, Emplaincourt PO, et al. Compendium of physical activities: An update of activity codes and MET intensities. *Med Sci Sports Exerc* 32: S498-S516, 2000.

Bardia A, Hartmann LC, Vachon CM, Vierkant RA, Wang AH, Olson JE, Sellers TA, Cerhan JR. Recreational physical activity and risk of postmenopausal breast cancer based on hormone receptor status. *Arch Intern Med* 166: 2478-2483, 2006.

Barnes DE, Yaffe K, Satariano WA, Tager IB. A longitudinal study of cardiorespiratory fitness and cognitive function in healthy older adults. *J Am Geriatr Soc* 51(4): 459-465, 2003.

Blair SN, Kohl HW, Barlow CE, Paffenbarger RS Jr, Gibbons LW, Macera CA. Changes in physical fitness and all-cause mortality: A prospective study of healthy and unhealthy men. *JAMA* 273: 1093-1098, 1995.

Blair SN, Kohl HW, Paffenbarger RS, Clark DG, Cooper KH, Gibbons LW. Physical fitness and all-cause mortality. *JAMA* 262: 2395-2401, 1989.

Brown M, Sinacore DR, Host HH. The relationship of strength to function in the older adult. *J Gerontol* 50A: 55-59, 1995.

Cady LD. Programs for increasing health and physical fitness of firefighters. *J Occup Med* 27: 110-114, 1985.

Carnethon MR, Gulati M, Greenland P. Prevalence and cardiovascular disease correlates of low cardiorespiratory fitness in adolescents and adults. *JAMA* 294: 2981-2988, 2005.

Caspersen CJ, Pereira MA, Curran KM. Changes in physical activity patterns in the United States, by sex and cross-sectional age. *Med Sci Sports Exerc* 32(9): 1601-1609, 2000.

Caspersen, CJ, Powell KE, Christenson GM. Physical activity, exercise, and physical fitness: Definitions and distinctions for health-related research. *Public Health Rep* 100: 126-131, 1985.

Chisholm DM, Collis ML, Kulak LL, Davenport W, Gruber N. Physical activity readiness. *British Columbia Medical Journal* 17: 375-378, 1975.

Church TS, Earnest CP, Skinner JS, Blair SN. Effects of different doses of physical activity on cardiorespiratory fitness among sedentary, overweight or obese postmenopausal women with elevated blood pressure: A randomized controlled trial. *JAMA* 297: 2081-2091, 2007.

Cobb LA, Weaver WD. Exercise: A risk for sudden death in patients with coronary heart disease. *J Am Coll Cardiol* 7: 215-219, 1986.

Courneya KS, Mackey JR, Jones LW. Coping with cancer: Can exercise help? *Phys Sportsmed* 28: 49-73, 2000.

Dallal CM, Sullivan-Halley J, Ross RK, Wang Y, Deapen D, Horn-Ross PL, Reynolds P, Stram DO, Clarke CA, Anton-Culver H, et al. Long-term recreational physical activity and risk of invasive and in situ breast cancer: The California teachers study. *Arch Intern Med* 167(4): 408-415, 2007.

Ehrman JK. General exercise prescription development. In: Ehrman JK, Gordon PM, Visich PS, Keteyian SJ, eds. *Clinical Exercise Physiology*. Champaign (IL): Human Kinetics, 2003.

Franco OH, de Laet C, Peeters A, Jonker J, Mackenbach J, Nusselder W. Effects of physical activity on life expectancy with cardiovascular disease. *Arch Intern Med* 165: 2355-2360, 2005.

Franklin BA, Conviser JM, Stewart B, Lasch J, Timmis GC. Sporadic exercise: A trigger for acute cardiovascular events? *Circulation* 102(suppl 2): II612, 2000.

Garcia AW, King AC. Predicting long-term adherence to aerobic exercise: A comparison of two models. *J Sport Exerc Psychol* 13: 394-410, 1991.

Gettman LR. Cost-benefit analysis of a corporate fitness program. *Fitness in Business* 1: 11-17, 1986.

Gettman LR. Economic benefits of physical activity. *PCPFS Research Digest* 2(7): 1-6, 1996.

Hahn RA, Teuesch SM, Rothenberg RB, Marks JS. Excess deaths from nine chronic diseases in the United States, 1986. *JAMA* 264(20): 2554-2559, 1998.

Haskell W, Lee IM, Pate RR, Powell KE, Blair SN, Franklin BA, Macera CA, Heath GW, Thompson PD, Bauman A. Physical activity and public health: Updated recommendation for adults. *Med Sci Sports Exerc* 39(8): 1423-1434, 2007.

Holt J, Holt LE, Pelham TW. Flexibility redefined. In: Bauer T, ed. *Biomechanics in sports XIII*. Thunder Bay (ON): Lakehead University. 170-174. 1996.

Iellamo F, Legramante JM, Massaro M, Raimondi G, Galante A. Effects of a residential exercise training on baroreflex sensitivity and heart rate variability in patients with coronary artery disease. A randomized, controlled study. *Circulation* 102: 2588-2592, 2000.

Kostka T, Berthouse SE, Lacour J, Bonnefoy M. The symptomatology of upper respiratory tract infections and exercise in elderly people. *Med Sci Sports Exerc* 32: 46-51, 2000.

Knudson DV, Magnusson P, McHugh M. Current issues in flexibility fitness. *PCPFS Research Digest* 3(10): 1-8, 2000.

Larson EB, Wang L, Bowen JD, McCormick WC, Teri L, Crane P, Kukull W. Exercise is associated with reduced risk for incident dementia among person 65 years of age and older. *Ann Intern Med* 144(2): 73-81, 2006.

Laukkanen JA, Lakka TA, Rauramaa R, Kuhanen R, Venalainen JM, Salonen R, Salonen JT. Cardiovascular fitness as a predictor of mortality in men. *Arch Intern Med* 161: 825-831, 2001.

Lautenschlager NT, Cox KL, Flicker L, Foster JK, van Bockxmeer FM, Xiao J, Greenop KR, Almeida OP. Effect of physical activity on cognitive function in older adults at risk for Alzheimer Disease. *JAMA* 300(9): 1027-1037, 2008.

Leitzmann MF, Rimm EB, Willett WC, Spiegelman D, Grodstein F, Stampfer MJ, Colditz GA, Giovannucci E. Recreational physical activity and the risk of cholecystectomy in women. *N Engl J Med* 341: 777-784, 1999.

Links. *Businessweek*. 44. 2008 March 17.

Manson JE, Skerrett PJ, Greenland P, VanItallie TB. The escalating pandemics of obesity and sedentary lifestyle. A call to action for clinicians. *Arch Intern Med* 164: 249-258, 2004.

Marcus BH, Albrecht AE, King TK, Parisi AF, Pinto BM, Roberts M, Niaura RS, Abrams DB. The efficacy of exercise as an aid for smoking cessation in women. A randomized controlled trial. *Arch Intern Med* 159: 1229-1234, 1999.

Matthews CD, Ockene IS, Freedson PS, Rosal MC, Herbert JR, Merriam PA. Moderate to vigorous physical activity and risk of upper-respiratory tract infection. *Med Sci Sports Exerc* 34(8): 1242-1248, 2002.

[MFMER] Mayo Foundation for Medical Education and Research. *Depression and anxiety: Exercise eases symptoms* [Internet]. [cited 2008 November 5]. Rochester (MN): Mayo Foundation for Medical Education and Research. Available from: www.mayoclinic.com/health/depression-and-exercise/MH00043. 2007 October 23.

McGinnis JM, Foege WH. Actual causes of death in the United States. *JAMA* 270(18): 207-212, 1993.

McTiernan A, Kooperberg C, White E, Wilcox S, Coates R, Adams-Campbell LL, Woods N, Ockene J. Recreational physical activity and the risk of breast cancer in postmenopausal women: The Women's Health Initiative Cohort Study. *JAMA* 290: 1331-1336, 2003.

Mittleman MA, Maclure M, Tofler GH, Sherwood JB, Goldberg RJ, Muller JE. Triggering of acute myocardial infarction by heavy physical exertion: protection against triggering by regular exertion: Determinants of Myocardial Infarction Onset Study Investigators. *N Engl J Med* 329: 1677-1683, 1993.

Mokdad AH, Marks JS, Stroup DF, Gerberding JL. Actual causes of death in the United States, 2000. *JAMA* 291(10): 1238, 1241, 2004.

[NCHC] National Coalition on Health Care. *Health Insurance Costs* [Internet]. [cited 2008 November 2]. Washington (DC): The National Coalition on Health Care. Available from: www.nchc.org/facts/cost.shtml. 2008 October 16.

[NSGA] National Sporting Goods Association. *Exercise walking remains No. 1* [Internet]. [cited 2008 April 13]. Mount Prospect (IL): National Sporting Goods Association. Available from: www.nsga.org/i4a/pages/index.cfm?pageid=3924. 2008 April.

Nieman DC. Does exercise alter immune function and respiratory infections? *PCPFS Research Digest*. 3(13): 1-8, 2001.

Nieman DC, Henson DA, Gusewitch G, Warren BJ, Dotson RC, Butterworth DE, Nehlsen-Cannarella SL. Physical activity and immune function in elderly women. *Med Sci Sports Exerc* 25: 823-831, 1993.

Nieman DC, Nehlsen-Cannarella SF, Henson DA, Koch AJ, Butterworth DE, Fagoaga OR, Utter A. Immune response to exercise training and/or energy restriction in obese females. *Med Sci Sports Exerc* 20: 679-686, 1998.

Nieman DC, Nehlsen-Cannarella SL, Markoff PA, Balk-Lamerton AJ, Yang H, Chritton DBW, Lee JW, Arabatzis K. The effects of moderate exercise training on natural killer cells and acute upper respiratory tract infection. *Int J Sports Med* 11: 467-473, 1990.

Opatz, JP, ed. *Economic impact of worksite health promotion*. Champaign (IL): Human Kinetics. 1994.

Paffenbarger RS, Hyde RT, Wing AL, Hsieh CC. Physical activity, all-cause mortality, and longevity of college alumni. *New Engl J Med* 314: 605-613, 1986.

Powell KE, Blair SN. The public health burdens of sedentary living habits: Theoretical but realistic estimates. *Med Sci Sports Exerc* 26: 851-856, 1994.

Pratt M, Macera CA, Wang G. Higher direct medical costs associated with physical inactivity. *Phys Sportsmed* 28: 63-70, 2000.

Pyron M. Never too late to start when it comes to exercise. *ACSM Fit Society Page* Summer: 1-4. 2003.

Rankinen T, Bouchard C. Dose-response issues concerning the relationship between regular physical activity and health. *PCPFS Research Digest* 3(18): 1-8, 2002.

Rockhill B, Willett WC, Hunter DJ, Manson JE, Hankinson SE, Spiegelman D, Colditz GA. A prospective study of recreational physical activity and breast cancer risk. *Arch Intern Med* 159: 2290-2296, 1999.

Sallis JF, Hovell MF, Hofstetter CR, Barrington E. Explanation of vigorous physical activity during two years using social learning variables. *Soc Sci Med* 34: 25-32, 1992.

Sallis JF, Hovell MF, Hofstetter CR, Elder JP, Caspersen CJ, Hackley M, Powell KE. Distance between homes and exercise facilities related to the frequency of exercise among San Diego residents. *Public Health Rep* 105: 179-185, 1990.

Sallis JF. Influences on physical activity of children, adolescents, and adults. *PCPFS Research Digest* 1(7): 1-7, 1994.

Shephard RJ. PAR-Q, Canadian home fitness test and exercise screening alternatives. *Sports Med* 5: 185-195, 1988.

Shephard RJ. Readiness for physical activity. *PCPFS Research Digest* 1(5): 1-9, 1994.

Shephard RJ. Worksite fitness and exercise programs: A review of methodology and health impact. *Am J Health Promot* 10: 436-452, 1996.

Shephard RJ, Shek PN. Cancer, immune function, and physical activity. *Can J Appl Physiol* 20: 1-25, 1995.

Siscovick DS, Ekelund LG, Johnson JL, Truong Y, Adler A. Sensitivity of exercise electrocardiography for acute cardiac events during moderate and strenuous physical activity: The Lipid Research Clinics Coronary Primary Prevention Trial. *Arch Intern Med* 151: 325-330, 1991.

Sisko A, Truffer C, Smith S, Keehan S, Cylus J, Poisal JA, Clemens MK, Lizonitz J. Health spending projections through 2018: Recession effects add uncertainty to the outlook. *Health Affairs* 28(2): w346-w357, 2009.

Smith JK. Exercise and atherogenesis. *Exerc Sport Sci Rev* 29(2): 49-53, 2001.

Spain CG, Franks BD. Healthy People 2010: Physical Activity and Fitness. *PCPFS Research Digest* 3(13): 1-16, 2001.

Stucky-Ropp RC, DiLorenzo TM. Determinants of exercise in children. *Prev Med* 22: 880-889, 1993.

Tappe MK, Duda JL, Ehrnwalk PM. Perceived barriers to exercise among adolescents. *J Sch Health* 59: 153-155, 1989.

Taylor WC, Whitt-Glover MC. Exercise design and implementation for specific populations. *ACSM Fit Society Page* Spring: 6. 2007.

Thompson PD, Franklin BA, Balady GJ, Blair SN, Corrado D, Estes NA 3rd, Fulton JE, Gordon NF, Haskell WL, Link MS, et al. Exercise and acute cardiovascular events: Placing the risks into perspective. *Med Sci Sports Exerc* 39(5): 886-897, 2007.

Thompson PD, Funk EJ, Carleton RA, Sturner WQ. Incidence of death during jogging in Rhode Island from 1975 through 1980. *JAMA* 247: 2535-2538, 1982.

U.S. Department of Health Education and Welfare (United States). *Surgeon General's Report on Health Promotion and Disease Prevention.* Washington (DC): Author. 1979.

[USDHHS] U.S. Department of Health and Human Services, Physical Activity Guidelines Advisory Committee (United States). *Physical Activity Guidelines Advisory Committee Report, 2008.* Washington (DC): Author. 2008.

[USDHHS] U.S. Department of Health and Human Services, Centers for Disease Control and Prevention, National Center for Chronic Disease Prevention and Health Promotion (United States). *Physical Activity and Health: A Report of the Surgeon General.* Atlanta: Author. 1996.

[USDHHS] U.S. Department of Health and Human Services (United States). *Dietary Guidelines for Americans* [Internet]. [cited 2008 July 9]. Available from: www.health.gov/dietaryguidelines/dga2005/report/. 2005.

Van Camp SP, Bloor CM, Mueller FO, Cantu FC, Olson HG. Nontraumatic sports death in high school and college athletes. *Med Sci Sports Exerc* 27: 6411-647, 1995.

Weuve J, Kang JH, Manson JE, Breteler MMB, Ware JH, Grodstein F. Physical activity, including walking and cognitive function in older women. *JAMA* 292: 1454-1461, 2004.

Whaley MH, ed. *ACSM's guidelines for exercise testing and prescription.* 7th ed. Philadelphia: Lippincott Williams & Wilkins. 2005.

Wilmore JH. Exercise, obesity, and weight control. *PCPFS Research Digest* 1(6): 1-11, 1994.

[WHO] World Health Organization. *Obesity and overweight.* Fact Sheet No. 311. Available from: Geneva. 2006 September.

Yaffe K, Barnes D, Nevitt M, Lui LY, Covinsky K. A prospective study of physical activity and cognitive decline in elderly women: Women who walk. *Arch Intern Med* 161: 1703-1708, 2001.

Chapter 2

Ainsworth BE. The compendium of physical activities. *PCPFS Research Digest* 4(2): 1-8, 2003.

Ainsworth BE, Haskell WL, Whitt MC, Irwin ML, Swartz AM, Strath SJ, O'Brien WL, Bassett DR Jr, Schmitz KH, Emplaincourt PO, et al. Compendium of physical activities: An update of activity codes and MET intensities. *Med Sci Sports Exerc* 32: S498-S516, 2000.

[ACSM] American College of Sports Medicine. The recommended quantity and quality of exercise for developing and maintaining cardiorespiratory and muscular fitness in healthy adults. *Med Sci Sports Exerc* 22: 265-274, 1990.

[ACSM] American College of Sports Medicine. The recommended quantity and quality of exercise for developing and maintaining cardiorespiratory and muscular fitness in healthy adults. *Med Sci Sports Exerc* 10: vii-x, 1978.

[ACSM] American College of Sports Medicine. The recommended quantity and quality of exercise for developing and maintaining cardiorespiratory and muscular fitness and flexibility in healthy adults. *Med Sci Sports Exerc* 30: 975-991, 1998.

[ACSM] American College of Sports Medicine. Position Stand: "Exercise for patients with coronary artery disease." *Med Sci Sports Exerc* 26(3): i-v, 1994.

Australian Government Department of Health and Ageing. *National Physical Activity Guidelines for Australians.* [Internet]. [cited 2009 November19]. Canberra: Author. http://www.health.gov.au/internet/main/publishing.nsf/Content/phd-physical-activity-adults-pdf-cnt.htm/$File/adults_phys.pdf

Brooks GA, Butte NF, Rand WM, Flatt J-P, Caballero B. Chronicle of the Institute of Medicine physical activity recommendation: How a physical activity recommendation came to be among dietary recommendations. *Am J Clin Nutr* 79(suppl): 921S-930S, 2004.

[CDHS] California Center for Physical Activity (United States). *Guidelines for physical activity across the lifespan* [Internet]. [cited 2008 September 26]. Available from: www.caphysicalactivity.org/facts_recomm1.html

[CDHS] California Department of Health Services, California Obesity Prevention Initiative Health Systems Workgroup (United States). *Obesity prevention for health care systems—Executive summary and literature review* [Internet]. [cited 2008 September 15]. Sacramento (CA) : California Department of Health Services. Available from: www.caphysicalactivity.com. 2002.

[CDC] Centers for Disease Control and Prevention and President's Council on Physical Fitness and Sports (United States). *Healthy People 2010* [Internet]. [cited 2008 February 26]. Atlanta (GA): Centers for Disease Control and Prevention. Available from: www.healthypeople.gov. 2000.

Feigenbaum MS, Pollock ML. Prescription of resistance training for health and disease. *Med Sci Sports Exerc* 31(1): 38-45, 1999.

Feury M. Walk off the weight. *Woman's Day* 74-78. 2000 September 12.

Fletcher GF, Blair SN, Blumenthal J, Caspersen C, Chaitman B, Epstein S, Falls H, Froelicher ES, Froelicher VF, Pina IL. Statement on exercise. Benefits and recommendations for physical activity programs for all Americans. A statement for health professionals by the Committee on Exercise and Cardiac Rehabilitation of the Council on Clinical Cardiology, American Heart Association. *Circulation* 86(1): 340-344, 1992.

Haskell W, Lee IM, Pate RR, Powell K, Blair SN, Franklin BA, Macera CA, Health GA, Thompson PD, Bauman A. Physical activity and public health: Updated recommendation for adults. *Med Sci Sports Exerc* 39(8): 1423-1434, 2007.

Haskell WL. Physical activity in the prevention and management of coronary heart disease. *PCPFS Research Digest* 2(1): 1-12, 1995.

Hellmich N. Journey to better fitness starts with 10,000 steps. *USA Today* 1. 1999 June 29.

[IOM] Institute of Medicine (United States). *Dietary Reference Intakes for energy, carbohydrates, fiber, fat, fatty acids, cholesterol, protein and amino acids* [Internet]. [cited 2008 May 12]. Washington (DC): The National Academies Press. Available from: www.nap.edu/books/0309085373/html. 2002.

Jakicic JM, Clark K, Coleman E, Donnelly JE, Foreyt J, Melanson E, Volek J, Volpe SL. American College of Sports Medicine Position Stand: Appropriate intervention strategies for weight loss and prevention of weight regain for adults. *Med Sci Sports Exerc* 33(12): 2145-2156, 2001.

Knudson DV, Magnusson P, McHugh M. Current issues in flexibility fitness. *PCPFS Research Digest* 3(10): 1-8, 2000.

Kruger J, Kohl HW. Centers for Disease Control. Trends in leisure-time physical inactivity by age, sex and race/ethnicity – United States, 1994-2004. *MMWR Morb Mortal Wkly Rep* 54: 991-994, 2005.

Kruger J, Kohl HW, Miles IJ. Prevalence of regular physical activity among adults – United States, 2001 and 2005. *MMWR Morb Mortal Wkly Rep* 56(46): 1209-1212, 2007.

Macera CA, Jones DA, Yore MM, Ham SA, Kohl HW, Kimset CD, Buchner D. Centers for Disease Control. Prevalence of physical activity, including lifestyle activities among adults – United States, 2000-2001. *MMWR Morb Mortal Wkly Rep* 52: 764-769, 2003.

[NIH] National Institutes of Health, National Heart, Lung and Blood Institute (United States). *Clinical guidelines on the identification, evaluation, and treatment of overweight and obesity in adults: The evidence report* [Internet]. [cited 2008 July 3]. Available from: www.nhlbi.nih.gov/guidelines/obesity. 1998.

[NIH] National Institutes of Health, National Heart, Lung and Blood Institute (United States). Strategies to increase physical activity. In: *Guidelines on overweight and obesity: Electronic textbook* [Internet]. [cited 2008 July 28]. Available from: www.nhlbi.nih.gov/guidelines/obesity/e_txtbk. 1998.

[NIH] National Institutes of Health, National Institute of Diabetes and Digestive and Kidney Diseases, Weight-control Information Network (United States). *Physical activity and weight control* [Internet]. [cited 2008 May 20]. Bethesda (MD): Weight-control Information Network. NIH Pub. No.: 03-4031. Available from :www.win.niddk.nih.gov/publications/physical.htm. 2003 March.

[NIH] National Institutes of Health, NIH Consensus Development Program (United States). *Physical Activity and Cardiovascular Health. NIH Consens Statement Online*

1995 December 18-20; 13(3): 1-33. [Internet]. [cited 2008 August 28]. Available from: http://consensus.nih.gov/1995/1995ActivityCardiovascularHealth101html.htm. 1995.

NIH Consensus Development Panel on Physical Activity and Cardiovascular Health. Physical activity and cardiovascular health. *JAMA* 276: 241-246, 1996.

[NSCA] National Strength and Conditioning Association. *NSCA position statement. Basic guidelines for the resistance training of athletes* [Internet]. [cited 2008 May 20]. Colorado Springs(CO): National Strength and Conditioning Association. Available from: www.nsca-lift.org http://www.nsca-lift.org/Publications/Basic%20guidelines%20for%20the%20resistance%20training%20of%20athletes.pdf

[NSCA] National Strength and Conditioning Association. *NSCA position statement. Health aspects of resistance exercise and training* [Internet]. [cited 2008 May 20]. Colorado Springs (CO): National Strength and Conditioning Association. Available from: http://www.nsca-lift.org/Publications/posstatements.shtml.

Pate RR, Pratt M, Blair SN, Haskell WL, Macera CA, Bouchard C, Buchner D, Ettinger W, Heath GW, King AC et al. Physical activity and pubic health. A recommendation from the Centers for Disease Control and Prevention and the American College of Sports Medicine. *JAMA* 273(5): 402-407, 1995.

Physical Activity Guidelines Advisory Committee. *Physical Activity Guidelines Advisory Committee report, 2008.* Washington (DC): U.S. Department of Health and Human Services. 2008.

Pollock ML, Franklin BA, Balady GJ, Chaitman BL, Fleg JL, Fletcher B, Limacher M, Piña IL, Stein RA, Williams M, et. al. Resistance exercise in individuals with and without cardiovascular disease. *Circulation* 101(7): 828-833, 2000.

The President's challenge [Internet]. [cited 2008 April 2]. Available from: www.presidentschallenge.org.

Public Health Agency of Canada (Canada). *Canada's physical activity guide to healthy active living* [Internet]. [cited 2008 August 25]. Ottawa, Ontario: Public Health Agency of Canada. Available from: www.paguide.com. 1998.

Quittner J. High-tech walking. *Time* 77, 2000 July 24.

Saris WHM, Blair SN, van Baak MA, Eaton SB, Davies PSW, DiPietro L, Fogelholm M, Rissanen A, Schoeller D, Swinburn B, et al. How much physical activity is enough to prevent unhealthy weight gain? Outcome of the IASO 1st Stock Conference and consensus statement. *Obesity Reviews* 4(2): 101-114, 2003.

Thompson PD, Buchner D, Pina IL, Balady GJ, Williams MA, Marcus BH, Berra K, Blair SN, Costa F, Franklin B, et al. Exercise and physical activity in the prevention and treatment of atherosclerotic cardiovascular disease: A statement from the Council on Clinical Cardiology (Subcommittee on Exercise, Rehabilitation , and Prevention) and the Council on Nutrition, Physical Activity, and Metabolism (Subcommittee on Physical Activity). *Circulation* 107(24): 3109-3116, 2003.

Tudor-Locke C. Taking steps towards increased physical activity: Using pedometers to measure and motivate. *PCPFS Research Digest* 3(17): 1-8, 2002a.

Tudor-Locke C, Ainsworth BE, Thompson RW, Matthews CE. Comparison of pedometer and accelerometer measures of free-living physical activity. *Med Sci Sports Exerc* 34(12): 2045-2051, 2002b.

[USDHHS] U.S. Department of Health and Human Services (United States). *2008 Physical Activity Guidelines for Americans* [Internet]. [cited 2008 October 7]. Washington(DC): USDHHS. Available from: www.health.gov/paguidelines. 2008.

[USDHHS] U.S. Department of Health and Human Services, Centers for Disease Control and Prevention, National Center for Chronic Disease Prevention and Health Promotion (United States). *Physical Activity and Health: A Report of the Surgeon General.* [Internet]. [cited 2008 February 3]. Atlanta: U.S. Department of Health and Human Services. Available from: www.cdc.gov/nccdphp/sgr/sgr.htm. 1996.

[USDHHS] U.S. Department of Health and Human Services, Centers for Disease Control and Prevention, National Center for Chronic Disease Prevention and Health Promotion, Division of Adult and Community Health (United States). *Behavioral Risk Factor Surveillance System.* [Internet]. [cited 2008 September 10]. Available from: http://apps.nccd.cdc.gov/brfss/index.asp.

[USDHHS] U.S. Department of Health and Human Services, National Institutes of Health, National Heart, Lung, and Blood Institute (United States). *Clinical guidelines on the identification, evaluation, and treatment of overweight and obesity in adults; evidence report.* Washington(DC): HHS, PHS. Pub. No.: 98-4083. 1998.

[USDHHS] U.S. Department of Health and Human Services (United States). *Dietary Guidelines for Americans* [Internet]. [cited 2008 July 9]. Available from: www.health.gov/dietaryguidelines/dga2005/report/. 2005.

[USDHHS] U.S. Department of Health and Human Services, National Institutes of Health, National Heart, Lung, and Blood Institute (United States). *The practical guide: Identification, evaluation and treatment of overweight and obesity in adults* [Internet]. [cited NIH Pub. No:. 00-4084. www.nhlbi.nih.gov/guidelines/obesity/prctgd_c.pdf, 2000 October.

Welk GJ, Differding JA, Thompson RW, Blair SN, Dziura J, Hart P. The utility of the Digi-walkier step counter to assess daily physical activity patterns. *Med Sci Sports Exerc.* 32(9 Suppl):S481-488, 2000.

Williams MA, Haskell WL, Ades PA, Amsterdam EA, Bittner V, Franklin BA, Gulanick M, Laing ST, Stewart KJ. Resistance exercise in individuals with and without cardiovascular disease: 2007 update. A scientific statement from the American Heart Association Council on Clinical Cardiology and Council on Nutrition, Physical Activity, and Metabolism. *Circulation* 116: 572-584, 2007.

[WHO] World Health Organization. *Global strategy on diet, physical activity and health.* Geneva, World Health

Organization. Available from: http://apps.who.int/gb/ebwha/pdf_files/WHA57/A57_R17-en.pdf. 2004.

[WHO] World Health Organization Global Strategy on Diet, Physical Activity and Health. Recommended amount of physical activity [Internet]. [cited 2008 September 9]. Geneva, World Health Organization. Available from: www.who.int/dietphysicalactivity/factsheet_recommendations/en/index.html.

[WHO] World Health Organization. *Global strategy on diet, physical activity and health.* Geneva, World Health Organization. Available from: www.who.int/dietphysical activity/strategy/eb11344/strategy_english_web.pdf. 2004.

Chapter 3

[ACE] American Council on Exercise. *Fit Facts: Three things every exercise program should have.* [Internet]. [cited 2008 July 7]. Available from: www.acefitness.org/fitfacts/fitfacts_display.aspx?itemid=2627 . 2001.

[ACSM] American College of Sports Medicine. General principles of exercise prescription In: *ACSM's Guidelines for Exercise Testing and Prescription.* 8th ed. Baltimore: Lippincott, Williams and Wilkins. 152-182. 2010.

Borg GAV. Psychophysical bases of perceived exhaustion. *Med Sci Sports Exerc* 14: 377-381, 1982.

Corbin CB, Lindsey R. *Fitness for Life.* 4th ed. Champaign (IL): Human Kinetics. 2002.

Crouter S. *Selecting and effectively using heart rate monitors* [Internet]. [cited 2008 April 28]. American College of Sports Medicine. Available from: www.acsm.org/AM/Template.cfm?Section=Brochures2&Template=/CM/ContentDisplay.cfm&ContentID=8110. 2005.

DeBusk RF, Stenestrand U, Sheehan M, Haskell WL. Training effects of long versus short bouts of exercise in healthy subjects. *Am J Cardiol* 65: 1010-1013, 1990.

Franks, BD. Personalizing physical activity prescription. *PCPFS Research Digest* 2(9): 1-10, 1997.

Hass CJ, Garzarella L, de Hoyos D, Pollock, ML. Single versus multiple sets in long-term recreational weightlifters. *Med Sci Sports Exerc* 32: 235-242, 2000.

Karvonen M, Kentala K, Mustala O. The effect of training on heart rate: a longitudinal study. *Ann Med Exp Biol Fenn* 35: 307-315, 1957.

Katzen M, Willett W. *Eat, drink, and weigh less: A flexible and delicious way to shrink your waist without going hungry.* New York: Hyperion. 2006.

Kerr D, Morton A, Dick I, Prince R. Exercise effects on bone mass in postmenopausal women are site-specific and load-dependent. *J Bone Miner Res* 11: 218-225, 1996.

Knudson DV, Magnusson P, McHugh M. Current issues in flexibility fitness. *PCPFS Research Digest* 3(10): 1-8, 2000.

Murphy MH, Hardman AE. Training effect of short and long bouts of brisk walking in sedentary women. *Med Sci Sports Exerc* 30: 152-157, 1998.

Persinger R, Foster C, Gibson M, Fater DCW, Porcari JP. Consistency of the talk test for exercise prescription. *Med Sci Sports Exerc* 36(9): 1632-1636, 2004.

Rhea MR, Alvar BA, Ball SD, Burkett LN. Three sets of weight training super to 1 set with equal intensity for eliciting strength. *J Strength Cond Res* 16(4): 525-529, 2002.

Starkey DB, Pollock ML, Ishida Y, Welsch MA, Brechue WF, Graves JE, Feigenbaum MS. Effect of resistance training volume on strength and muscle thickness. *Med Sci Sports Exerc* 28: 1311-1320, 1996.

Utter AC, Kang J, Robertson RJ. *ACSM current comment: Perceived exertion.* Indianapolis (IN): American College of Sports Medicine. 2000. Available from: www.acsm.org/cc.

Vincent KR, Braith RW. Resistance exercise and bone turnover in elderly men and women. *Med Sci Sports Exerc* 34(1): 17-23, 2002.

Whaley MH, ed. General principles of exercise prescription. In: *ACSM's guidelines for exercise testing and prescription.* 7th ed. Baltimore: Lippincott, Williams and Wilkins. 133-173. 2006.

Chapter 4

[AAP] American Academy of Pediatrics, Committee on Sports Medicine and Fitness. Fitness, activity, and sports participation in the preschool child. *Pediatrics* 90(6): 1002-1004, 1992.

[AAP] American Academy of Pediatrics, Committee on Sports Medicine and Fitness. Infant exercise programs. *Pediatrics* 82(5): 800, 1988.

[AAP] American Academy of Pediatrics, Council on Sports Medicine and Fitness and Council on School Health. Active healthy living: Prevention of childhood obesity through increased physical activity. *Pediatrics* 117(5): 1834-1842, 2006. [Internet]. [cited 2008 September 2]. Available from: www.pediatrics.org/cgi/doi/10.1542/peds.2006-0472.

American Heart Association. "Exercise (Physical Activity) and Children" [Internet] www.americanheart.org/presenter.jhtml?identifier=4596 [cited 14 June 2008].

Australian Government Department of Health and Ageing. *Australia's physical activity recommendations for 5-12 year olds.* Canberra: Department of Health and Ageing. 2004. Available from: www.health.gov.au/internet/main/publishing.nsf/Content/9D7D393564FA0C42CA256F970014A5D4/$File/kids_phys.pdf.

California Center for Physical Activity. "The California Center for Physical Activity's Guidelines for Physical Activity Across the Lifespan: Preschool Children" 2002. [Internet] www.caphysicalactivity.org/facts_recomm1.html [cited September 28 2008]

[CDHS] California Department of Health Services, California Obesity Prevention Initiative Health Systems Workgroup (United States). *Obesity prevention for health care systems—Executive summary and literature review* [Internet]. [cited 2008 September 15]. California Department of

Health Services. Available from: www.caphysicalactivity. com. 2002.

Carrel AL, Bernhardt DT. Exercise prescription for the prevention of obesity in adolescents. *Curr Sports Med Rep* 3: 330-336, 2004.

Clements R. Encouraging kids to be active: the importance of play. *ACSM Fit Society Page* Late Summer/Early Fall: 3, 2007.

Malina RM. Building an active mindset in children. *ACSM Fit Society Page* Late Summer/Early Fall: 4, 2007.

[NASPE] National Association for Sport and Physical Education. *Active start: A statement of physical activity guidelines for children from birth to age 5*. Reston (VA): NASPE. 2004.

[NASPE] National Association for Sport and Physical Education. *Physical activity and children: A statement of guidelines 2003*. Reston (VA): NASPE. 2003.

[PCPFS] President's Council on Physical Fitness and Sports, National Association for Sport and Physical Education, and the Kellogg Company (United States). *Kids in action: Fitness for children birth to age five*. Available from: fitness. gov/funfit/Kidsinactionbook.pdf. 2003.

[USDHHS] U.S. Department of Health and Human Services (United States). *2008 Physical Activity Guidelines for Americans*. [Internet]. [cited 2008 October 11]. Washington (DC): U.S. Department of Health and Human Services. Available from: www.health.gov/paguidelines, 2008.

[USDHHS] U.S. Department of Health and Human Services (United States). *The Surgeon General's call to action to prevent and decrease overweight and obesity*. Washington (DC): U.S. Department of Health and Human Services. 2001.

Chapter 5

[AAP] American Academy of Pediatrics Committee on Nutrition. Prevention of pediatric overweight and obesity. *Pediatrics* 112(2): 424-430, 2003.

[AAP] American Academy of Pediatrics Council on Sports Medicine and Fitness and Council on School Health. Physical fitness and activity in schools. *Pediatrics* 105(5): 1156-1157, 2000.

[AAP] American Academy of Pediatrics Council on Sports Medicine and Fitness and Council on School Health. Active healthy living: Prevention of childhood obesity through increased physical activity. *Pediatrics* 117(5): 1834-1842, 2006.

[ACSM] American College of Sports Medicine. Exercise prescription for healthy populations. In: *ACSM's Guidelines for Exercise Testing and Prescription*. 8th ed. Baltimore: Lippincott, Williams and Wilkins. 187-189. 2010.

[ACSM] American College of Sports Medicine. *Pre-participation physical exams* [Internet]. [cited 2008 August 21]. Indianapolis, IN: American College of Sports Medicine, 2002. Available from: www.acsm.org/AM/TextTemplate. cfm?Section=Search&section=Brochures&template=/CM/ ContentDisplay.cfm&ContentFileID=262.

[ACE] American Council on Exercise. *Fit facts: Strength training for kids: A guide for parents and teachers* [Internet]. [cited 2008 July 7]. San Diego,: American Council on Exercise. Available from: www.acefitness.org/fitfacts. 2001.

Armstrong N, Balding J, Gentle P, Kirby S. Patterns of physical activity among 11-16 year-old British children. *BMJ* 301: 203-205, 1990.

Armstrong N, Bray S. Physical activity patterns defied by continuous heart rate monitoring. *Arch Disease Children* 66: 245-247, 1991.

Associated Press. Diabetes in children set to soar. *MSNBC*. 2003 June 16.

Australian Government Department of Health and Ageing. *Australia's physical activity recommendations for 12-18 year olds*. Canberra: Department of Health and Ageing. 2004. [Internet] [cited 2009 October 4] www.getmoving.tas. gov.au/RelatedFiles/Physicalguidlinesfor12-18yrolds.pdf

Australian Government Department of Health and Ageing. *Australia's physical activity recommendations for 5-12 year olds*. Canberra: Department of Health and Ageing. 2004. [Internet] [cited 2009 October 4] www.health.gov.au/internet/ main/publishing.nsf/Content/9D7D393564FA0C42CA2 56F970014A5D4/$File/kids_phys.pdf

Bar-Or O. Health benefits of physical activity during childhood and adolescence. *PCPFS Research Digest* 2(4): 1-6, 1995.

Baumert PW Jr, Henderson JM, Thompson NJ. Health risk behaviors of adolescent participants in organized sports. *J Adolesc Health* 22: 460-465, 1998.

Bernhardt DT, Gomez J, Johnson MD, Martin TJ, Rowland TW, Small E, LeBlanc C, Malina R, Krein C, Young JC, et al. Committee on Sports Medicine and Fitness. Strength training by children and adolescents. *Pediatrics*. 107(6): 1470-2, 2001.

Biddle S, Sallis JF, Cavill N, eds. *Young and active? Young people and health-enhancing physical activity – evidence and implications*. London: Health Education Authority. 1998.

Bonjour J-P. *Invest in your bones* 2001[Internet]. [cited 2008 November 11]. Nyon, Switzerland: International Osteoporosis Foundation. Available from: www.iofbonehealth. org/download/osteofound/filemanager/publications/pdf/ invest_in_your_bones.pdf.

Brenner JS and the Council on Sports Medicine and Fitness. Overuse injuries, overtraining, and burnout in child and adolescent athletes. *Pediatrics* 119: 1242-1245, 2007.

California Department of Education [Internet]. [cited 2008 July 15]. www.cde.ca.gov/cyfsbranch/lsp/health/pecommunications.htm. 2002

[CDHS] California Department of Health Services, California Obesity Prevention Initiative Health Systems Workgroup (United States). *Obesity prevention for health care systems—Executive summary and literature review* [Internet]. [cited 2008 September 15] . Sacramento, CA: California Department of Health Services. Available from: www. caphysicalactivity.com. 2002.

[CDHS] California Department of Health Services (United States). "The California Center for Physical Activity's Guidelines for Physical Activity Across the Lifespan: Children (5-12 Years)". [Internet]. [cited 2008 September 15]. Sacramento(CA): California Department of Health Services. Available from: www.caphysicalactivity.org/facts_recomm1.html. 2002.

[CDHS] California Department of Health Services, "The California Center for Physical Activity's Guidelines for Physical Activity Across the Lifespan: Youth (13-17 Years)" (United States). [Internet]. [cited 2008 September 15]. Sacramento, CA: California Department of Health Services. Available from: www.caphysicalactivity.org/facts_recomm1.html. 2002.

Carrel Al, Bernhardt DT. Exercise prescription for the prevention of obesity in adolescents. *Curr Sports Med Rep* 3: 330-336, 2004.

Caspersen CJ, Pereira MA, Curran KM. Changes in physical activity patterns in the United States, by sex and cross-sectional age. *Med Sci Sports Exerc* 32(9): 1601-1609, 2000.

[CDC] Centers for Disease Control & Prevention and American College of Sports Medicine. Summary statement. Workshop of physical activity and public health. *Sports Medicine Bulletin* 28(4): 7, 1994.

[CDC] Centers for Disease Control and Prevention. Guidelines for school and community programs to promote lifelong physical activity among young people. *MMWR Morb Mortal Wkly Rep* 46(RR-6): 1-36, 1997. [Internet] [cited 2008 October 5] www.cdc.gov/healthyyouth/physicalactivity/guidelines

[CDC] Centers for Disease Control and Prevention and President's Council on Physical Fitness and Sports (United States). *Healthy People 2010* [Internet]. [cited 2008 February 26]: Centers for Disease Control and Prevention, Atlanta, GA. Available from: www.healthypeople.gov. 2000.

[CDC] Centers for Disease Control and Prevention (United States). *Physical activity and good nutrition: Essential elements to prevent chronic diseases and obesity.* Atlanta, GA: Centers for Disease Control and Prevention. 2003.

[CDC] Centers for Disease Control and Prevention and President's Council on Physical Fitness and Sports (United States). *Healthy People 2010* "Physical activity in children and adolescents" [Internet]. [cited 2008 February 26]: Centers for Disease Control and Prevention, Atlanta, GA. Available from: www.healthypeople.gov. 2000.

[CDC] Centers for Disease Control and Prevention. Youth risk behavior surveillance – United States, 1997. *MMWR Morb Mortal Wkly Rep* 47(55-3): 1-89, 1998.

[CDC] Centers for Disease Control and Prevention. Youth risk behavior surveillance – United States, 2007. *MMWR Morb Mortal Wkly Rep* 57(SS-4):1-131 , 2008.

CDC. Increasing physical activity: A Report on recommendations of the task force on community preventive services. *Morbidity and Mortality Weekly Report* 50(RR-18): 1-16, 200Corbin CG, Pangrazi RP. Physical activity for youth:

How much is enough? In: *FITNESSGRAM/ACTIVITY-GRAM Reference Guide.* Dallas: Cooper Institute. 2008.

Corbin CG, Pangrazi RP, Welk GJ. Toward an understanding of appropriate physical activity levels for youth. *PCPFS Research Digest* 1(8): 1-12, 1994.

Dietz WH. Health consequences of obesity in youth: Childhood predictors of adult disease. *Pediatrics* 101(3)Supp: 518-525, 1998.

Dietz WH. Physical activity recommendations: Where do we go from here? *J Pediatr* 146: 719-720, 2005.

Dowda M. Ainsworth BE, Addy CL, Saunders R, Riner W. Environmental influences, physical activity, and weight status in 8-to 16-year olds. *Arch Pediatr Adolesc Med* 155: 711-717, 2001.

Escobedo LG, Marcus SE, Holtzman D, Giovino GA. Sports participation, age at smoking initiation and the risk of smoking among US high school students. *JAMA* 269: 1391-1395, 1993.

Faigenbaum AD. Youth resistance training. *PCPFS Research Digest* 4(3): 1-8, 2003.

Faigenbaum AD, Kraemer WJ, Cahill B, Chandler J, Dziados J, Elfrink LD, Forman E, Gaudiose M, Micheli L, Nitka M, and Roberts S. . National Strength & Conditioning Association. Youth resistance training: Position statement paper and literature review. *J Strength Cond Res* 18(6): 63-75, 1996.

Faigenbaum AD, Westcott WL. Resistance training for obese children and adolescents. *PCPFS Research Digest* 8(3): 1-8, 2007.

Gilson ND, Cooke CB, Mahoney CA. A comparison of adolescent moderate-t-vigorous physical activity participation in relation to a sustained or accumulated criterion. *Health Educ Res* 16(3): 335-341, 2001.

Grunbaum JA, Kann L, Kinchen S, Williams B, Ross JG, Lowry R, Kolbe L. Youth risk behavior surveillance – United States, 2001. *MMWR Morb Mortal Wkly Rep* 51(SS-4): 1-64, 2002.

Haskell, W.L. Health consequences of physical activity: Understanding and challenges regarding dose response. *Medicine and Science in Sports and Exercise, 26(6),* 649–660, 1994.

Health Education Authority, United Kingdom. *Young and active? Young people and health-enhancing physical activity – evidence and implications.* London: Health Education Authority. 1998.

[IOM] Institute of Medicine (United States). *Dietary Reference Intakes for energy, carbohydrate, fiber, fat, fatty acids, cholesterol, protein and amino acids* [Internet]. [cited 2008 May 12]. Washington (DC): The National Academies Press. Available from: www.nap.edu/books/0309085373/html. 2002.

Kann L, Kinchen SA, Williams BI, Ross JG, Lowry R, Grunbaum JA, Kolbe LJ. Youth Risk Behavior Surveillance--United States, 1999. State and local YRBSS Coordinators. J Sch Health. 2000 Sep;70(7):271-85, 2000.

Kaplan JP, Liverman CT, Kraak VA, eds. *Preventing childhood obesity: Health in the balance*. Washington (DC): The National Academies Press. 1-20; 237-284. 2005. [Internet] [cited 2008 November 28 www.nap.edu/catalog.php?record_id=11015]

Kavey RW, Daniels SR, Lauer RM, Atkins DL, Hayman LL, Taubert K. American Heart Association "Primary Prevention of Atherosclerotic Cardiovascular Disease Beginning in Childhood". *Circulation* 107: 1562-1566, 2003.

Keays J, Allison R. The effects of regular moderate to vigorous physical activity on student outcomes: A review. *Can J Public Health* 86: 62-66, 1995.

Kohrt WM, Bloomfield SA, Little KD, Nelson ME, Yingling VR. Physical activity and bone health. *Med Sci Sports Exerc* 36(11): 1985-1996, 2004

Kulig K, Brener ND, McManus T. Sexual activity and substance use among adolescents by category of physical activity plus team sports participation. *Arch Pediatr Adolesc Med* 157: 905-912, 2003.

Malina RM. Physical fitness of children and adolescents in the United States: status and secular change. Medicine and sport science. 50:67-90, 2007.

Malina RM. Growth and maturation: Do regular physical activity and training for sport have a significant influence? In: Armstrong N, van Mechelen W, eds. *Paediatric exercise science and medicine*. Oxford: Oxford University Press. 95-106. 2000.

Malina RM. Growth and maturation of young athletes – Is training for sport a factor? In: Chan K-M, Michele LJ, eds. *Sports and Children*. Hong Kong: Williams & Wilkins. 133-161. 1998.

Minne HW, Pfeifer M. (International Osteoporosis Foundation, Nyon Switzerland). *Invest in your bones. Move it or lose it* [Internet]. [cited 2008 June 26]. IOF, Nyon, Switzerland: International Osteoporosis Foundation. Available from: www.iofbonehealth.org/publications/move-it-or-lose-it.html. 2005.

[NCHS] National Center for Health Statistics. *Prevalence of overweight among children and adolescents: United States, 1999-2002* [Internet]. [cited 2008 April 6]. Hyattsville, MD: National Center for Health Statistics. Available from: www.cdc.gov/nchs/products/pubs/pubd/hestats/overwght99.htm.

[NASPE] National Association for Sport and Physical Education. *Comprehensive school physical activity programs*. Reston (VA): National Association for Sport and Physical Education. 2008. [Internet] [cited 2009 October 4] www.aahperd.org/naspe/standards/upload/Comprehensive-School-Physical-Activity-Programs2-2008.pdf

[NASPE] National Association for Sport and Physical Education. *Guidelines for after school physical activity and intramural sport programs*. Reston (VA): National Association for Sport and Physical Education. 2001.

[NASPE] National Association for Sport and Physical Education. *Physical activity and fitness recommendations for physical activity professionals*. Reston (VA): National Association for Sport and Physical Education. 2002.

[NASPE] National Association for Sport and Physical Education. *Physical activity for children: A statement of guidelines for children 5-12 (2nd edition)*. Reston (VA): NASPE Publications. 2004.

National Association of State Boards of Education. *Fit, healthy and ready to learn: A school health policy guide. Part 1: Physical activity, healthy eating and tobacco use prevention*. Alexandria (VA): National Association of State Boards of Education. 2000.

[NSCA] National Strength and Conditioning Association. *Youth resistance training* [Internet]. [cited 2008 May 20]. Colorado Springs, CO: National Strength and Conditioning Association. Available from: http://www.nsca-lift.org/youthpositionpaper/Youth_Pos_Paper_200902.pdf, 2009.

Nelson MC, Gordon-Larsen P. Physical activity and sedentary behavior patterns are associated with selected adolescent health risk behaviors. *Pediatrics* 117: 1281-1290, 2006.

Opatz JP. *Economic impact of worksite health promotion*. Champaign (IL): Human Kinetics. 1994.

Pate RR, Davis MG, Robinson TN, Stone EJ, McKenzie TL, Young JC. AHA scientific statement: Promoting physical activity in children and youth. *Circulation* 114: 1214-1224, 2006. [Internet] [cited 2009 October 4] circ.ahajournals.org/cgi/content/full/114/11/1214

Pate RR, Heath GW, Dowda M, Trost SG. Associations between physical activity and other health behaviors in a representative sample of US adolescents. *Am J Public Health* 86: 1577-1581, 1996.

Pate RR, Trost SG, Levin S, Dowda M. Sports participation and health-related behavior among US youth. *Arch Pediatr Adolesc Med* 154: 904-911, 2000.

[PAGAC] Physical Activity Guidelines Advisory Committee (United States). *Physical Activity Guidelines Advisory Committee Report, 2008*. Washington (DC): U.S. Department of Health and Human Services. 2008.

Physical Best/NASPE. *Physical Education for Lifelong Learning*. Champaign (IL): Human Kinetics. 2005.

Powell KE. Dysinger W. Childhood participation in organized school sports and physical education as precursors of adult physical activity. *Am J Prev Med* 3(5): 276-281, 1987.

[PCPFS] President's Council on Physical Fitness and Sports, National Association for Sport and Physical Education, and the Kellogg Company (United States). *Kids in action: Fitness for children birth to age five*. Washington, D.C.: U.S. Department of Health and Human Services. 2003.

[PCPFS] "The President's Challenge" [Internet]. [cited 2008 April 2]. Available from: www.presidentschallenge.org.

Public Health Agency of Canada (Canada). *Canada's Physical Activity Guide for Youth* [Internet]. [cited 2008 August 25]. Ottawa, Ontario : Public Health Agency of Canada. Available from: www.paguide.com. 2002.

Public Health Agency of Canada (Canada). *Canada's Physical Activity Guide to Healthy Active Living: Family Guide to Physical Activity for Children (6-9 Years of Age)*

[Internet]. [cited 2008 August 25]. Ottawa, Ontario: Public Health Agency of Canada. Available from: www.phac-aspc.gc.ca/pau-uap/paguide/child_youth/pdf/kids_family_guide_e.pdf. 2002.

Public Health Agency of Canada (Canada). *Canada's Physical Activity Guide to Healthy Active Living: Family Guide to Physical Activity for Youth 10-14 Years of Age* [Internet]. [cited 2008 August 25]. Ottawa, Ontario: Public Health Agency of Canada. Available from: www.phac-aspc.gc.ca/pau-uap/paguide/child_youth/pdf/yth_family_guide_e.pdf. 2002.

Pyron M. Preparticipation physical exams. *ACSM Fit Society Page* Late Summer/Early Fall: 6, 2007.

Robertson K, Adolfsson P, Riddell MC, Scheiner G, Hanas R. Exercise in children and adolescents with diabetes. *Pediatr Diabetes* 9: 65-77, 2008.

Rowland TW. Aerobic response to endurance training in prepubescent children: A critical analysis. *Med Sci Sports Exerc* 17: 493-497, 1985.

Rowland TW. Growth and Exercise. In: *Children's Exercise Physiology*. 2nd ed. Human Kinetics, Champaign, IL. 22-41, xx. 2005.

Sabo DF, Miller KE, Farrel MP, Melnick MJ, Barnes GM. High school athletic participation, sexual behavior and adolescent pregnancy: A regional study. *J Adolesc Health* 25: 207-216, 1999.

Sady SP. Cardiorespiratory exercise training in children. *Clin Sports Med*: 5:493-514, 1986.

Sallis JF. Age-related decline in physical activity: A synthesis of human and animal studies. *Med Sci Sports Exerc* 32(9): 1598-1600, 2000.

Sallis JF. Influences on physical activity of children, adolescents, and adults. *PCPFS Research Digest* 1(7): 7, 1994.

Sallis JF, McKenzie TL, Kolody B, Lewis M, Marshall S, Rosengard P. Effects of health-related physical education on academic achievement: Project SPARK. *Research Quarterly for Exercise and Sport* 70: 127-134, 1999.

Sallis JF, Patrick K, Long BL. Physical activity guidelines for adolescents: Consensus statement. *Ped Exerc Sci* 6(4): 299-301, 1994.

Saris WHM, Blair SN, van Baak MA, Eaton SB, Davies PSW, DiPietro L, Fogelholm M, Rissanen A, Schoeller D, Swinburn B, et al. How much physical activity is enough to prevent unhealthy weight gain? Outcome of the IASO 1st Stock Conference and consensus statement. *Obesity Reviews* 4(2): 101-114, 2003.

Shephard RJ. Curricular physical activity and academic performance. *Pediatr Exerc Sci* 9: 113-126, 1997.

Shephard RJ. Habitual physical activity and academic performance. *Nutrition Reviews* 54(4S): S32-S36, 1996.

Shephard RJ, Volle M, Lavalee M, LaBarre R, Jequier JC, Rajic M. Required physical activity and academic grades: A controlled longitudinal study. In: Limarinen and Valimaki, eds. *Children and Sport*. Berlin: Springer Verlag. 58-63. 1984.

Sleap M, Waburton P. Physical activity levels of 5-11 year-old children in England as determined by continuous observation. *Res Q Exerc Sport* 63(3): 238-245, 1992.

Spinks AB, Macpherson AK, Bain C, McClure RJ. Compliance with the Australian national physical activity guidelines for children: Relationship to overweight status. *J Sci Med Sport* 10(3): 156-163, 2007.

Strong WB, Malina RM, Blimkie CJ, Daniels SR, Dishman RK, Gutin B, Hergenroeder AC, Must A, Nixon PA, Pivarnik JM, , et al. Evidence based physical activity for school-age youth. *Journal of Pediatrics* 146: 732-737, 2005.

Suitor CW, Kraak VI. *Adequacy of evidence for physical activity guidelines development: Workshop summary*. Washington (DC): National Academies Press. 2007.

Symons CW, Cinelli B, James TC, Groff P. Bridging student health risks and academic achievement through comprehensive school health program. *J Sch Health* 67(6): 220-227, 1997.

Taras H. Physical activity and student performance at school. *J Sch Health* 75:214-218, 2005.

The Cooper Institute for Aerobics Research. *FITNESSGRAM*. Champaign (IL): Human Kinetics. 1999.

Trost SG, Pate RR, Sallis JF, Freedson PS, Taylor WC, Dowda M, Sirard J. Age and gender differences in objectively measured physical activity in youth. *Med Sci Sports Exerc* 34(2): 350-355, 2002.

[USDHHS] U.S. Department of Health and Human Services, National Institutes of Health, National Heart, Lung, and Blood Institute (United States). *The practical guide: Identification, evaluation and treatment of overweight and obesity in adults* [Internet]. [cited NIH Pub. No:. 00-4084. www.nhlbi.nih.gov/guidelines/obesity/prctgd_c.pdf, 2000 October.

[USDHHS] U.S. Department of Health and Human Services (United States). *Dietary Guidelines for Americans* [Internet]. [cited 2008 July 9]. Available from: www.health.gov/dietaryguidelines/dga2005/report/. 2005.

[USDHHS] U.S. Department of Health and Human Services (United States). *2008 Physical Activity Guidelines for Americans* Washington, D.C. U.S. Department of Health and Human Services [Internet]. [cited 2008 October 7]. Available from: www.health.gov/paguidelines. 2008.

Valerio G, Spagnuolo M, Lombardi F, Spadaro R, Siano M, Franzese A. Physical activity and sports participation in children and adolescents with type 1 diabetes mellitus. *Nutr Metab Cardiovasc Dis* 17(5): 376-382, 2003.

Vincent SD, Pangrazi RP. An examination of the activity patterns of elementary school children. *Pediatr Exerc Sci* 14(4): 432-452, 2002.

Washington RL, Bernhardt DT, Gomez J, Johnson MD, Martin TJ, Rowland TW, Small E, LeBlanc C, Krein C, Malina R et al. American Academy of Pediatrics Committee on Sports Medicine and Fitness strength training by children and adolescents. *Pediatrics* 107(6): 1470-1472, 2001.

Whaley MH, ed. Exercise testing and prescription for children and elderly. In: *ACSM's Guidelines for Exercise Testing and Prescription.* 7th ed. Baltimore: Lippincott, Williams and Wilkins. 237-245. 2005.

[WHO] World Health Organization. Global Strategy on Diet, Physical Activity and Health. *Recommended Amount of Physical Activity* [Internet]. [cited 2008 September 9]. World Health Organization. Available from: www.who.int/dietphysicalactivity/factsheet_recommendations/en/index.html.

[WHO] World Health Organization. Global Strategy on Diet, Physical Activity and Health. *Physical Activity and Young People.* [Internet]. [cited 2008 September 9]. Geneva. Available from: www.who.int/dietphysicalactivity/factsheet_young_people/en/index.html. 2008.

Chapter 6

Alleyne J, Pettica P. *Position Statement: Exercise and Pregnancy Discussion Paper.* [Internet] http://www.casm-acms.org/documents/PregnancyDiscussionPaper.pdf [cited 2009 8 October]

[ACOG] American College of Obstetricians and Gynecologists: ACOG Committee opinion. Number 267, January 2002: Exercise during pregnancy and the postpartum period. *Obstet Gynecol* 99: 171-173, 2002.

[ACSM] American College of Sports Medicine. Exercise prescription for healthy populations: Pregnancy. In: *ACSM's Guidelines for Exercise Testing and Prescription.* 8th ed. Baltimore: Lippincott, Williams and Wilkins. 183-187. 2010.

[ACE] American Council on Exercise. Fit Facts: Exercise and Pregnancy, 2001. [Internet] [cited 2009 October 8] http://www.acefitness.org/FITFACTS/pdfs/fitfacts/itemid_44.pdf

American Pregnancy Association Exercise Guidelines During Pregnancy. [Internet]. [cited 2009 October 8]. www.americanpregnancy.org/pregnancyhealth/exercise-guidelines.html

Artal R, Clapp JF, Vigil DV. *ACSM current comment: Exercise during pregnancy.* Indianapolis (IN): American College of Sports Medicine. 2000. Available from: www.acsm.org/AM/Template.cfm?Section=current_comments1&Template=/CM/ContentDisplay.cfm&ContentID=8638

Artal R, Masaki DI, Khodiguian N, Romem Y. Exercise prescription in pregnancy: weight-bearing versus non-weight-bearing exercise. *Am J Obstet Gynecol* 161: 1464, 1989.

Artal R, O'Toole M. Guidelines of the American College of Obstetricians and Gynecologists for exercise during pregnancy and the postpartum period. *Br J Sports Med* 37: 6-12, 2003.

Artal R, Romen Y, Paul RH, Wiswell R. Fetal bradycardia induced by maternal exercise. *Lancet* 2: 258-260, 1984.

Artal R. Lockwood CJ, Barss VA. UpToDate. *Anatomical and physiological changes of pregnancy and exercise* [Internet]. [cited 2008 May 12]. Available from: www.uptodate.com.

Artal R. Lockwood CJ, Barss VA. UpToDate. *Recommendations for exercise during pregnancy and the postpartum period* [Internet]. [cited 2008 May 12]. Available from: www.uptodate.com

Bell BB, Dooley MMP. Royal College of Obstetricians and Gynaecologists statement number 4: Exercise in pregnancy. 1-7, 2006.

Bell BB, Dorset, Dooley MMP [RCOG] Royal College of Obstetrics and Gynecology. Exercise in Pregnancy (RCOG Statement 4) January 1, 2006, London [Internet] [cited 2009 October 8] www.rcog.org.uk/womens-health/clinical-guidance/exercise-pregnancy.

Bessinger RC, McMurray RG, Hackney AC. Substrate utilization and hormonal responses to moderate intensity exercise during pregnancy and after delivery. *Am J Obstet Gynecol* 186: 757-764, 2002.

Boissoneault JS, Blashack ML. Incidence of diastasis recti abdominis during the childbearing year. *Phys Ther* 68: 1082-1086, 1998.

Bopp MF, Lovelady CA, Hunter CP, Kinsella TC. Maternal diet and exercise: Effects on long-chain polyunsaturated fatty acid concentrations in breast milk. *J Amer Diet Assoc* 105: 1098-1103, 2005.

Borodulin KM, Evenson KR, Wen F, Herring AH, Benson AM . Physical activity patterns during pregnancy. *Med Sci Sports Exerc* 40(11): 1901-1908, 2008.

Bung P, Artal R. Gestational diabetes and exercise: a survey. *Semin Perinatol* 20: 328-333, 1996.

Carpenter MW, Sady SP, Hoegsberg B, Sady MA. Fetal heart rate response to maternal exertion. *JAMA* 259: 3006-3009, 1988.

Clapp J. Exercise–risk and benefits. In: Queenan JT, Hobbins JC, Spong CY, eds. *Protocols for High-Risk Pregnancies.* 4th ed. Ames (IA): Blackwell Publishing. 45-53. 2005.

Clapp JF III, Kim H, Burciu B, Lopez B. Beginning regular exercise in early pregnancy: effect on fetoplacental growth. *Am J Obstet Gynecol.* 183:1484-1488, 2000.

Clapp JF. Fetal heart rate response to running in mid-pregnancy and late pregnancy. *Am J Obstet Gynecol* 153: 251-252, 1985.

Clapp JF, Capeless El. Neonatal morphometrics after endurance exercise during pregnancy. *Am J Obstet Gynecol.*163: 1805-1811, 1990.

Class JF III. Exercise during pregnancy. A clinical update. *Clin Sports Med* 19: 273-286, 2000.

Collings CA, Curet LB, Mullin JP. Maternal and fetal responses to a maternal aerobic exercise program. *Am J Obstet Gynecol.* 145: 702-707, 1983.

Davies GA, Wolfe LA, Mottola MF, MacKinnon C , Arsenault MY, Bartellas E, Cargill Y, Gleason T, Iglesias S, Klein M, et al. SOGC Clinical Practice Obstetrics Committee, Canadian Society for Exercise Physiology Board of Directors. Exercise in pregnancy and the postpartum period. *J Obstet Gynaecol Can* 25: 516-529, 2003a.

Davies G, Wolfe LA, Mottola MF. Joint SOGC/CSEP Clinical Practice Guidelines: Exercise in pregnancy and the postpartum period. *Can J Appl Physiol* 28: 329-341, 2003b.

Dempsey JC, Butler CL, Williams MA. No need for a pregnant pause: physical activity may reduce the occurrence of gestational diabetes mellitus and preeclampsia. *Exerc Sport Sci Rev* 33(3): 141-149, 2005.

Dewey KG, Lovelady CA, Nommsen-Rivers LA, McCrory MA, Lonnerdal B. A randomized study of the effects of aerobic exercise by lactating women on breast-milk volume and composition. *N Engl J Med* 330: 449-453, 1994.

Duncombe D, Skouteris H, Wertheim, EH, Kelly L, Fraser V, Paxton SJ . Vigorous exercise and birth outcomes in a sample of recreational exercisers: a prospective study across pregnancy. *Aust N Z J Obstet Gynaecol* 46: 288-292, 2006.

Entin PL, Munhall KM. Recommendations regarding exercise during pregnancy made by private/small group practice obstetricians in the USA. *J Sports Sci Med* 5: 449-458, 2006.

Evenson KR, Savitz DA, Huston SL. Leisure-time physical activity among pregnant women in the US. *Paediatr Perinat Epidemiol* 18(6): 400-407, 2004.

Gregory RL, Wallace JP, Fell LD, Marks J, King BA. Effect of exercise on milk immunoglobulin A. *Med Sci Sports Exerc* 29: 1596-1601, 1997.

Grisso JA, Main DM, Chiu G, Snyder ES. Effects of physical activity and lifestyle factors on uterine contraction frequency. *Am J Perinatol* 9: 489, 1992.

Hale RW, Milne L. The elite athlete and exercise in pregnancy. *Semin Perinatol* 20: 277-284, 1996. Hall DC, Kaufmann DA. Effects of aerobic and strength conditioning on pregnancy outcomes. *Am J Obstet Gynecol* 157: 1199-1203, 1987.

Hatch MC, Shu XO, McLean DE, Levin B, Begg M, Reuss L, Susser M. Maternal exercise during pregnancy, physical fitness, and fetal growth. *Am J Epidemiol* 137: 1105-1114, 1993.

Huch R. Physical activity at altitude in pregnancy. *Semin Perinatol*. 20(4): 303-314, 1996.

Kettles MA, Cole C, Wright B. Women's Health and Fitness Guide. Human Kinetics, Champaign, IL, 2006.

KidsHealth. Exercising During Pregnancy. [Internet] [cited 2009 October 8] http://kidshealth.org/parent/pregnancy_newborn/pregnancy/exercising_pregnancy.html

Klebanoff, MA, Shiono PH, Carey JC. The effect of physical activity during pregnancy on preterm delivery and birth weight. *Am J Obstet Gynecol* 163: 1450-1456, 1990.

Kramer MS. Aerobic exercise for women during pregnancy. *Cochrane Database Syst Rev* :CD000180, 2002.

Krandel KR, Kase T. Training in pregnancy women: effects on fetal development and birth. *Am J Obstet Gynecol* 178: 280-286, 1998.

Larsson L, Lindqvist PG. Low-impact exercise during pregnancy–a study of safety.

Acta Obstet Gynecol Scand. 84(1): 34-38, 2005

Lokey EA, Tran ZV, Wells CL, Myers BC, Tran AC. Effects of physical exercise on pregnancy outcomes: A meta-analytic review. *Med Sci Sports Exerc* 23: 1234-1239, 1991.

Lovelady CA, Hunter CP, Geigerman C. Effect of exercise on immunological factors in breast milk. *Pediatrics* 111: e148-e152, 2003.

Lovelady CA, Lonnerdal B, Dewey KG. Lactation performance of exercising women. *Am J Clin Nutr* 52: 103-109, 1990.

Lovelady CA, Nommsen-Rivers LA, McCrory MA, Dewey KG. Effects of exercise on plasma lipids and metabolism of lactating women. *Med Sci Sports Exerc* 27: 22-28, 1995.

Lovelady CA, Williams JP, Garner KE, Moreno KL, Taylor ML, Leklem JE. Effects of energy restriction and exercise on vitamin B-6 status of women during lactation. *Med Sci Sports Exerc* 33: 512-518, 2001.

Marcoux S, Brisson J, Fabia J. The effect of leisure time physical activity on the risk of preeclampsia and gestational hypertension. *J Epidemiol Community Health* 43: 147-152, 1989.

Marquez-Sterling S, Perry AC, Kaplan TA, Halberstein RA, Signorile JF. Physical and psychological changes with vigorous exercise in sedentary primigravidae. *Med Sci Sports Exerc* 32: 58-62, 2000.

McMurray RG, Mottola MF, Wolfe LA, Artal R. Recent advances in understanding maternal and fetal responses to exercise. *Med Sci Sports Exerc* 25: 1305-1321, 1993.

Morkved S, Bo K. The effect of post-natal exercises to strengthen the pelvic floor muscles. *Acta Obstet Gynecol Scand* 75: 382-385, 1996.

Mottola MF. Exercise in the postpartum period: Practical applications. *Curr Sports Med Rep* 1: 362-368, 2002.

Naeye RL, Peters EC. Working during pregnancy: Effects on the fetus. *Pediatrics* 69: 724, 1982.

Paisley TS, Joy EA, Price RJ. Exercise during pregnancy: A practical approach. *Curr Sports Med Rep* 2: 325-330, 2003.

Perkins CD, Pivarnik JM, Reeves MJ, Feltz DL, Womack CJ. Maternal physical activity and birth-weight: A meta-analysis. *Med Sci Sports Exerc* 27(Suppl 5): S177, 2005.

[PAGAC] Physical Activity Guidelines Advisory Committee. *Physical Activity Guidelines Advisory Committee Report, 2008*. Washington (DC): U.S. Department of Health and Human Services. 2008.

Pivarnik JM, Chambliss HO, Clapp JF, Dugan SA, Hatch MC, Lovelady CA, Mottola MF, Williams MA. Impact of physical activity during pregnancy and postpartum on chronic disease risk. *Med Sci Sports Exerc* 38: 989-1006, 2006.

Rooney BL, Schauberger CW, Mathiason MA. Impact of perinatal weight change on long-term obesity and obesity-related illnesses. *Obstet Gynecol* 106(6): 1349-1356, 2005.

Saftlas AF, Logsden-Sackett N, Wang W, Woolson R Bracken MB. Work, leisure-time physical activity, and risk of preeclampsia and gestational hypertension. *Am J Epidemiol* 160(8): 758-765, 2004.

Sampselle CM, Miller J, Mims BL, DeLancey JOL, Ashton-Miller JA. Antonakos CL. Effect of pelvic muscle exercise on transient incontinence during pregnancy and after birth. *Obstet Gynecol* 91: 406-412, 1998.

Sampselle CM, Seng J, Yeo SA, Killion C, Oakley D. Physical activity and postpartum well-being. *JOGN Nurs* 28: 41-49, 1999.

Scott S. Exercise during pregnancy. *ACSM's Health & Fitness Journal* 10(2): 37-39, 2006.

Scott S. Exercise in the postpartum period. *ACSM's Health & Fitness Journal.* 10(4): 40-41, 2006.

Smith BJ, Cheung NW, Bauman AE, Zehle K, McLean M. Postpartum physical activity and related psychosocial factors among women with recent gestational diabetes mellitus. *Diabetes Care* 28: 2650-2654, 2005.

Sorensen TK, Williams MA, Lee I-M. Recreational physical activity during pregnancy and risk of preeclampsia. *Hypertension* 41: 1273-1280, 2003.

Sternfeld B, Quesenberry CP Jr, Eskenazi B, Newman LA. Exercise during pregnancy and pregnancy outcome. *Med Sci Sports Exerc* 27: 634, 1995.

Suitor CW, Kraak VI. *Adequacy of evidence for physical activity guidelines development: Workshop summary.* Washington (DC): National Academies Press. 2007.

[USDHHS] U.S. Department of Health and Human Services (United States). *2008 physical activity guidelines for Americans* [Internet]. [cited 2008 October 7]. Washington, D.C. USDHHS. Available from: www.health.gov/paguidelines. 2008.

[USDHHS] U.S. Department of Health and Human Services (United States). *Dietary Guidelines for Americans* [Internet]. [cited 2008 July 9]. Available from: www.health.gov/dietaryguidelines/dga2005/report/. 2005.

Wang TW, Apgar BS. Exercise during pregnancy. *Am Fam Physician* 57(8): 1846-1852, 1998.

Whaley MH, ed. Other clinical conditions influencing exercise prescription. In: *ACSM's Guidelines for exercise testing and prescription.* 7th ed. Baltimore: Lippincott, Williams and Wilkins. 205-236. 2005.

Wolfe LA, Lowe-Wyldem SJ, Tranmer JE, McGrath MJ. Fetal heart rate during maternal static exercise. *Can J Sport Sci* 13: 95, 1988.

Wolfe LA, Mottola MF. *Parmed-X for pregnancy: Physical activity readiness medical examination.* Ottawa (ON): Can Soc Exerc Physiol. 1996.

Wright KS, Quinn TJ, Carey GB. Infant acceptance of breast milk after maternal exercise. Pediatrics;109:585-589, 2002.

Chapter 7

Abbott RD, White LR, Ross GW, Masaki KH, Curb JD, Petrovitch H. Walking and dementia in physically capable elderly men. *JAMA* 292(12): 1447-1453, 2004.

Adams KJ. *ACSM current comment: Strength, power, and the baby boomer* [Internet]. [cited 2008 July 2]. Indianapolis (IN): American College of Sports Medicine. Available from: www.acsm.org/AM/Template. cfm?Section=Current_Comments1&Template=/CM/ContentDisplay.cfm&ContentID=8653.

[ACSM] American College of Sports Medicine. Exercise prescription for healthy populations: Elderly People. In: *ACSM's Guidelines for exercise testing and prescription.* 8th ed. Baltimore: Lippincott, Williams and Wilkins. 190-194. 2010.

Australian Government (Department of Health and Ageing) (1999) *National Physical Activity Guidelines for Australians*, Canberra, Australia, 1999. [Internet] [cited 2009 October 9] http://fulltext.ausport.gov.au/fulltext/1999/feddep/physguide.pdf

Batty GD. Physical activity and coronary heart disease in older adults. A systematic review of epidemiological studies. *Eur J Public Health* 12(3): 171-176, 2002.

Bayles C. Frailty. In: Durstine JL, Moore GE, eds. *ACSM's Exercise management for persons with chronic diseases and disabilities.* 2nd ed. Champaign (IL): Human Kinetics. 157-163. 2003.

Bixby WR. Spalding TW, Haufler AJ, Deeny SP, Mahlow PT, Zimmerman JB, Hatfield BD. The unique relation of physical activity to executive function in older men and women. *Med Sci Sports Exerc* 39(8): 1408-1416, 2007.

Butler RN, Davis R, Lewis CB, Nelson ME Strauss E. Physical fitness: Benefits of exercising for the older patient. *Geriatrics* 53(10): 46-62, 1998.

California Center for Physical Activity: Guidelines for Physical Activity Across the Lifespan: Older Adult (>60 years), Sacramento, CA [Internet]. [cited 2009 October 8]. www.caphysicalactivity.org/facts_recomm1d.html.

[CDHS] California Department of Health Services, California Obesity Prevention Initiative Health Systems Workgroup (United States). *Obesity prevention for health care systems—Executive summary and literature review* [Internet]. [cited 2008 September 15]. Sacramento, CA: California Department of Health Services. Available from: www.caphysicalactivity.com. 2002.

[CDC] Centers for Disease Control and Prevention and The Merck Company Foundation. *The state of aging and health in America 2007* [Internet]. Whitehouse Station (NJ): The Merck Company Foundation; [cited 2008 July 25]. Available from: www.cdc.gov/aging. 2007.

Chodzko-Zajko W, Proctor DN, Fiatarone Singh MA, Minson CT, Nigg CR, Salem GJ Skinner JS. Exercise and physical acivity for older adults. *Med Sci Sports Exerc* 41(7): 1510-1530, 2009 July.

Department of Health (United Kingdom). *At least five a week: Evidence on the impact of physical activity and its relationship to health* [Internet]. [cited 2008 November 1]. London: Department of Health. Available from: www.dh.gov.uk/dr_consum_dh/groups/dh_digitalassets/@dh/@en/documents/digitalasset/dh_4080981.pdf . 2005.

Dutta C, Guralnik J, Blair SN, Buchner D, Chodzko-Zajko W, King AC, Milner C, Ory M, Pahor M, Prohaska T et al. *Exercise & Physical Activity: Your everyday guide from the National Institute on Aging.* Baltimore: National Institutes of Health National Institute on Aging. 2008. [Internet] [cited 2009 October 8] www.nia.nih.gov/Healthinformation/Publications/ExerciseGuide/

Fiatarone MA, Marks EC, Ryan ND, Meredith CN, Lipsitz LA, Evans WJ. High-intensity strength training in nonagenarians. Effects on skeletal muscle. *JAMA* 263: 3029-3034, 1990.

Fletcher GF, Balady GJ, Amsterdam EA, Chaitman B, Eckel R, Fleg J, Froelicher VF, Leon AS, Pina IL, Rodney R et al. Exercise standards for testing and training: A statement for health care professionals from the American Heart Association. *Circulation* 104: 1694-1740, 2001.

Frontera WF, Meredith CN. O'Reilly KP, Evans WJ. Strength conditioning in older men: Skeletal muscle hypertrophy and improved function. *J Appl Physiol* 64: 1038-1044, 1988.

[IOM] Institute of Medicine. *Retooling for an aging American: Building the health care workforce.* Washington (DC): National Academies Press. 2008.

Keenan TA. Physical activity survey, 2006. *AARP Bulletin* [Internet]. [cited 2008 July 2]. Washington, DC. Available from: www.research.aarp.org. 2006.

Klein DA. Flexibility in aging: Stretching to mend the bend. *ACSM Fit Society Page.* Summer: 1-11. 2003.

Leenders NYJM. The Elderly. In: Ehrman JK, Gordon PM, Visich PS, Keteyian SJ Eds. *Clinical Exercise Physiology.* Champaign (IL): Human Kinetics. 571-587. 2003.

Mazzeo RS. *ACSM current comment: Exercise and the older adult.* [Internet]. [cited 2008 July 2]. Indianapolis (IN): American College of Sports Medicine. Available from: www.acsm.org/cc.

Mazzeo RS, Cavanagh P, Evans W, Fiatarone M, Hagberg J, McAuley E, Startzell JK. ACSM position stand: Exercise and physical activity for older adults. *Med Sci Sports Exerc* 30(6): 992-1008, 1998.

Morganti CM, Nelson ME, Fiatarone MA, Dallal GE, Economos CD, Crawford BM, Evans WJ. Strength improvements with 1 yr of progressive resistance training in older women. *Med Sci Sports Exerc* 27(6): 906-912, 1995.

Nelson ME, Rejeski WJ, Blair SN, Duncan PW, Judge JO, King AC, Macera CA, Castaneda-Sceppa C. Physical activity and public health in older adults: Recommendation from the American College of Sports Medicine and the American Heart Association. *Med Sci Sports Exerc* 39(8): 1435-1445, 2007.

Nelson ME, Rejeski WJ, Blair SN, Duncan PW, Judge JO, King AC, Macera CA, Castaneda-Sceppa C American College of Sports Medicine; American Heart Association. Physical activity and public health in older adults: recommendations from the American College of Sports Medicine and the American Heart Association. *Circulation.* 2007;116(9):1094-1105.

[PAGAC] Physical Activity Guidelines Advisory Committee. *Physical Activity Guidelines Advisory Committee Report, 2008.* Washington (DC): U.S. Department of Health and Human Services. 2008. Available from: www.health.gov/paguidelines.

Potteiger, JA (ed.). Exercise and the older adult. *ACSM Fit Society Page.* Summer: 1-11, 2003.

Public Health Agency of Canada/Canadian Society for Exercise Physiology. *Canada's Physical Activity Guide to Healthy Active Living for Older Adults* [Internet]. [cited 2008 November 1]. Ottawa, Ontario: Public Health Agency of Canada. Available from: www.phac-aspc.gc.ca/pau-uap/paguide/older/phys_guide.html. 2002.

Robert Wood Johnson Foundation. *National blueprint: Increasing physical activity among adults 50 and older* [Internet]. [cited 2008 November 6]. Princeton, NJ: Robert Wood Johnson Foundation. Available from: www.agingblueprint.org/PDFs/Final_Blueprint_Doc.pdf. 2001.

Rogers ME. (National Center on Physical Activity and Disability, Chicago IL). *First steps to active health: Balance and flexibility exercises for older adults* [Internet]. [cited 2008 July 28]. Chicago: National Center on Physical Activity and Disability. Available from: www.ncpad.org. 2006.

Sallis JF. Age-related decline in physical activity: A synthesis of human and animal studies. *Med Sci Sports Exerc* 32(9): 1598-1600, 2000.

Sevick MA. Bradham DD, Muender M, Chen GJ, Enarson C, Dailey M, Ettinger WH. Cost-effectiveness of aerobic and resistance exercise in seniors with knee osteoarthritis. *Med Sci Sports Exerc* 32(9): 1534-1540, 2000.

[USDHHS] U.S. Department of Health and Human Services (United States). Centers for Disease Control and Prevention. *Promoting active lifestyles among older adults* [Internet]. [cited 2008 June 27]. Atlanta, GA. Available from: www.mhqp.org/guidelines/preventivePDF/CDC_AdultAcvt.pdf.

[USDHHS] U.S. Department of Health and Human Services (United States), Dietary Guidelines for Americans, 2005. Washington, D.C. [Internet] [cited 2008 August 9] www.health.gov/dietaryguidelines/dga2005/report/

[USDHHS] U.S. Department of Health and Human Services (United States). *2008 physical activity guidelines for Americans* [Internet]. [cited 2008 October 7]. Washington, DC, U.S. Department of Health and Human Services. Available from: www.health.gov/paguidelines. 2008.

Weuve J, Kang JH, Manson JE Breteler MM, Ware JH, Grodstein F. Physical activity, including walking, and cognitive function in older women. *JAMA* 292(12): 1454-1461, 2004.

Whaley MH, ed. Other clinical conditions influencing exercise prescription. In: *ACSM's guidelines for exercise testing and prescription.* 7th ed. Baltimore: Lippincott, Williams and Wilkins. 246-251. 2005.

[WHO] World Health Organization. *Keep fit for life. Meeting the nutritional needs of older persons.* Geneva: World Health Organization. 2002. [Internet] [cited 2009 October 8] www.who.int/nutrition/publications/olderpersons/en/index.html

[WHO] World Health Organization. Global Strategy on Diet, Physical Activity and Health. *Recommended amount of physical activity* [Internet]. [cited 2008 September 9]. Geneva, Switzerland: World Health Organization. Available from: www.who.int/dietphysicalactivity/factsheet_old-eradults/en/index.html. 2008.

[WHO] World Health Organization, Ageing and Health Programme. *Growing older–staying well: Ageing and physical activity in everyday life.* Geneva: World Health Organization. Geneva, Switzerland. 1998.

[WHO] Physical activity through transport as part of daily activities. World Health Organization Regional Office for Europe, Copenhagen, Netherlands, 2002. [Internet] [cited 2009 October 8] http://www.euro.who.int/document/Trt/Booklet.pdf.

Chapter 8

[ACSM] American College of Sports Medicine. Exercise prescription for other clinical populations. In: *ACSM's Guidelines for Exercise Testing and Prescription.* 8th ed. Baltimore: Lippincott, Williams and Wilkins. 250-253. 2010.

[ADA] American Diabetes Association. Executive Summary: Standards of Medical Care in Diabetes—2008 *Diabetes Care* 31:S5-S11, 2008.

[ADA] American Diabetes Association Diagnosis and Classification of Diabetes Mellitus *Diabetes Care* S12-S54, 2008.

Bardia A, Hartmann LC, Vachon CM, Vierkant RA, Wang AH, Olson JE, Sellers TA, Cerhan JR. Recreational physical activity and risk of postmenopausal breast cancer based on hormone receptor status. *Arch Intern Med* 166: 2478-2483, 2006.

Byers T, Nestle M, McTiernan A, Doyle C, Currie-Williams A, Gansler T, Thun M . American Cancer Society guidelines on nutrition and physical activity for cancer prevention: Reducing the risk of cancer with healthy food choices and physical activity. *CA Cancer J Clin* 52: 92-119, 2002.

Calle EE, Murphy TK, Rodriguez D, Thun MJ, Heath CW Jr. Diabetes mellitus and pancreatic cancer mortality in a prospective cohort of United States adults. *Cancer Causes Control* 9: 402-410, 1998.

Chiasson JL, Josse RG, Gomis R, Hanefeld M, Karasik A, Laakso M. Acarbose for prevention of type 2 diabetes mellitus: The STOP-NIDDM randomial trial. *Lancet* 359: 2072-2077, 2002.

Dallal CM, Sullivan-Halley J, Ross RK, Wang Y, Deapen D, Horn-Ross PL, Reynolds P, Stram DO, Clarke CA, Anton-Culver H, et al. Long-term recreational physical activity and risk of invasive and in situ breast cancer: The California teachers study. *Arch Intern Med.* 167: 408, 2007.

Frezza EE, Wachtel MS, Chiriva-Internati M. The influence of obesity on the risk of developing colon caner. *Gut* 55(2): 285-291, 2005.

Friedenreich CM, Cust AE Physical activity and breast cancer risk: impact of timing, type and dose of activity and population subgroup effects *Br J Sports Med*; 42: 636 – 647, 2008.

Frisch RE, Wyshak G, Albright NL, Albright TE, Schiff I, Witschi J, Marguglio M. Lower prevalence of breast cancer

and cancers of the reproductive system among former college athletes compared to non-athletes. *Br J Cancer* 52: 885-891, 1985.

Gerstein HC, Yusuf S, Bosch J, Pogue J, Sheridan P, Dinccag N, Hanefeld M, Hoogwerf B, Laakso M, Mohan V, et al. Effect of rosiglitazone on the frequency of diabetes in patients with impaired glucose tolerance or impaired fasting glucose: A randomized controlled trial. *Lancet* 368: 1096-1105, 2006.

Grundy SM, Cleeman JI, Daniels SR, Donato KA, Eckel RH, Franklin BA, Gordon DJ, Krauss RM, Savage PJ, Smith SC Jr, et al. Diagnosis and management of the metabolic syndrome: An American Heart Association/National Heart, Lung, and Blood Institute scientific statement: Executive summary. *Circulation* 112: e285-e290, 2005. [Internet] [cited 2009 October 9] www.circ.ahajournals.org/cgi/content/full/112/17/e285.

International Diabetes Federation. *Diabetes Prevalence* [Internet]. [cited 2008 August 19]. Brussels, Belgium: International Diabetes Federation. Available from: www.idf.org.

International Diabetes Federation. *Did you know?* [Internet]. [cited 2008 August 19]. Brussels, Belgium: International Diabetes Federation. Available from: www.idf.org.

Knowler WC, Barrett- Connor E, Fowler SE, Hamman RF, Lachin JM, Walker EA, Nathan DM. Reduction in the incidence of type 2 diabetes with lifestyle intervention or metformin. *N Engl J Med* 346: 393-403, 2002.

Kriska A. Physical activity and the prevention of Type II (non-insulin-dependent) diabetes. *PCPFS Research Digest* 2(10): 1-12, 1997.

Kushi LH, Byers T, Doyle C, Bandera EV, McCullough M, Gansler T, Andrews KS. American Cancer Society guidelines on nutrition and physical activity for cancer prevention: Reducing the risk of cancer with healthy food choices and physical activity. *CA Cancer J Clin* 56: 254-281, 2006.

Laaksonen DE, Lakka HM, Salonen JT, Niskanen LK, Rauramaa R, Lakka TA . Low levels of leisure-time physical activity and cardiorespiratory fitness predict development of the metabolic syndrome. *Diabetes Care* 25(9): 1612-1618, 2002.

Lee I-M. Physical activity and cancer. *PCPFS Research Digest* 2(2): 1-9, 1995.

Levi F, La Vecchia C, Negri E, Franceschi S. Selected physical activities and risk of endometrial cancer. *B J Cancer* 67: 846-851, 1993.

Lindblad P, Chow WH, Chan J, Bergström A, Wolk A, Gridley G, McLaughlin JK, Nyren O, Adami HO. The role of diabetes mellitus in the aetiology of renal cell cancer. Diabetologia. 42: 107-112, 1999.

Martinez ME, Giovannucci E, Speigelman D, Hunter DJ, Willett WC, Colditz GA . Leisure-time physical activity, body size, and colon cancer in women. Nurses' Health Study Research Group. *J Natl Cancer Inst.* 89: 948-955, 1997.

Maruti SS, Willett WC, Feskanich D, Rosner B, Colditz GA. A prospective study of age-specific physical activity and premenopausal breast cancer. *J Natl Cancer Inst* 100(10): 728-737, 2008.

Mazzeo RS. The influence of exercise and aging on immune function. *Med Sci Sports Exerc* 26: 586-592, 1994.

McGinnis JM, Foege WH. Actual Causes of Death in the United States. Journal of the American Medical Association, 270(18): 2207–2212, 1993.

McTiernan A, Kooperberg C, White E, Wilcox S; Coates R; Adams-Campbell LL; Woods N; Ockene J. Recreational physical activity and the risk of breast cancer in postmenopausal women: the Women's Health Initiative Cohort Study. *JAMA* 290: 1331-1336, 2003.

McTiernan A, Tworoger SS, Ulrich CM, Yasui Y, Irwin ML, Rajan KB, Sorensen B, Rudolph RE, Bowen D, Stanczyk FZ, et al. Effect of exercise on serum estrogens in postmenopausal women: 12 month randomized clinical trial. *Cancer Res* 64(8): 2923-2928, 2004.

McTiernan A, Ulrich C, Slate S, Potter, J. Physical activity and cancer etiology: Associations and mechanisms. *Cancer Causes Control* 9: 487-509. 1998.

Mokdad AH, Marks JS, Stroup DF, Gerberding JL. Actual causes of death in the United States, 2000. *JAMA* 291(10): 1238, 1241, 2004.

[NCI] National Cancer Institute. *Questions and answers: Physical activity and cancer* [Internet]. [cited 2008 May 20]. Bethesda, Maryland: National Cancer Institute. Available from: www.cancer.gov/cancertopics/factsheet/prevention/physicalactivity.

[NCEP] National Cholesterol Education Program, Expert Panel on Detection, Evaluation, and Treatment of High Blood Cholesterol in Adults (Adult Treatment Panel III). Third Report of the National Cholesterol Education Program (NCEP) Expert Panel on Detection, Evaluation, and Treatment of High Blood Cholesterol in Adults (Adult Treatment Panel III) final report. *Circulation* 106: 3143-3421, 2002.

Pan XR, Li GW, Hu YH, Wang JX, Yang WY, An ZX, Hu ZX, Lin J, Xiao JZ, Cao HB, et al. Effects of diet and exercise in preventing NIDDM in people with impaired glucose tolerance. The Da Qing IGT and Diabetes Study. *Diabetes Care.* 20: 537-544, 1997.

[PAGAC] Physical Activity Guidelines Advisory Committee. *Physical Activity Guidelines Advisory Committee Report, 2008*. Washington (DC): U.S. Department of Health and Human Services. 2008.

Pronck N. The metabolic syndrome–the "syndrome X factor"–at work. *ACSM's Heath & Fitness Journal* 10(5): 38-42, 2006.

Ramachandran A, Snehalatha C, Mary S, Selvam S, Kumar CK, Seeli AC, Shetty AS. The Indian Diabetes Prevention Programme shows that lifestyle modification and metformin prevent type 2 diabetes in Asian Indian subjects with impaired glucose tolerance (IDPP-1). *Diabetologia* 49: 289-297, 2006.

Rapp K, Schroeder J, Klenk J, Ulmer H, Concin H, Diem G, Oberaigner W, Weiland SK. Obesity and incidence of cancer: A large cohort study of over 145,000 adults in Austria. *Br J Cancer* 93: 1062-1067, 2005.

Rockhill B, Willett WC, Hunter DJ, Manson JE, Hankinson SE, Spiegelman D, and Colditz GA . A prospective study of recreational physical activity and breast cancer risk. *Arch Intern Med* 159: 2290-2296, 1999.

Shephard RJ, Rhind S, Shek PN. The impact of exercise on the immune system: NK cells interleukins 1 and 2, and related responses. *Exerc Sport Sci Rev* 23: 215-241, 1995(a).

Shephard RJ, Shek PN. Associations between physical activity and susceptibility to cancer: Possible mechanisms. *Sports Med* 26(5): 293-315, 1998.

Shephard RJ, Shek PN. Cancer, immune function, and physical activity. *Can J Appl Physiol* 20: 1-25, 1995(b).

Slattery ML, Edwards SL, Ma KN, Friedman GF, Potter JD. Physical activity and colon cancer: A public health perspective. *Ann Epidemiol* 7: 137-145. 1997.

Slattery ML. Physical activity and colorectal cancer. *Sports Med* 34(4): 239-252. 2004.

Tuomilehto J, Lindstrom J, Eriksson JG, Valle TT, Hamalainen H, Ilanne-Parikka P, Keinanen-Kiukaanniemi S, Laakso M, Louheranta A, Rastas M, et al. Prevention of type 2 diabetes mellitus by changes in lifestyle among subjects with impaired glucose tolerance. *N Engl J Med* 344: 1343-1350, 2001.

United Kingdom Testicular Cancer Study Group. Aetiology of testicular caner: Association with congenital abnormalities, age at puberty, infertility, and exercise. *Br Med J* 308: 1393-1399, 1994.

[USDHHS] U.S. Department of Health and Human Services (United States). *2008 physical activity guidelines for Americans* [Internet]. [cited 2008 October 7]. Washington, D.C. [USDHHS] U.S. Department of Health and Human Services. Available from: www.health.gov/paguidelines. 2008.

Vainio H, Bianchini F, eds. *Weight control and physical activity, IARC handbooks of cancer prevention, volume 6. Lyon, France, IARC Press*: IARC Press. 2002.

Vihko RK, Apter DL. The epidemiology and endocrinology of the menarche in relation to breast cancer. *Cancer Surv* 5: 561-571, 1986.

Will JC, Galuska DA, Vinicor F, Calle EE. Colorectal cancer: Another complication of diabetes mellitus? *Am J Epidemiol* 147: 816-825, 1998.

Wolk A, Gridley G, Svensson M, Nyren O, McLaughlin JK, Fraumeni JF, Adami HO. A prospective study of obesity and cancer risk (Sweden). *Cancer Causes Control* 12(1): 1321, 2001.

World Cancer Research Fund/American Institute for Cancer Research. *Food, nutrition, physical activity, and the prevention of cancer: A global perspective*. Washington (DC): AICR. 2007. [Internet] [cited 2009 October 8] www.dietandcancerreport.org.

Chapter 9

Ahmed RL, Thomas W, Yee D, Schmitz KH. Randomized controlled trial of weight training and lymphedema in breast cancer survivors. *J Clin Oncol* 24(18): 2765-2772, 2006.

[ACS] American Cancer Society. *Nutrition and physical activity during and after cancer treatment: An American Cancer Society guide for informed choices* [Internet]. [cited 2008 April 18]. Atlanta, Georgia. Available from: http://caonline.amcancersoc.org/cgi/content/full/56/6/323. 2006.

[ACS] American Cancer Society Physical activity and the cancer patient. Atlanta, Georgia [Internet]. [cited 2008 April 30]. www.cancer.org/docroot/mit/content/mit_2_3x_physical_activity_and_the_cancer_patient.asp

[ACSM] American College of Sports Medicine. Exercise prescription for other clinical populations. In: *ACSM's guidelines for exercise testing and prescription*. 8th ed. Baltimore: Lippincott, Williams and Wilkins. 228-232. 2010.

Bailar JC, Smith EM. Progress against cancer? *N Engl J Med* 314(19): 1226-1232, 1986.

Cho OH, Yoo YS, Kim NC. Efficacy of comprehensive group rehabilitation for women with early breast cancer in South Korea. *Nurs Health Sci* 8(3): 140-146, 2006.

Courneya KS. Physical exercise and quality of life in postsurgical colorectal cancer patients. *Psychol Health Med* 4: 181-187, 2004.

Courneya KS. A randomized trial of exercise and quality of life in colorectal cancer survivors. *Eur J Cancer Care* 12: 347-357, 2003.

Culos-Reed SN, Carlson LE, Daroux LM, Hately-Aldous S. A pilot study of yoga for breast cancer survivors: Physical and psychological benefits. *Psychooncology* 15(10): 891-897, 2006.

Daley AJ, Crank H, Saxton JM, Mutrie N, Coleman R, Roalfe A. Randomized trial of exercise therapy in women treated for breast cancer. *J Clin Oncol* 25(13): 1713-1721, 2007.

Doyle C, Kushi LH, Byers T, Courneya KS, Demark-Wahnefried W, Grant B, McTiernan A, Rock CL, Thompson C, Gansler T, Andrews KS and the 2006 Nutrition, Physical Activity and Cancer Survivorship Advisory Committee . Nutrition and physical activity during and after cancer treatment: An American Cancer Society Guide for informed choices. *CA Cancer J Clin* 56: 323-353, 2006.

Enger SM. Exercise activity, body size and premenopausal breast cancer survival. *Br J Cancer*, 90: 2138-2141, 2004

Herrero F, San Juan AF, Fleck SJ, Foster C, Lucia A. Combined aerobic and resistance training in breast cancer survivors: A randomized controlled pilot trial. *Int J Sports Med* 27(7): 573-580, 2006.

Holmes MD. Physical activity and survival after breast cancer diagnosis. *JAMA* 293: 2479-2486, 2005.

Hutnick NA, Williams NI, Kraemer WJ, Orsega-Smith E, Dixon RH, Bleznak AD, Mastro AM. Exercise and lymphocyte activation following chemotherapy for breast cancer. *Med Sci Sports Exerc* 37(11): 1827-1835, 2005.

McKenzie DC, Kalda AL. Effect of upper extremity exercise on secondary lymphedema in breast cancer patients: A pilot study. *J Clin Oncol* 21(3): 463-466, 2003.

Meloni G. Obesity and autologous stem cell transplantation in acute myeloid leukemia. *Bone Marrow Transplant* 28: 365-367, 2001.

Meyerhardt, JA. Physical activity and survival after colorectal cancer diagnosis. *J Clin Oncol* 24: 3527-3534, 2006.

Meyerhardt, JA. Impact of physical activity on cancer recurrence and survival in patients with stage III colon cancer: Findings from CALGB 89803. *J Clin Oncol* 24: 3535-3541, 2006.

Mustian KM, Katula JA, Zhao H. A pilot study to assess the influence of tai chi chuan on functional capacity among breast cancer survivors. *J Support Oncol* 4(3): 139-145, 2006.

NCI National Cancer Institute. Cancer Trends Progress Report, Bethesda, Maryland, 2007. [Internet]. [cited 2008 April 30]. http://progressreport.cancer.gov/.

National Cancer Institute. *DevCan: Probability of developing or dying of cancer software* [Computer program]. Version 6.2.1. Bethesda (MD): Statistical Research and Applications Branch. [cited 2008 November 13]. Available from: www.srab.cancer.gov/devcan. 2007.

[PAGAC] Physical Activity Guidelines Advisory Committee. *Physical Activity Guidelines Advisory Committee report, 2008*. Washington (DC): U.S. Department of Health and Human Services. 2008.

Pinto BM, Frierson GM, Rabin C, Trunzo JJ, Marcus BH. Home-based physical activity intervention for breast cancer patient. *J Clin Oncol* 23(15): 3577-3587, 2005.

Sandel SL, Judge JO, Landry N, Faria L, Ouellette R, Majczak M. Dance and movement program improves quality-of-life measures in breast cancer survivors. *Cancer Nurs* 28(4): 301-309, 2005.

Schairer JR, Keteyian SJ. Cancer. In: Ehrman JK, Gordon PM, Visich PS, Keteyian SJ, eds. *Clinical exercise physiology*. Champaign (IL): Human Kinetics. 403-418. 2003.

Schmitz KH, Holtzman J, Courneya KS, Masse LC, Duval S, Kane R. Controlled physical activity trials in cancer survivors: A systematic review and meta-analysis. *Cancer Epidemiol Biomarkers Prev* 14(7): 1588-1595. 2005..

Schwartz AL. Cancer. In: Durstine JL and Moore GE, eds. *ACSM's exercise management for persons with chronic disease and disabilities*. 2nd ed. Champaign (IL): Human Kinetics. 166-172. 2003.

Schwartz AL. *Cancer fitness: Exercise programs for patients and survivors*. New York: Fireside. 2004.

Segal RJ. Resistance exercise in men receiving androgen deprivation therapy for prostate cancer. *J Clin Oncol* 21: 1653-1659, 2003.

Tartter PI, Gajdos C, Rosenbaum Smith S, Estabrook A, Rademaker AW. The prognostic significance of Gail model risk factors for women with breast cancer. *Am J Surg*. 184(1): 11-15, 2002.

Tartter PI. Cholesterol, weight, height, Quetelet's index and colon cancer recurrence. *J Surg Oncol* 20: 3302-3316, 2002.

[USDHHS] U.S. Department of Health and Human Services (United States). *2008 physical activity guidelines for Americans* [Internet]. [cited 2008 October 7]. Washington, DC. [USDHHS] U.S. Department of Health and Human Services. Available from: www.health.gov/paguidelines. 2008.

Windsor PM. A randomized, controlled trial of aerobic exercise for treatment-related fatigue in men receiving radical external beam radiotherapy for localized prostate carcinoma. *Cancer* 101: 550-557, 2004.

Chapter 10

Ades PA, Gunther P, Meacham CP, Handy MA, LeWinter MM. Hypertension, exercise, and beta-adrenergic blockade. *Ann Intern Med* 109: 629-634, 1988.

[AACVPR] American Association of Cardiovascular and Pulmonary Rehabilitation. *Guidelines for cardiac rehabilitation and secondary prevention programs*. 4th ed. Champaign (IL): Human Kinetics. 2004.

[ACSM] American College of Sports Medicine. Exercise prescription for other clinical populations. In: *ACSM's Guidelines for Exercise Testing and Prescription*. 8th ed. Baltimore: Lippincott, Williams and Wilkins. 248-250. 2010.

[ACSM] American College of Sports Medicine. Exercise prescription for patients with cardiovascular disease. In: *ACSM's Guidelines for Exercise Testing and Prescription*. 8th ed. Baltimore: Lippincott, Williams and Wilkins. 211-222. 2010.

[ACSM] American College of Sports Medicine. Position stand: Physical activity, physical fitness, and hypertension. *Med Sci Sports Exerc* 25(10): i-x, 1993.

[ACSM] American College of Sports Medicine. Position stand: Exercise for Patients with Coronary Artery Disease. *Med Sci Sports Exerc* 26(3):i-v, 1994.

[ACE] American Council on Exercise. Exercising with heart disease. *Fit Facts* [Internet]. [cited 2008 September 16]. San Diego, California: American Council on Exercise. Available from: www.acefitness.org/fitfacts. 2001.

[AHA] American Heart Association. *Heart disease and stroke statistics–2009 update* 2009 [Internet]. [cited 2008 December 16]. Dallas, Texas: American Heart Association. Available from: www.americanheart.org/presenter. jhtml?identifier=1200026.

Cleroux J, Feldman RD, Petrella RJ. Recommendations on physical exercise training. *CMAJ* 160(9 Suppl): S21-S28, 1999.

Contractor AS, Gordon NF. Hypertension. In: Ehrman JK, Gordon PM, Visich PS, Keteyian SJ, eds. *Clinical exercise physiology*. Champaign (IL): Human Kinetics. 281-296. 2003.

Ehrman JK. Myocardial infarction. In: Ehrman JK, Gordon PM, Visich PS, Keteyian SJ, eds. *Clinical exercise physiology*. Champaign (IL): Human Kinetics. 201-226. 2003.

Franklin BA. Coronary artery bypass graft surgery and percutaneous transluminal coronary angioplasty. In: Durstine JL, Moore GE, eds. *ACSM's exercise management for persons with chronic diseases and disabilities*. 2nd ed. Champaign (IL): Human Kinetics. 32-39. 2003.

Franklin BA. Myocardial infarction. In: Durstine JL, Moore GE, eds. *ACSM's exercise management for person with chronic diseases and disabilities*. 2nd ed. Champaign (IL): Human Kinetics. 24-31. 2003.

Franklin BA. Coronary Artery Bypass Graft Surgery and Percutaneous Transluminal Coronary Angioplasty. In: Durstine JL, Moore GE, eds. *ACSM's exercise management for person with chronic diseases and disabilities*. 2nd ed. Champaign (IL): Human Kinetics. 32-39. 2003.

Franklin BA, Balady GJ, Berra K, Gordon NF, Pollock ML. *ACSM current comment: Exercise for persons with cardiovascular disease* [Internet]. [cited 2008 June 12]. Indianapolis (IN): American College of Sports Medicine. Available from: www.acsm.org/cc. or http://www.acsm.org/AM/Template.cfm?Section=current_comments1&Template=/CM/ContentDisplay.cfm&ContentID=8639

Gordon NF. Hypertension. In: Durstine JL, Moore GE, eds. *ACSM's Exercise management for person with chronic diseases and disabilities*. 2nd ed. Champaign (IL): Human Kinetics. 76-80. 2003.

Hagberg JM. (American College of Sports Medicine). *Public information brochure. Exercise your way to lower blood pressure* 2005 [Internet]. [cited 2008 June 12]. Indianapolis (IN): American College of Sports Medicine. Available from: www.acsm.org/AM/Template.cfm?Section=brochures2&Template=/CM/ContentDisplay.cfm&ContentID=1733.

Hagberg JM, Park J-J, Brown MD. The role of exercise training in the treatment of hypertension: An update. *Sports Med* 30(3): 193-206, 2000.

Kelley GA, Kelley KS. Progressive resistance exercise and resting blood pressure: A meta-analysis of randomized controlled trials. *Hypertension*. 35: 838-843, 2000.

Marzolini S, Oh PI, Thomas SG, Goodman JM. Aerobic and resistance training in coronary disease: Single versus multiple sets. *Med Sci Sports Exerc* 40(9): 1557-1564, 2008.

McDonnell TR, Laubach CA Jr. Revascularization of the Heart. In: Ehrman JK, Gordon PM, Visich PS, Keteyian SJ, eds. *Clinical exercise physiology*. Champaign (IL): Human Kinetics. 227-242. 2003.

[NHBPEP] National High Blood Pressure Education Program (United States). *The seventh report of the joint national committee on prevention, detection, evaluation, and treatment of high blood pressure (JNC 7)* [Internet]. [cited 2008 August 6]. Bethesda (MD): National Heart, Lung, and Blood Institute. NIH Publication No.: 03-5233. Available from: www.nhlbi.nih.gov/guidelines/hypertension. 2003.

Pescatello LS, Franklin BA, Fagard R, Farquhar WB, Kelley GA, Ray CA. Exercise and hypertension. *Med Sci Sports Exerc* 36(3): 533-553, 2004.

Pollock ML, Franklin BA. Balady GJ, Chaitman BL, Fleg JL, Fletcher B, Limacher M, Piña IL, Stein RA, Williams M, Bazzarre T. Resistance exercise in individuals with and without cardiovascular disease. Benefits, rationale, safety, and prescription. An advisory from the Committee on Exercise, Rehabilitation and Prevention, Council on Clinical Cardiology, American Heart Association. *Circulation* 101: 828-833, 2000.

Vasan RS, Beiser A, Seshadri S, Larson MG, Kannel WB, D'Agostino RB, Levy D. Residual lifetime risk for developing hypertension in middle-aged women and men: The Framingham Heart Study. *JAMA* 287: 1003-1010, 2002.

Whelton SP, Chin A, Xin X, He J. Effect of aerobic exercise on blood pressure: A meta-analysis of randomized, controlled trials. *Ann Intern Med* 136: 493-503, 2002.

Williams B, Poulter NR, Brown MJ, Davis M, McInnes GT, Potter JF, Sever PS, McG Thom S . Guidelines for management of hypertension: Report of the fourth working party of the British Hypertension Society, 2004–BHS IV. *J Hum Hypertens* 18: 139-185, 2004. [Internet]. [cited 2008 December 1]. www.bhsoc.org/pdfs/BHS_IV_Guidelines.pdf

Williams MA, Haskell WL, Ades PA, Amsterdam EA, Bittner V, Franklin BA, Gulanick M, Laing ST, Stewart KJ; American Heart Association Council on Clinical Cardiology; American Heart Association Council on Nutrition, Physical Activity, and Metabolism. Resistance exercise in individuals with and without cardiovascular disease: 2007 update. A scientific statement from the American Heart Association Council on Clinical Cardiology and Council on Nutrition, Physical Activity, and Metabolism. *Circulation* 116: 572-584, 2007.

[WHO] World Health Organization, International Society of Hypertension Writing Group. 2003 World Health Organization (WHO)/International Society of Hypertension (ISH) statement on management of hypertension. *J Hypertens* 21: 1983-1992, 2003.

Chapter 11

[ACE] American Council on Exercise. Exercise and arthritis. *Fit Facts* [Internet]. [cited 2009 October 10]. San Diego, California: American Council on Exercise. Available from: www.acefitness.org/fitfacts. 2009.

[ACE] American Council on Exercise. Reduce Your Risk for Osteoporosis Now. *Fit Facts* [Internet]. [cited 2009 Octoberr 10]. San Diego, California: American Council on Exercise. Available from: www.acefitness.org/fitfacts. 2009.

[ACE] American Council on Exercise. Prevent osteoporosis now. *Fit Facts*[Internet]. [cited 2008 June 28]. San Diego, California: American Council on Exercise. Available from: www.acefitness.org/fitfacts. 2001.

[ACR] American College of Rheumatology. *Exercise and arthritis* [Internet]. [cited 2010 February 19]. Atlanta, Georgia: American College of Rheumatology: American College of Rheumatology. Available from: www.rheumatology.org/practice/clinical/patients/diseases_and_conditions/exercise.asp. 2006.

[ACR] American College of Rheumatology, subcommittee on osteoarthritis guidelines. Recommendations for the medical management of osteoarthritis of the hip and knee.

Arthritis Rheum 43(9): 1905-1915, 2000. [Internet]. [cited 2010 February 19]. www.rheumatology.org/practice/clinical/guidelines/oa-mgmt/oa-mgmt.asp

[ACSM] American College of Sports Medicine. Exercise prescription for other clinical populations. In: *ACSM's guidelines for exercise testing and prescription*. 8th ed. Baltimore: Lippincott, Williams and Wilkins. 225-228, 256-258. 2010.

American Geriatrics Society Panel on Exercise and Osteoarthritis. Exercise prescription for older adults with osteoarthritis pain: Consensus practice recommendations. *J Am Geriatr Soc* 49: 808-823, 2001.

Arthritis Foundation. *Top three types of exercise* [Internet]. [cited 2008 April 6]. Atlanta, Georgia: Arthritis Foundation. Available from: www.arthritis.org/types-exercise.php. 2007.

Bloomfield SA, Smith SS. Osteoporosis. In: *ACSM's exercise management for person with chronic diseases and disabilities*. 2nd ed. Durstine JL, Moore GE, eds. Champaign (IL): Human Kinetics. 222-229. 2003.

Brown JP, Josse RG, for the Scientific Advisory Council of the Osteoporosis Society of Canada. 2002 clinical practice guidelines for the diagnosis and management of osteoporosis.

CMAJ 2002;167(10 Suppl):S1-34.

[CDC] Centers for Disease Control and Prevention and The Merck Company Foundation. *The state of aging and health in America 2007* [Internet]. [cited 2008 July 25]. Whitehouse Station (NJ): The Merck Company Foundation. Available from: www.cdc.gov/aging. 2007.

Ettinger WH Jr, Burns R, Messier SP, Applegate W, Rejeski WJ, Morgan T, Shumaker S, Berry MJ, O'Toole M, Monu J, et al. A randomized trial comparing aerobic exercise and resistance exercise with a health education program in older adults with knee osteoarthritis. The Fitness Arthritis and Seniors Trial (FAST). *JAMA* 277: 25-31, 1997.

Felson DT, Lawrence RC, Hochberg MC, McAlindon T, Dieppe PA, Minor MA, Blair SN, Berman BM, Fries JF, Weinberger M et al. Osteoarthritis: new insights. Part 2: Treatment approaches. *Ann Intern Med* 133(9): 727-737, 2000.

Hakkinen A, Sokka T, Kotaniemi A, Kautiainen H, Jappinen I, Laitinen L, Hannonen P. Dynamic strength training in patients with early rheumatoid arthritis increases muscle strength but not bone density. *J Rheumatol* 26: 1257-1263, 1999.

[IOF] International Osteoporosis Foundation. *Facts and statistics about osteoporosis and its impact* [Internet]. [cited 2008 September 9]. Nyon, Switzerland: International Osteoporosis Foundation. Available from: www.iofbonehealth.org/facts-and-statistics.html. 2007

Iversen MD, Liang MH, Bae SC. *Exercise in rehabilitation medicine: Selected arthritides: Rheumatoid arthritis, osteoarthritis, spondylarthropathies, systemic lupus erythematosus, polymyositis/dermatomyositis, and systemic sclerosis*. Champaign (IL): Human Kinetics. 1999.

[JHAC] Johns Hopkins Arthritis Center. *ACR clinical classification criteria for rheumatoid arthritis* [Internet]. [cited 2008

October 16]. Baltimore, Maryland: Johns Hopkins Arthritis Center. Available from: www.hopkins-arthritis.org/physician-corner/education/acr/acr.html#class_rheum.

Kanis JA, Johnell O, Oden A, Dawson A, De Laet C, Jonsson B. (2000) Long-term risk of osteoporotic fracture in Malmo. *Osteoporos Int* 11:669-674.

Kelley GA. Aerobic exercise and lumbar spine bone mineral density in postmenopausal women: A meta-analysis. *J Am Geriatr Soc* 46(2): 143-152, 1998.

Kelley GA, Kelley KS, Tran ZV. Resistance training and bone mineral density in women: A meta-analysis of controlled trials. *Am J Phys Med Rehabil* 80(1): 65-77, 2001.

Kohrt WM, Bloomfield SA, Little KD, Nelson ME, Yingling VR. Physical activity and bone health. *Med Sci Sports Exerc* 36(11): 1985-1996, 2004

Lawrence RC, Everett DF, Hochberg MC. *Arthritis. In Health status and well-being of the elderly: National Health and Nutrition Examination I epidemiologic follow-up survey.* New York: Oxford University Press. 1990.

Martyn-St James M, Carroll S. High intensity resistance training and postmenopausal bone loss: A meta analysis. *Osteoporosis Int* 17(8): 1225-1240, 2006.

Martyn-St James M, Carroll S. Progressive high-intensity resistance training and bone mineral density changes among premenopausal women: Evidence of discordant site-specific skeletal effects. *Sports Med* 36(8): 683-704, 2006.

Melton LJ 3rd, Chrischilles EA, Cooper C, Lane AW, Riggs BL. Perspective. How many women have osteoporosis? *J Bone Miner Res.* 7(9): 1005-1010, 1992.

Minne HW, Pfeifer M. (International Osteoporosis Foundation, Nyon Switzerland). *Invest in your bones. Move it or lose it* [Internet]. [cited 2008 June 26]. Nyon, Switzerland: International Osteoporosis Foundation. Available from: www.iofbonehealth.org/publications/move-it-or-lose-it.html. 2005.

Minor MA, Hewett JE, Webel RR, Anderson SK; Kay DR. Efficacy of physical conditioning exercise in patients with rheumatoid arthritis and osteoarthritis. *Arthritis Rheum* 32: 1396-1405, 1989.

Minor MA, Kay DR. Arthritis. In: *ACSM's exercise management for person with chronic diseases and disabilities.* 2nd ed. Durstine JL, Moore GE, eds. Champaign (IL): Human Kinetics. 210-216. 2003.

[NCPAD] National Center on Physical Activity and Disability. *Osteoarthritis and exercise* [Internet]. [cited 2008 July 2]. Chicago, Illinois: National Center on Physical Activity and Disability. Available from: www.ncpad.org/disability/fact_sheet.php?sheet=120. 2007a.

[NCPAD] National Center on Physical Activity and Disability. *Rheumatoid arthritis and exercise* [Internet]. [cited 2008 July 2]. Chicago, Illinois: National Center on Physical Activity and Disability. Available from: www.ncpad.org/disability/fact_sheet.php?sheet=131. 2007b.

[NIH] National Institutes of Health; National Institute of Arthritis and Musculoskeletal and Skin Diseases *Handout on Health: Osteoarthritis.* National Institute of Arthritis and Musculoskeletal and Skin Diseases, Bethesda, Maryland, 2002 (updated May 2006). [Internet] [cited 2008 December 3] www.niams.nih.gov/Health_Info/Osteoarthritis/default.asp

[NOF] National Osteoporosis Foundation. *Prevention: Exercise for healthy bones* [Internet]. [cited 2008 July 22]. Washington, D.C.: National Osteoporosis Foundation. Available from: www.nof.org/prevention/exercise.htm. 2008.

[NOF] National Osteoporosis Foundation. *Fast facts on osteoporosis* [Internet]. [cited 2008 July 22]. Washington, D.C.: National Osteoporosis Foundation. Available from: www.nof.org/osteoporosis/diseasefacts.htm. 2008.

Nelson ME. Fiatarone MA, Morganti Cm, Trice I, Greenberg RA, Evans WJ. Effects of high intensity strength training on multiple risk factors for osteoporotic fractures. A randomized controlled trial. *JAMA.* 272(24): 1909-1914, 1994.

Nieman DC. Exercise soothes arthritis joint effects. *ACSM's Health and Fitness Journal* 4: 20-27, 2000.

O'Reilly SC, Muir KR, Doherty M. Effectiveness of home exercise on pain and disability from osteoarthritis of the knee: A randomized controlled trial. *Ann Rheum Dis* 58: 15-19, 1999.

Osteoporosis prevention, diagnosis, and therapy. *NIH Consensus Statement 2000 March 27-29* 17(1): 1-36, 2000. [Internet]. [cited 2009 October 9]. www.consensus.nih.gov/2000/2000Osteoporosis111html.htm

Ottawa panel evidence-based clinical practice guidelines for therapeutic exercises and manual therapy in the management of osteoarthritis. *Phys Ther* 85(9):907-971, 2005.

Ries MD, Philbin EF, Groff GD. Relationship between severity of gonoarthrosis and cardiovascular fitness. *Clin Orthop* 313: 169-176, 1995.

Semble EL, Loeser RF, Wise CM. Therapeutic exercise for rheumatoid arthritis and osteoarthritis. *Semin Arthritis Rheum* 20(1): 32-40, 1990.

Sisto SA, Malanga G. Osteoarthritis and therapeutic exercise. *Am J Phys Med Rehabil* 85(supple): S69-S78, 2006.

Slemenda C, Brandt KD, Heilman DK, Mazzuca S, Braunstein EM, Katz BP, Wolinsky FD. Quadriceps weakness and osteoarthritis of the knee. *Ann Intern Med* 127: 97-104, 1997.

Stenstrom CH. Home exercise in rheumatoid arthritis Functional class II: Goal setting versus pain attention. *J Rheumatol* 21: 627-634, 1994.

[USDHHS] U.S. Department of Health and Human Services, Public Health Service, Office of the Surgeon General (United States). *Bone health and osteoporosis: A report of the Surgeon General. Executive summary* [Internet]. [cited 2008 April 30]. Rockville (MD): U.S. Department of Health and Human Services. Available from: www.surgeongeneral.gov/library/bonehealth. 2004.

[USDHHS] U.S. Department of Health and Human Services (United States). *2008 physical activity guidelines for Americans* [Internet]. [cited 2008 October 7]. U.S. Department of Health and Human Services Washington, D.C. Available from: www.health.gov/paguidelines. 2008.

Van den Ende CH, Hazes JM, le Cessie S, Mulder WJ, Belfor DG, Breedveld FC, Dijkmans BA. Comparison of high and low intensity training in well controlled rheumatoid arthritis. *Ann Rheum Dis* 55(11): 798-805, 1996.

Verbrugge LM. Women, men and osteoarthritis. *Arthritis Care Res* 6: 212-220, 1995.

[WHO] World Health Organization. Assessment of fracture risk and its application to screening for postmenopausal osteoporosis. *World Health Organ Tech Rep Ser* 843, 1994.

Chapter 12

Albright A, Franz M, Hornsby G, Kriska A, Marrero D, Ullrich I, Verity LS. American College of Sports Medicine position stand: exercise and type 2 diabetes. *Med Sci Sports Exerc* 32: 1345-1360, 2001.

[ACE] American Council on Exercise. Exercise and type 1 diabetes. *Fit Facts* [Internet]. [cited 2008 June 28]. San Diego, California: American Council on Exercise. Available from www.acefitness.org/fitfacts. 2001.

[ACE] American Council on Exercise. Exercise and type 2 diabetes. *Fit Facts* [Internet]. [cited 2008 June 28]. San Diego,: American Council on Exercise. Available from www.acefitness.org/fitfacts. 2001.

[ACSM] American College of Sports Medicine. Exercise prescription for other clinical populations. In: *ACSM's Guidelines for Exercise Testing and Prescription* 8th ed. Baltimore: Lippincott, Williams and Wilkins. 232-237. 2010.

[ADA] American Diabetes Association/American College of Sports Medicine. Joint statement: Diabetes mellitus and exercise. *Med Sci Sports Exerc* 29(12): i-iv 1997.

[ADA] American Diabetes Association. Diabetes mellitus and exercise. *Diabetes Care* 24(suppl 1): S51-S55, 2001.

[ADA] American Diabetes Association. Diagnosis and classification of diabetes mellitus. *Diabetes Care* 31(S1): S55-S78, 2008a.

[ADA] American Diabetes Association. Physical activity/exercise and diabetes. *Diabetes Care* 26: S73-S77, 2003.

[ADA] American Diabetes Association. Executive Summary: Standards of medical care in diabetes—2008 *Diabetes Care* 31(S1):S5-S11, 2008c.

[ADA] American Diabetes Association. Standards of medical care in diabetes–2008. *Diabetes Care* 31(S1): S12-S54, 2008b.

[ADA] American Diabetes Association. *Total prevalence of diabetes and pre-diabetes* [Internet]. [cited 2008 June 8]. Alexandria,: American Diabetes Association. Available from: www.diabetes.org/diabetes-statistics-prevalence.jsp.

Annuzzi G, Riccardi G, Capaldo B, Kaijser L. Increased insulin-stimulated glucose uptake by exercised human muscles one day after prolonged physical exercise. *Eur J Clin Invest* 21: 6-12, 1991.

Bax JJ, Young LH, Frye RL, Bonow RO, Steinberg HO, Barrett EJ. Screening for coronary artery disease in patient with diabetes. *Diabetes Care* 30: 2729-2736, 2007.

Boule' NG, Haddad E, Kenny GP, Wells GA, Sigal RJ. Effects of exercise on glycemic control and body mass index in type 2 diabetes mellitus: A meta-analysis of controlled clinical trials. *JAMA* 286: 1218-1227, 2001.

Boule' NG, Kenny GP, Haddad E, Wells GA, Sigal RJ. Meta-analysis of the effect of structured exercise training on cardiorespiratory fitness in type 2 diabetes mellitus. *Diabetologia* 46: 1071-1081, 2003.

[CDC] Centers for Disease Control and Prevention. *National diabetes fact sheet* [Internet]. Centers for Disease Control and Prevention Atlanta, Georgia. Available from: www.cdc.gov/diabetes/pubs/factsheet07.htm. 2007.

Church, TS, LaMonte MJ, Barlow CE, Blair SN. Cardiorespiratory fitness and body mass index as predictors of cardiovascular disease mortality among men with diabetes. *Arch Intern Med* 165:2114-2120, 2005.

Did you Know? International Diabetes Federation. Brussels, Belgium., [Internet] [cited 2008 August 19] http://www.idf.org/sound_bites .

Durstine JL, Moore GE, eds. *ACSM's exercise management for persons with chronic diseases and disabilities*. 2nd ed. Champaign (IL): Human Kinetics. 2003.

Expert Committee on the Diagnosis and Classification of Diabetes Mellitus. Follow-up report on the diagnosis of diabetes mellitus. *Diabetes Care* 26: 3160-3167, 2003.

Ford ES, DeStefano F. Risk factors for mortality from all causes and from coronary heart disease among persons with diabetes. Findings from the National Health and Nutrition Examination Survey I Epidemiologic Follow-up Study. *Am J Epidemiol* 133: 1220-1230, 1991.

Gilmer TP, O'Connor PJ, Rush WA, Crain AL, Whitebird RR, Hanson AM, Solberg LI. Predictors of health care costs in adults with diabetes. *Diabetes Care* 28: 59-64, 2005.

Gregg EW, Gerzoff RB, Caspersen CJ, Williamson DF, Narayan KM. Relationship of walking to mortality among US adults with diabetes. *Arch Intern Med* 163: 1440-1447, 2003.

Haffner SM, Lehto S, Ronnemaa T, Pyörälä K, Kallio V, Laakso M. Mortality from coronary heart disease in subjects with type 2 diabetes and in nondiabetic subjects with and without prior myocardial infarction. *N Engl J Med* 339: 229-234, 1998.

Hornsby WG, Albright AL. Diabetes. In: Durstine JL, Moore GE, eds. *ACSM's Exercise management for person with chronic diseases and disabilities*. 2nd ed. Champaign (IL): Human Kinetics. 133-141, 2003.

Hu FB, Stampfer MJ, Solomon C, Liu S, Colditz GA, Speizer FE, Willett WC, Manson JE. Physical activity and risk for cardiovascular events in diabetic women. *Ann Intern Med* 134: 96-105, 2001.

Ivy JL. Role of exercise training in the prevention and treatment of insulin resistance and non-insulin-dependent diabetes mellitus. *Sports Med* 24: 321-336, 1997.

Kriska A. Physical activity and the prevention of type II (non-insulin dependent) diabetes. *PCPFS Research Digest* 2(10): 1-12, 1997

Lehmann R, Vokac A, Niedermann K, Agosti K, Spinas GA. Loss of abdominal fat and improvement of the cardiovascular risk profile by regular moderate exercise training in patients with NIDDM. *Diabetologica* 38: 1313-1319, 1995.

Mayer-Davis DJ, D'Agostino R Jr, Karter AJ, Haffner SM, Rewers MJ, Saad M, Bergman RN. Intensity and amount of physical activity in relation to insulin sensitivity: the Insulin Resistance Atherosclerosis Study. *JAMA* 279: 669-674, 1998.

McDonald MS. Postexercise late-onset hypoglycemia in insulin-dependent diabetic patients. *Diabetes Care* 10: 584-588, 1987.

Roberts L, Jones TW, Fournier PA. Exercise training and glycemic control in adolescents with poorly controlled type 1 diabetes mellitus. *J Pediatr Endocrinol Metab* 15: 621-627, 2002.

Robertson K, Adolfsson P, Riddell MC, Scheiner G, Hanas R. Exercise in children and adolescents with diabetes. *Pediatr Diabetes* 9: 65-77, 2008.

Sigal RJ, Kenny GP, Boule' NG, Wells GA, Prud'homme D, Fortier M, Reid RD, Tulloch H, Coyle D, Phillips P, et al. Effects of aerobic training, resistance training, or both on glycemic control in type 2 diabetes: A randomized trial. *Ann Intern Med* 147: 357-369, 2007.

Sigal RJ, Kenny GP, Wasserman DH, Castaneda-Sceppa C, White RD. Physical activity/exercise and type 2 diabetes. A consensus statement from the American Diabetes Association. *Diabetes Care* 29(6): 1433-1438, 2006.

Stewart KJ. Exercise training and the cardiovascular consequences of type 2 diabetes and hypertension. Plausible mechanisms for improving cardiovascular health. *JAMA* 288: 1622-1631, 2002.

Unger J. Introduction to diabetes. In: Unger J, ed. *Diabetes Management in Primary Care*. Philadelphia: Lippincott Williams & Wilkins. 1-42. 2007.

[USDHHS] U.S. Department of Health and Human Services (United States). *2008 physical activity guidelines for Americans* [Internet]. [cited 2008 October 7]. U.S. Department of Health and Human Services Washington, D.C. Available from: www.health.gov/paguidelines. 2008.

Valensi P, Sachs RN, Harfouche B, Lormeau B, Paries J, Cosson E, Paycha F, Leutenegger M, Attali JR. Predictive value of cardiac autonomic neuropathy in diabetic patients with or without silent myocardial ischemia. *Diabetes Care* 24: 339-343, 2001.

Valerio G, Spagnuolo M, Lombardi F, Spadaro R, Siano M, Franzese A. Physical activity and sports participation in children and adolescents with type 1 diabetes mellitus. *Nutr Metab Cardiovasc Dis* 17(5): 376-382, 2003.

Wackers FJ, Young LH, Inzucchi, SE, Chyun DA, Davey JA, Barrett EJ, Taillefer R, Wittlin SD, Heller GV, Filipchuk N, and others. Detection of silent myocardial ischemia in asymptomatic diabetic subjects: The DIAD study. *Diabetes Care* 27: 1954-1961, 2004.

Wei M, Gibbons LW, Kampert JB, Nichaman MZ; Blair SN. Low cardiorespiratory fitness and physical inactivity as predictors of mortality in men with type 2 diabetes. *Ann Intern Med* 132: 605-611, 2000.

Yamanouchi K, Shinozaki T, Chikada K, Nishikawa T, Ito K, Shimizu S, Ozawa N, Suzuki Y, Maeno H, Kato K, et al. Daily walking combined with diet therapy is a useful means for obese NIDDM patients not only to reduce body weight but also to improve insulin sensitivity. *Diabetes Care* 18: 775-778, 1995.

Chapter 13

[ACSM] American College of Sports Medicine. Exercise prescription for other clinical populations. In: *ACSM's Guidelines for Exercise Testing and Prescription*. 8th ed. Baltimore: Lippincott, Williams and Wilkins. 237-244. 2010.

Andersson C, Grooten W, Hellsten M, Kaping K, Mattsson E. Adults with cerebral palsy: Walking ability after progressive strength training. *Dev Med Child Neurol* 45(4): 220-228, 2003.

Blanchard Y, Darrah J. *ASCM current comment: Health-related fitness for children and adolescents with cerebral palsy* [Internet]. [cited 2008 June 12]. Indianapolis (IN): American College of Sports Medicine. Available from: www.acsm.org/cc.

[CDC] Centers for Disease Control and Prevention National Center on Birth Defects and Developmental Disabilities (Centers for Disease Control and Prevention). Disability and health state chartbook, 2006 : profiles of health for adults with disabilities. Atlanta, GA: U.S. Dept. of Health & Human Services, Centers for Disease Control and Prevention, National Center on Birth Defects and Developmental Disabilities, 2006.

[CDC] U.S. Department of Health and Human Services Centers for Disease Control and Prevention. Disability and Health in the United States, 2001-2005. Publication 2008-1035, U.S. Department of Health and Human Services, Hyattsville, Maryland, 2008.

Dodd KJ, Taylor NF, Damiano DL. A systematic review of the effectiveness of strength-training programs for people with cerebral palsy. *Arch Phys Med Rehabil*. 83(8): 1157-1164, 2002.

Ferrini AF, Ferrini RL. *Health in the Later Years*. San Francisco: McGraw-Hill. 2000

Figoni SF. Spinal cord disabilities: Paraplegia and tetraplegia. In: Durstine JL, Moore GE, eds. *ACSM's Exercise management for persons with chronic diseases and disabilities*. 2nd ed. Champaign (IL): Human Kinetics. 247-253. 2003.

Formisano R, Pratesi L, Modarelli F, Bonefati V, Meco G. Rehabilitation and Parkinson's disease. *Scand J Rehabil Med* 24(3): 157-160, 1992.

Goetz C, Thelen J, MacLeod C, Carvey P, Bartley E, Stebbins G. Blood levodopa levels and unified Parkinson's disease rating scale function: With and without exercise. *Neurology* 43(5): 1040-1042, 1993.

Gordon NF, Gulanick M, Costa F, Fletcher G, Franklin BA, Roth EJ, Shephard T. Physical activity and exercise recommendations for stroke survivors: an American Heart Association scientific statement from the Council on Clinical Cardiology, Subcommittee on Exercise, Cardiac Rehabilitation, and Prevention; the Council on Cardiovascular Nursing; the Council on Nutrition, Physical Activity, and Metabolism; and the Stroke Council. *Circulation* 109(16): 2031-2041, 2004.

Kohl HW III, Powell KE, Gordon NF, Blair SN Paffenbarger RS Jr. Physical activity, physical fitness, and sudden cardiac death. *Epidemiol Rev* 14:37-58, 1992.

Laskin JJ. Cerebral palsy. In: Durstine JL, Moore GE, eds. *ACSM's Exercise management for person with chronic diseases and disabilities.* 2nd ed. Champaign (IL): Human Kinetics. 288-294, 2003.

Macko RF, DeSouza CA, Tretter LD, Silver KH, Smith GV, Anderson PA, Tomoyasu N, Gorman P, Dengel DR. Treadmill aerobic exercise training reduces the energy expenditure and cardiovascular demands of hemiparetic gait in chronic stroke patients: A preliminary report. *Stroke* 28: 326-330, 1997.

Mulcare JA. Multiple Sclerosis. In: Durstine JL, Moore GE, eds. *ACSM's Exercise management for persons with chronic diseases and disabilities.* 2nd ed. Champaign (IL): Human Kinetics. 267-272. 2003.

[NCPAD] National Center on Physical Activity and Disability. *Duchenne Muscular Dystrophy and Exercise* [Internet]. [cited 2008 July 2]. Chicago, Illinois: National Center on Physical Activity and Disability. Available from: www.ncpad.org/disability/fact_sheet.php?sheet=142, 2007.

[NCPAD] National Center on Physical Activity and Disability. *Exercise/Fitness: Resistance Training for Persons with Physical Disabilities* [Internet]. [cited 2008 March 25]. Chicago, Illinois: National Center on Physical Activity and Disability. Available from: www.ncpad.org/exercise/fact_sheet.php?sheet=107, 2005.

[NCPAD] National Center on Physical Activity and Disability. *Multiple Sclerosis and Exercise* [Internet]. [cited 2009 October 10]. Chicago, Illinois: National Center on Physical Activity and Disability. Available from: www.ncpad.org/disability/fact_sheet.php?sheet=186, 2009.

[NCPAD] National Center on Physical Activity and Disability. *Multiple Sclerosis: Designing an Exercise Program* [Internet]. [cited 2008 July 2]. Chicago, Illinois: National Center on Physical Activity and Disability. Available from: www.ncpad.org/disability/fact_sheet.php?sheet=187, 2007.

[NCPAD] National Center on Physical Activity and Disability. *Muscular Dystrophy* [Internet]. [cited 2008 July 2]. Chicago, Illinois: National Center on Physical Activity and Disability. Available from: www.ncpad.org/disability/fact_sheet.php?sheet=73&view=all , 2007.

[NCPAD] National Center on Physical Activity and Disability. *Parkinson's Disease and Exercise* [Internet]. [cited 2009 October 10]. Chicago, Illinois: National Center on Physical Activity and Disability. Available from: www.ncpad.org/disability/fact_sheet.php?sheet=59, 2009.

[NCPAD] National Center on Physical Activity and Disability. *Spinal Cord Injury and Exercise* [Internet]. [cited 2008 July 2]. Chicago, Illinois: National Center on Physical Activity and Disability. Available from: www.ncpad.org/disability/fact_sheet.php?sheet=130, 2007.

Palmer S, Mortiner J, Webster D, Bistevino R, Dickman G. *Exercise therapy for Parkinson's disease.* Arch Phys Med Rehabil, 67(10), 741-745, 1986.

Palmer-McLean K, Harbst KB. Stroke and Brain Injury. In: Durstine JL, Moore GE, eds. *ACSM's Exercise management for persons with chronic diseases and disabilities.* 2nd ed. Champaign (IL): Human Kinetics. 238-246, 2003.

Petajan JH, White AT. Recommendations for physical activity in patients with MS. *Sports Med* 27(3): 179-191, 1999.

Protas EJ, Stanley RK. Parkinson's Disease. In: Durstine JL, Moore GE, eds. *ACSM's Exercise management for persons with chronic diseases and disabilities.* 2nd ed. Champaign (IL): Human Kinetics. 295-302. 2003.

Reuter I, Engelhardt M, Stecker K, Baas H. Therapeutic value of exercise training in Parkinson's disease. *Med Sci Sports Exerc* 31(11): 1544-1549, 1999.

Rimmer JH. Promoting inclusive physical activity communities for people with disabilities. *PCPFS Research Digest* 9(2): 1-8, 2008.

Schenkman M, Cutson T, Kuchibhatla M, Chandler J, Pieper C, Ray L, Laub K. Exercise to improve spinal flexibility and function for people with Parkinson's disease: A randomized, controlled trial. *J Am Geriatr Soc* 46: 1207-1216, 1998.

Seaman JA. Physical activity & fitness for persons with disabilities: A paradigm shift. *PCPFS Research Digest* 1-12, March 1999.

Shephard R. *Fitness in Special Populations.* Champaign (IL): Human Kinetics. 1990.

Tarnopolsky MA. Muscular Dystrophy. In: Durstine JL, Moore GE, eds. *ACSM's Exercise management for person with chronic diseases and disabilities.* 2nd ed. Champaign (IL): Human Kinetics. 254-261, 2003.

Thompson PD, Klocke FJ, Levine BD, Van Camp SP. 26th Bethesda conference: Recommendations for determining eligibility for competition in athletes with cardiovascular abnormalities: Task Force 5: Coronary artery disease. *Med Sci Sports Exerc* 26(suppl): S271-S275, 1994.

[UCP] United Cerebral Palsy. *Exercise principles and guidelines for persons with cerebral palsy and neuromuscular disorders* [Internet]. [cited 2008 July 2]. Washington, D.C.: United Cerebral Palsy. Available from: www.ucp.org/ucp_channeldoc.cfm/1/15/11500/11500-11500/639, 2008a.

[UCP] United Cerebral Palsy. *Exercise principles and guidelines for persons with cerebral palsy and neuromuscular disorders: Components of an exercise session* [Internet]. [cited 2008 July 2]. Washington, D.C.: United Cerebral Palsy http://www.ucp.org/ucp_channeldoc.cfm/1/15/15/15-15/643, 2008b.

[UCP] United Cerebral Palsy. *Exercise & Fitness: General exercise guidelines* [Internet]. [cited 2008 July 2]. Washington, D.C.: United Cerebral Palsy www.ucp.org/ucp_channeldoc.cfm/1/15/11500/11500-11500/3178, 2008c.

[USDHHS] U.S. Department of Health and Human Services. *Healthy People 2010.* 2nd ed. With Understanding and Improving Health and Objectives for Improving Health. 2 vols. Washington, DC: U.S. Government Printing Office, November 2000.

[USDHHS] U.S. Department of Health and Human Services (United States). *2008 physical activity guidelines for Americans* [Internet]. [cited 2008 October 7]. U.S. Department of Health and Human Services Washington, D.C.. Available from: www.health.gov/paguidelines. 2008.

Chapter 14

[AAAAI] American Academy of Allergy Asthma & Immunology. *Tips to Remember: Exercise-induced asthma* [Internet]. [cited 2008 June 30]. Milwaukee, Wisconsin: American Academy of Allergy Asthma & Immunology. Available from: www.aaaai.org/patients/publicedmat/tips/exerciseinducedasthma.stm. 2009.

[AAAAI] American Academy of Allergy Asthma & Immunology. *Topic of the month: February 2006: Exercising with asthma* [Internet]. [cited 2008 June 30]. Milwaukee, Wisconsin: American Academy of Allergy Asthma & Immunology. Available from: www.aaaai.org/patients/topicofthemonth/0206. 2006.

[AAAAI] American Academy of Allergy Asthma & Immunology. *Topic of the month: February 2007: Get exercise-induced asthma under control* [Internet]. [cited 2008 June 30]. Milwaukee, Wisconsin: American Academy of Allergy Asthma & Immunology. Available from: www.aaaai.org/patients/topicofthemonth/0207. 2007.

[ACSM] American College of Sports Medicine. Exercise prescription for other clinical populations. In: *ACSM's guidelines for exercise testing and prescription.* 8th ed. Baltimore: Lippincott, Williams and Wilkins. 260-264. 2010.

[ACE] American Council on Exercise. Exercise and asthma. *Fit facts* [Internet]. [cited 2008 June 28]. San Diego, California: American Council on Exercise. Available from www.acefitness.org/fitfacts/fitfacts_display.aspx?itemid=2594. 2001.

Clark, CJ. Asthma. In: Durstine JL and Moore GE eds. *ACSM's exercise management for persons with chronic diseases and disabilities.* 2nd ed. Champaign (IL): Human Kinetics. 105-110. 2003.

Cooper, CB. Chronic Obstructive Pulmonary Disease In: Durstine JL and Moore GE eds. *ACSM's exercise management for persons with chronic diseases and disabilities.* 2nd ed. Champaign (IL): Human Kinetics. 92-98. 2003.

Cypcar D, Lemanske DF. Asthma and exercise. *Clin Chest Med* 15: 351-368. 1994.

Mayers LB, Rundell KW. *ACSM current comments: Exercise-induced asthma* [Internet]. [cited 2008 April 30]. Available from: www.acsm.org. 2000.

[NHLBI] National Heart, Lung, and Blood Institute. National Asthma Education and Prevention Program. *Breathing difficulties related to physical activity for students with asthma: Exercise-induced asthma: Information for physical educators, coaches and trainers* [Internet]. [cited 2008 July 23]. Bethesda (MD): National Heart, Lung, and Blood Institute. Available from: www.nhlbi.nih.gov/health/prof/lung/asthma/exercise_induced.pdf. 2005.

[NHLBI] National Heart, Lung, and Blood Institute. National Asthma Education and Prevention Program, School Asthma Education Subcommittee. *Asthma and physical activity in the school* [Internet]. [cited 2008 July 7] Bethesda (MD): National Heart, Lung, and Blood Institute. Available from: www.nhlbi.nih.gov/health/public/lung/asthma/phy_asth.pdf. 2004.

[NHLBI] National Heart, Lung, and Blood Institute. National asthma education and prevention program. *Expert Panel Report 3: Guidelines for the Diagnosis and Management of Asthma* [Internet]. [cited 2008 July 7]. Bethesda (MD): National Heart, Lung, and Blood Institute. NIH No.: 08-5846. Available from: www.nhlbi.nih.gov/guidelines/asthma. 2007.

Van Gent R. Van der Ent CK, Van Essen-Zandvliet LEM, Rovers MM, Kimpen JLL, de Meer G van der Ent CK. No differences in physical activity in (un)diagnosed asthma and healthy controls. *Pediatr Pulmonol* 42(11): 1018-1023, 2007.

Chapter 15

[ACSM] American College of Sports Medicine. Health-Related Physical Fitness Testing and Interpretation. In: *ACSM's Guidelines for Exercise Testing and Prescription.* 8th ed. Baltimore: Lippincott, Williams and Wilkins. 71-72. 2010.

Chang R-KR, Gurvitz M, Rodriguez S, Hong E, Klitzner TS. Current practice of exercise stress testing among pediatric cardiology and pulmonology centers in the United States. *Pediatr Cardiol* 27: 110-116. 2006.

Corbin CB, Pangrazi RP, Frank BD. Definitions: Health, fitness, and physical activity. *PCPFS Research Digest.* March 2000.

Hoffman J. Aerobic power and endurance. In: *Norms for fitness, performance, and health.* Champaign (IL): Human Kinetics. 72. 2006.

Jurca R, Jackson A, LaMonte M, Morrow J Jr., Blair S, Wareham N, Haskell W, van Mechelen W, Church T, Jakicic J. Assessing Cardiorespiratory Fitness Without Performing Exercise Testing *Am J Prev Med* 29(3): 185-193, 2005.

Knudson DV, Magnusson P, McHugh M. Current issues in flexibility fitness. *PCPFS Research Digest* 3(10): 1-8, 2000.

Stephens P, Paridon SM. Exercise testing in pediatrics. *Pediatr Clin North Am* 51(6): 1569-1587, 2004.

Tomassoni TL. Introduction: The role of exercise in the diagnosis and management of chronic disease in children and youth. *Med Sci Sports Exerc* 28(4): 403-405, 1996.

Welk GJ, Meredith MD, eds. *Fitnessgram/Activitygram Reference Guide.* Dallas: The Cooper Institute. 2008.

Whaley MH. *ACSM's guidelines for exercise testing and prescription*. 7th ed. Baltimore: Lippincott, Williams & Wilkins. 2005.

Chapter 16

Aktas MK, Ozduran V, Pothier CE, Lang R, Lauer MS. Global risk scores and exercise testing for predicting all-cause mortality in a preventive medicine program. *JAMA* 292: 1462-1468, 2004.

Blair SN, Kampert JB, Kohl HW 3rd, Barlow CE; Macera CA; Paffenbarger RS Jr; Gibbons LW. Influences of cardiorespiratory fitness and other precursors on cardiovascular disease and all-cause mortality in men and women. *JAMA* 276: 205-210, 1996.

Gibbons LW, Mitchell TL, Wei M, Blair SN, Cooper KH. Maximal exercise test as a predictor of risk for mortality from coronary heart disease in asymptomatic men. *Am J Cardiol* 86: 53-58, 2000.

Gibbons RJ, Balady GJ, Bricker JT, Chaitman BR, Fletcher GF, Froelicher VF, Mark DB, McCallister BD, Mooss AN, O'Reilly MG, et al. American College of Cardiology/American Heart Association Task Force on Practice Guidelines. Committee to Update the 1997 Exercise Testing Guidelines. ACC/AHA 2002 guideline update for exercise testing: summary article. A report of the American College of Cardiology/American Heart Association Task Force on Practice Guidelines (Committee to Update the 1997 Exercise Testing Guidelines). *Circulation*. 106:1883-1892, 2002.

Goraya TY, Jacobsen SJ, Pellikka PA, Miller TD, Khan A, Weston SA, Gersh BJ, Roger VL . Prognostic value of treadmill exercise testing in elderly persons. *Ann Intern Med* 132: 862-870, 2000.

Jurca R, Jackson AS, LaMonte MJ, Morrow Jr JR, Blair SN, Wareham NJ, Haskell WL, van Mechelen W, Church TS, Jakicic JM, Lauukkanen R. Assessing cardiorespiratory fitness without performing exercise testing. *Am J Prev Med* 29(3): 185-193, 2005.

Lauer M, Froelicher ES, William M, Kligfield P. Exercise testing in asymptomatic adults: A statement for professionals from the American Heart Association Council on Clinical Cardiology, Subcommittee on Exercise, Cardiac Rehabilitation, and Prevention. *Circulation* 112: 771-776, 2005.

Mark DB, Shaw L, Harrell FE Jr, Hlatky MA, Lee KL, Bengtson JR, McCants CB, Califf RM, Pryor DB. Prognostic value of a treadmill exercise score in outpatients with suspected coronary artery disease. *N Engl J Med* 325: 849-853, 1991.

Mora S, Redberg RF, Cui Y, Whiteman MK, Flaws JA, Sharrett AR, Blumenthal RS. Ability of exercise testing to predict cardiovascular and all-cause death in asymptomatic women: A 20-year follow-up of the lipid research clinics prevalence study. *JAMA* 290: 1600-1607, 2003.

Myers J, Parkash M, Froelicher V, Do D, Partington S, Atwood JE. Exercise capacity and mortality among men referred for exercise testing. *N Engl J Med* 346: 793-801, 2002.

O'Rourke RA, Brundage BH, Froelicher VF, Greenland P, Grundy SM, Hachamovitch R, Pohost GM, Shaw LJ, Weintraub WS, Winters WL., Jr. American College of Cardiology/American Heart Association Expert Consensus document on electron-beam computed tomography for the diagnosis and prognosis of coronary artery disease. *Circulation* 102: 126-140, 2000.

Shephard RJ. Readiness for physical activity. *PCPFS Research Digest* 1(5): 1-8, 1994.

U.S. Preventive Services Task Force. Screening for coronary heart disease: recommendation statement. *Ann Intern Med* 140: 569-572, 2004.

Vanhees L, Fagard R, Thijs L, Staessen J, Amery A. Prognostic significance of peak exercise capacity in patients with coronary artery disease. *J Am Coll Cardiol* 23: 358-363, 1994.

Visich PS. Graded exercise testing. In: Ehrman JK, Gordon PM, Visich PS, Keteyian SJ eds. *Clinical exercise physiology*. Champaign (IL): Human Kinetics. 75-101. 2004.

Wei M, Kampert JB, Barlow CE, Nichaman MZ, Gibbons LW, Paffenbarger RS Jr., Blair SN. Relationship between low cardiorespiratory fitness and mortality in normal-weight, overweight, and obese men. *JAMA* 282: 1547-1553, 1999.

Whaley MH. *ACSM's guidelines for exercise testing and prescription*. 7th ed. Baltimore: Lippincott, Williams & Wilkins. 2005.

Chapter 17

[ACE] American Council on Exercise. Healthy hydration. *Fit Facts* [Internet]. [2008 June 28]. San Diego, California: American Council on Exercise. Available from: www.acefitness.org/fitfacts. 2008.

[ACE] American Council on Exercise. Reduce your risk for osteoporosis now. [Internet] [cited 2010 February 18] www.acefitness.org/fitfacts/fitfacts_display.aspx?itemid=2609. 2009.

[ACE] American Council on Exercise. Scaling the new pyramid. *Fit Facts* [Internet]. [2008 June 28]. San Diego, California: American Council on Exercise. Available from: www.acefitness.org/fitfacts. 2001.

[ACE] American Council on Exercise. Supplements: Too much of a good thing? *Fit Facts* [Internet]. [2008 June 28]. San Diego, California: American Council on Exercise. Available from: www.acefitness.org/fitfacts. 2001.

[ACSM] American College of Sports Medicine. Exercise prescription for other clinical populations: diabetes mellitus In: *ACSM's Guidelines for Exercise Testing and Prescription*. 8th ed. Baltimore: Lippincott, Williams and Wilkins. 232-237, 2010

[ACSM] [ADA] [DC] American College of Sports Medicine. American Dietetic Association, Dieticians of Canada. Joint Position Statement. Nutrition and Athletic Performance. *Med Sci Sports Exerc* 41(3):709-731, 2009.

[ACSM] American College of Sports Medicine. Position stand on exercise and fluid replacement. *Med Sci Sports Exerc* 28(1): i-vii, 1996.

[ADA] American Dietetic Association. Position of the American Dietetic Association, Dietitians of Canada, and the American College of Sports Medicine: nutrition and athletic performance. *J Am Diet Assoc* 100:1543–56, 2000.

Ames BN, Gold LS, Willett WC. The causes and prevention of cancer. *Proc Natl Acad Sci U S A* 2: 5258-5265. 1995.

Applegate E, Clark K (American College of Sports Medicine). *Selecting and effectively using sports drinks, carbohydrate gels and energy bars* [Internet]. [cited 2008 August 21]. Indianapolis, Indiana: American College of Sports Medicine. Available from: www.acsm.org/AM/Template.cfm?Section=brochures2&Template=/CM/ContentDisplay.cfm&ContentID=12036. 2005.

Autier P, Gandini S. Vitamin D supplementation and total mortality. *Arch Intern Med* 167(16): 1730-1737, 2007.

Blackwood AM, Sagnella GA, Cook DG, Cappuccio FP. Urinary calcium excretion, sodium intake and blood pressure in a multi-ethnic population: Results of the Wandsworth Heart and Stroke Study. *J Hum Hypertens.* 15(4): 229-237, 2001.

Canada's Food Guide [Internet]. Health Canada. [cited 2008 October 4]. Available from: www.hc-sc.gc.ca/fn-an/food-guide-aliment/index-eng.php. 2007.

Casa DJ, Armstrong LE, Hillman SK, et al. National Athletic Trainers' Association position statement: Fluid replacement for athletes. *J Athl Train* 35(2): 212-224, 2000.

Dewey KG, Lovelady CA, Nommsen-Rivers LA, et al. A randomized study of the effects of aerobic exercise by lactating women on breast-milk volume and composition. *N Engl J Med* 330: 449, 1994.

Doyle C, Kushi LH, Byers T, et al. Nutrition and physical activity during and after cancer treatment: An American Cancer Society guide for informed choices. *CA Cancer J Clin* 56: 323-353, 2006.

Flegal KM, Graubard BI, Williamson DF, Gail MH. Excess deaths associated with underweight, overweight, and obesity *JAMA* 293(15): 1861-1867, 2005.

Holick MF. Vitamin D deficiency. *N Engl J Med* 357: 266-81, 2007.

Horvath PJ, Eagen CK, Ryer-Calvin SD, Pendergast DR. The effects of varying dietary fat on the nutrient intake of male and female runner. *J Am Coll Nutr* 19(1): 42-51, 2000a.

Horvath PJ, Eagen CK, Fisher NM, Leddy JJ, Pendergast DR. The effects of varying dietary fat on performance and metabolism in trained male and female runners. *J Am Coll Nutr* 19(1): 52-60, 2000b.

[IOM] Institute of Medicine. *Dietary reference intakes for energy, carbohydrate, fiber, fat, protein and amino acids* [Internet]. [cited 2008 April 6]. Washington (DC): National Academies Press: Available from: www.nap.edu/books/0309085373/html. 2002.

[IOM] Institute of Medicine. *Dietary reference intakes for water, potassium, sodium, chloride, and sulfate* [Internet]. [cited 2008 April 6]. Washington (DC): National Academies Press. Available from: www.nap.edu/catalog.php?record_id=10925. 2004.

Jones and Bartlett. *Recommended Dietary Allowances.* National Research Council (1989). Recommended Dietary Allowances, 10th edition. Washington, DC: National Academy Press.

Lemon PWR. Effects of exercise on dietary protein requirements. *Int J Sport Nutr* 8: 426-447, 1998.

Lichtenstein AH, Appel LJ, Brands M, et al. Diet and lifestyle recommendations revision 2006: A scientific statement from the American Heart Association Nutrition Committee. *Circulation* 114: 82-96, 2006.

Manore MM. Nutrition and physical activity: Fueling the active individual. *PCPFS Research Digest* 5(1): 1-8, 2004. Available online at www.fitness.gov/publications/digests/digest-march2004.pdf.

Mottola MF. Exercise in the postpartum period: Practical applications. *Curr Sports Med Rep* 1: 362-368, 2002.

Osteoporosis prevention, diagnosis, and therapy. *NIH Consensus Statement.* 17(1):1-36, 2000.

Robertson K, Adolfsson P, Riddell MC, Scheiner G, Hanas R. Exercise in children and adolescents with diabetes. *Pediatr Diabetes.* 9(1):65-77, 2008.

Sawka MN, Burke LM, Eichner ER, et al. Exercise and fluid replacement. *Med Sci Sports Med* 39(2): 377-390, 2007.

Sobal J, Marquart LF. Vitamin/mineral supplement use among athletes: A review of the literature. *Int J Sport Nutr* 4: 320-334, 1994.

Steinmetz KA, Potter JD. Vegetables, fruit, and cancer. I. Epidemiology. *Cancer Causes Control* 2: 325-357. 1991.

[USDA] My pyramid [Internet]. Alexandria (VA): U.S. Department of Agriculture, Food and Nutrition Service. [cited 2008 March 25]. Available from: www.mypyramid.gov. 2005 September.

[USDA] My pyramid for kids [Internet]. Alexandria (VA): U.S. Department of Agriculture, Food and Nutrition Service. [cited 2008 March 25]. Available from: www.mypyramid.gov/kids/index.html. 2005 September.

[USDHHS] U.S. Department of Health and Human Services and National Institutes of Health, National Heart, Lung, and Blood Institute. *National Cholesterol Education Program: ATP III guidelines at-a-glance quick desk reference* [Internet]. [cited 2008 June 21]. Bethesda, Maryland: National Institutes of Health. Available from: www.nhlbi.nih.gov/guidelines/cholesterol/atglance.pdf. 2001.

U.S. Department of Health and Human Services and U.S. Department of Agriculture. Dietary Guidelines for Americans, 2005. 6th Edition, Washington, DC: U.S. Government Printing Office, January 2005. [Internet]. [cited 2008 April 6]. U.S. Department of Health and Human Services. Available from: www.health.gov/dietaryguidelines/dga2005/document/pdf/DGA2005.pdf. 2005.

Willett WC. Micronutrients and cancer risk. *Am J Clin Nutr* 59: 1162-1165, 1994.

[WCRF] World Cancer Research Fund/American Institute of Cancer Research WCRF/AICR Expert Report, Food, Nutrition, Physical Activity and the Prevention of Cancer: a Global Perspective London, UK and Washington, D.C. [Internet] [cited 2008 September 28] http://www.dietand-cancerreport.org/ 2007

[WHO] World Health Organization. *Global strategy on diet, physical activity and health* [Internet]. [cited 2008 April 30]. Geneva: World Health Organization. Report No.: WHA57.17. Available from: www.who.int/dietphysicalac-tivity/strategy/eb11344/strategy_english_web.pdf. 2004.

[WHO] World Health Organization *Keep fit for life. Meeting the nutritional needs of older persons.* Geneva, World Health Organization, 2002.

Chapter 18

[ACSM] American College of Sports Medicine. *Selecting and effectively using a home treadmill* [Internet]. [cited 2008 April 28]. Indianapolis, Indiana: American College of Sports Medicine. Available from: www.acsm.org/AM/Template.cfm?Section=brochures2. 2005.

[ACSM] American College of Sports Medicine. *Selecting and effectively using a stationary bicycle* [Internet]. [cited 2008 April 21]. Indianapolis, Indiana: American College of Sports Medicine. Available from: www.acsm.org/AM/Template.cfm?Section=brochures2. 2005.

[ACSM] American College of Sports Medicine. *Selecting and effectively using free weights* [Internet]. [cited 2008 April 21]. Indianapolis, Indiana: American College of Sports Medicine. Available from: www.acsm.org/AM/Template.cfm?Section=brochures2. 2005.

[ACSM] American College of Sports Medicine. *Selecting and effectively using home weights* [Internet]. [cited 2008 April 21]. Indianapolis, Indiana: American College of Sports Medicine. Available from: www.acsm.org/AM/Template.cfm?Section=brochures2. 2005.

[ACE] American Council on Exercise. How to choose a health club. *Fit Facts* [Internet]. [cited 2009 October 10]. San Diego, California: American Council on Exercise. Available from: www.acefitness.org/fitfacts. 2009.

[ACE] American Council on Exercise. How to choose an exercise video. *Fit Facts* [Internet]. [cited 2009 October 10]. San Diego, California: American Council on Exercise. Available from: www.acefitness.org/fitfacts. 2009.

[ACE] American Council on Exercise. How to design your own home gym. *Fit Facts* [Internet]. [cited 2009 October 10]. San Diego, California: American Council on Exercise. Available from: www.acefitness.org/fitfacts. 2009.

[ACE] American Council on Exercise. What you need to know to purchase a treadmill. *Fit Facts* [Internet]. [cited 2009 October 10]. San Diego, California: American Council on Exercise. Available from: www.acefitness.org/fitfacts. 2009.

Crouter S. *Selecting and effectively using heart rate monitors* [Internet]. [cited 2008 April 28]. Indianapolis, Indiana: American College of Sports Medicine. Available from: www.acsm.org/AM/Template.cfm?Section=brochures2. 2005.

Hosea T. *Selecting and effectively using a rowing machine* [Internet]. [cited 2008 April 28]. Indianapolis, Indiana: American College of Sports Medicine. Available from: www.acsm.org/AM/Template.cfm?Section=brochures2. 2005.

[ICAA] International Council on Active Aging. *How to select an age-friendly fitness facility* [Internet]. [cited 2008 September 9]. Vancouver (BC): International Council on Active Aging. Available from: www.icaa.cc/facilitylocator/ICAAFacilityTest.pdf. 2008.

[NCPAD] National Center on Physical Activity and Disability. Choosing a fitness center [Internet]. [cited 2008 July 28]. Chicago, Illinois: National Center on Physical Activity and Disability. Available from: www.ncpad.org/exercise. 2006.

[NCPAD] National Center on Physical Activity and Disability. *Exercise video list* [Internet]. [cited 2008 July 28]. Chicago, Illinois: National Center on Physical Activity and Disability. Available from: www.ncpad.org/exercise. 2007.

Schneider P. *Selecting and effectively using a pedometer* [Internet]. [cited 2008 April 28]. Indianapolis, Indiana: American College of Sports Medicine. Available from: www.acsm.org/AM/Template.cfm?Section=brochures2. 2005.

Williford H, Olson M. *Selecting and effectively using a health/fitness facility* [Internet]. [cited 2008 April 28]. Indianapolis, Indiana: American College of Sports Medicine. Available from: www.acsm.org/AM/Template.cfm?Section=brochures2. 2005.

Williford H, Olson M. *Selecting and effectively using a stair stepper/climber* [Internet]. [cited 2008 April 28]. Indianapolis, Indiana: American College of Sports Medicine. Available from: www.acsm.org/AM/Template.cfm?Section=brochures2. 2005.

Williford H, Olson M. *Selecting and effectively using an elliptical trainer* [Internet]. [cited 2008 April 28]. Indianapolis, Indiana: American College of Sports Medicine. Available from: www.acsm.org/AM/Template.cfm?Section=brochures2. 2005.

Zeni AI, Hoffman MD, Clifford PS. Energy expenditure with indoor exercise machines. *JAMA* 276(8): 604-606, 1996.

INDEX

Note: The italicized *f* and *t* following page numbers refer to figures and tables, respectively.

A

AAP. *See* American Academy of Pediatrics
AAP Council on Sports Medicine and Fitness 98
AARP. *See* American Association of Retired Persons
abdominal fat distribution 151-152
absenteeism expenses 11
academic performance, role of physical education in 81-82
ACSM. *See* American College of Sports Medicine
active aerobic pursuits 56
Active Lifestyle Program 34, 89
Active Start 63-64
activities of daily living (ADLs) 18
Activitygram 57, 58*f*, 88
acute myocardial infarction 12
adolescents. *See also* children; infants
AAP Council on Sports Medicine and Fitness 98
Active Healthy Living 79
age-appropriate physical activity guidelines 92
bone mineral accrual in 95
dietary reference intakes 88
exercise for the prevention of obesity 94-95
obesity 68, 94
physical activity pyramid 55-56, 56*f*
preparticipation physical exam 99-101
prevention of atherosclerotic CVD 85
proper weight for 264-265
resistance training guidelines 78, 99
adults. *See also* older adults
ACSM definition for overweight or obese adults 27-28
aerobic exercise for healthy adults 17-18

aerobic testing 249-250
body composition testing 251-252, 251*t*
cardiorespiratory fitness 22, 27
dietary guidelines for 35
flexibility testing 250-251
flexibility training 27
improving quality of life 30
muscular strengthening and endurance 18-19
polyunsaturated fats 265-267
The President's Challenge 34-36
prevention of CVD 272
resistance training 27
strength testing 250
without CVD 32
adults, older
ACSM guidelines for 124-127
aerobic benefits 122
aerobic exercise 126
aerobic training 139
balance exercise 129, 133, 139
cardiorespiratory fitness 130
contraindications to exercise 141
endurance exercise 129, 133
fall risk 137
flexibility exercise 139
flexibility training 130, 133
frailty 134-138
injury-related deaths 137
National Blueprint guidelines for 128, 130-132
neurological benefits 123-124
neuromuscular training 140
resistance training 130, 139
stability and flexibility benefits 123
strength benefits 122-123
strength training 129, 133
who are frail or at risk for falling 134-141, 138*t*
aerobic endurance training 122, 139
for people with arthritis 185
for people with RA 182
aerobic exercise
ACSM guidelines for 27

benefits of 8, 122
Canadian guidelines for 38
cardiorespiratory fitness 47-48
definition of 6
effect of on CAD 9
for healthy adults 17-18, 23
musculoskeletal adaptations to 8
for older adults 122, 126
for postpartum women 119
for pregnant women 113
aerobic exercise machines 293-297
Aerobics (Cooper) 31
Aerobics Center Longitudinal Study 6
aerobic testing 249-250
after-school physical activity 87
age-appropriate physical activity guidelines 92
age-related loss of strength 10
AHA. *See* American Heart Association
AHA nutrition committee 272
alcoholic beverages 270-271
alcoholic consumption 289
alpha-linolenic acid 267, 280, 283
Alzheimer's disease 10
American Academy of Pediatrics (AAP)
guidelines for infants 61
position statement on obesity in children 77-78
American Academy of Pediatrics (AAP)
prevention of obesity 77
school and community health guidelines 85-87
American Association of Retired Persons (AARP) 122
American Cancer Society (ACS) 147, 157-158, 286
American College of Cardiology (ACC) 255
American College of Obstetricians and Gynecologists (ACOG) 103

American College of Rheumatology (ACR) 184
American College of Sports Medicine (ACSM)
    cardiorespiratory exercise 49
    cardiorespiratory fitness 22
    on fitness and health benefits 27-28
    flexibility exercise 50
    focus of original guidelines 15
    history and development of guidelines 4-5
    individualized exercise prescription guidelines 48-49
    original guidelines for physical fitness 17
    resistance exercise 49-50
    significant historical guidelines 24
    on weight maintenance 27-28
American Council on Exercise (ACE) 187
    recommendations for activity duration 55
American Dietetic Association 281-282
American Geriatrics Society (AGS) 190
American Heart Association (AHA)
    managing risk factors for CVD 86
    original guidelines for physical fitness 17
    prevention of atherosclerotic cardiovascular disease 85-87
    prevention of CVD 271
    on resistance exercise 31
    significant historical guidelines 23-24
American Institute for Cancer Research 149-150, 286-287
American Pregnancy Association 110
amino acids 278
anatomical changes during pregnancy 105
anthropometric methods 251
antioxidant supplementation 278
appropriate reasons for testing 256
arthritis 182
    general guidelines 185-187
    osteoarthritis 183t
    rheumatoid 181-183, 183t
Arthritis Foundation 187
asthma
    in children 244-246
    Dutch study on 246
    exercise induced 239-240

Fit Fact sheet on 241
    management plan 244
atheroprotective cytokines 8
Australian Government Department of Health and Ageing 38-39, 66, 91, 136
Australia's Recommendations for 5-12 Year Olds 91
autonomy 73

B
babies. See infants
balance exercise 129, 133, 139
barriers to activity, perceived 13
BECOME FIT 69
Behavioral Risk Factor Surveillance System (BRFSS) 104
Bernhardt, D.T. 67
bicycle ergometer 259
bioelectrical impedance analysis (BIA) 251
Blair, S.N. 9
blood cell counts 164
BMI-for-age percentiles 265, 266f
body composition norms 252t
body composition standards 252t
body composition testing 251-252, 251t
body mass index (BMI)
    ACSM definition for overweight or obese adults 27-28
    calculating 264
bone density 194
bone health 95-96
Borg scale 131-132
bradycardia 226
brain injury 235-236
breast cancer 163
breast-feeding women 115
    benefits of exercise 117-118
British Hypertension Society (BHS) 172
bronchospasm 239
Bruce protocol 259
burnout 100

C
CAD. See coronary artery disease
caffeine in gels 276
calcium 267, 278, 280, 283
    osteoporosis 288
calcium excretion 289
California Center for Physical Activity 36-37, 94-95
California Department of Education 73
California Department of Health Services (CDHS) 36-37, 65-66

California Obesity Prevention Initiative 36-37
Canada's food guide 273-274
Canada's Physical Activity Guide to Healthy Active Living 90-92
Canadian Home Fitness Test 13
Canadian Medical Association Journal 172
Canadian Pediatric Society 90
Canadian Society for Exercise Physiology (CSEP) 38, 90-91, 109
cancer
    American Cancer Society 157-158
    benefits for survivors 157
    benefits of activity during treatment 156-157
    blood cell counts 164-165
    breast 150
    colorectal 150
    contraindications to exercise 164-165
    deaths in America 145
    endometrial 146-147, 150
    energy expenditure 146
    general guidelines 147-150
    hormone levels 146-147
    immune system 147
    lymphedema 164
    metabolic syndrome 150-151, 151f
    pancreatic 150
    prevention 285-286
    prevention and treatment 9
    public health recommendations 287
    reducing the risk of 286
    skin conditions 164
    specific cancers 163-164
    survival 285-286
    testicular 147
carbidopa 226
carbohydrate gels 276-277
carbohydrates
    athletes intake 279
    DRIs 267
    in fluids 276
    RDA 283
    recommendations for 270
    required by pregnant women 285
cardiac contraindications to exercise 175
cardiac death, sudden 12
cardiac drift 46
cardiac exercise testing 258
    appropriate reasons for 256
    candidates for 255-256

contraindications to 257
other methods of 260-261
procedures for 259-260
protocols for 258-259
sensitivity of 257
utility of information acquired 257-258
cardiac rehabilitation 10-11
cardiometabolic health 150-151
cardiometabolic risk 151
cardiorespiratory activity 27
cardiorespiratory conditioning 51
cardiorespiratory fitness 130
   aerobic exercise 47-48
   California guidelines for 37
   emphasis on 31
   guidelines for 22
cardiovascular disease (CVD)
   preventing CVD 272
      physical inactivity as a risk factor for 15
      preactivity screening and clearance 19
      risk factors of 9
Carrel, D.T. 67
Centers for Disease Control and Prevention (CDC)
   medical expenses 11
   significant historical guidelines 24
central fat distribution 151-152
cerebral palsy
   guidelines for 220-224
certification of personal trainers 304
certifying organizations 304
CHD. *See* coronary heart disease
childhood obesity 67-69
children. *See also* adolescents; infants
   2001 Youth Risk Behavior Survey 72
   AAP Council on Sports Medicine and Fitness 98
   Active Healthy Living 79
   with asthma 244-246
   atherosclerotic cardiovascular disease 85-87
   Australia's Recommendations for 91
   benefits of exercise testing 252-253
   BMI-for-age percentiles 266*f*
   bone health 96
   bone mineral accrual in 95
   changes in physical activity needs 73-74, 73*f*
   C-LPAM 74
   decline in physical activity 72
   dietary reference intakes 88

energy needs 265
establishment of earliest guidelines for 74-76
food pyramid 280
lifetime physical activity model 74
manifestations of too much activity 99
medical evaluation before participation 98-99
minimum activity standard 75
MyActivity Pyramid
MyPyramid 280, 282
noncompliance with the electronic media guideline 87
obesity 67-69, 77-79
optimal functioning standard 75
polyunsaturated fats 265-267
The President's Challenge 89-90
proper weight for 264-265
recommendations for activity and nutrition 69
resistance training 96-98
resistance training for 99
Circulation 86, 126, 272
Clinical Exercise Physiology (Ehrman) 158, 160, 176
cognition 10
collagen fibers 123
College of Family Physicians of Canada 90
colorectal cancer 150
Committee on Exercise, Cardiac Rehabilitation, and Prevention 23, 31-32
Comprehensive School Physical Activity Programs 84
computed tomography 261
concerns for pregnant competitive athletes 107-108
Constella Group 78
Consumer Reports 293-294
contraindications to exercise 141, 164-165, 175
contraindications to exercise testing 257
Cooper, Kenneth 31
   Aerobics 31
Cooper Aerobics Center 252
Cooper Institute 57, 81, 252
Corbin, C.B. 74
coronary artery calcification 261
coronary artery disease (CAD)
   aerobic exercise guidelines 173-177
   diagnosis of 258
   nutrition and activity 288
   physical inactivity as a risk for 4
coronary heart disease (CHD)

risk factors for 23-24
coronary perfusion 12
Council on Clinical Cardiology 23, 31-32
Council on Sports Medicine and Fitness 61
creatine supplements 278
Current Comments 103
Current Sports Medicine Reports 68, 114, 119
CVD. *See* cardiovascular disease

D
DASH (Dietary approaches to stop hypertension) 172, 268, 270
dehydration 275-276, 275*t*
demographics 12-13
densitometry 251
Department of Agriculture. *See* U.S. Department of Agriculture
Department of Education, California 73
Department of Health, United Kingdom 135
Department of Health and Ageing, Australian Government 38
depression 237
diabetes
   complications of 205
   food pyramid 289-290
   gestational 104, 117
   hyperglycemia 208
   hypoglycemia 208
   maintaining glucose levels 290*t*
   neuropathy 209
   overview 205
   physical activity and 9, 207-208
   preactivity screening and clearance 19
   premeal insulin bolus 216*t*
   preventing and managing 206-297
   protein guidelines 290
   type 1 213-218
   type 2 151-152
diet. *See also* nutrition
   achieving and maintaining a healthy weight 33-34
   for children and adolescents 88-89
   guidelines for adults 35
   nutrient intake 32-34
   physical activities for health 4-5
   Physical Activity Guidelines for Americans 19
Dietary Reference Intakes (DRIs) 265-267
dietary sodium 289

Dietary Supplement Health and Education Act 277
dietary supplements 277-278
Dietitians of Canada 281-282
disability expenses 11
diseases, dietary guidelines for cancer 285-288
disease severity 258
dobutamine stress echocardiography 261
dopamine 224
dose-response relationship 7-8, 7f, 19
doubly labeled water analyses 33
duration of activity 47
Dutch study on asthma 246
dynamic flexibility 7, 27
dyslipidemia 9

E
ECG abnormalities 257
effort-to-benefit ration 74
Ehrman, J.K. 158
Eisenhower, Dwight D. 4
electrocardiogram (ECG) monitoring 161
electrolytes 275-277
electron beam computed tomography 261
electronic media guideline 87
elliptical trainers 295-296
endometrial cancer 146-147, 150
endorphins 10
endurance 133
    benefits of improvement in 48
    muscular strength and 18-19
endurance training 129
energy balance 263-264
energy bars 276-277
ergometer 259
essential fatty acids (EFAs) 280
exercise
    aerobic 6
    contraindications for 175
    contraindications to 141, 164-165
    gravity-reduced 222
    hemodynamic response to exercise 260
    high-impact 198
    measurement of intensity 45
    physical activity *versus* 6
    protection against cancer 9
    smoking cessation effects of 10
    target heart rate range 45t
exercise echocardiography 260
exercise intensity. *See* intensity
exercise programs
    components of 6-7, 6t

exercise videos, CDs and DVDs 300
exertional myocardial ischemia 257

F
facioscapulohumeral M.D. 226-228
fall risks 137
fat 268
fatigue 156, 237
fat intake 279-280, 283
fetal blood flow 107
fiber 267, 283
Fit Facts
    asthma 241
    postpartum health 118
    strength training 99
Fit Facts information sheets (ACE)
    health maintenance 55
    on hydration 275
fitness and health benefits 27-28
Fitnessgram 81, 88, 250, 252t
Fit Society Page (ACSM) 127-128
FITT principle
    definitions 44-45
    for people with arthritis 188
    recommendations for pregnant women 108-109
flexibility exercise 139
    benefits for older adults 123
    for older adults 126-127
    for people with arthritis 185
    for people with RA 182-183
    physical activity pyramid 56
    recommendations for 50
flexibility training
    ACSM guidelines 27
    benefits of improvement in 27
    Canadian guidelines for 38
    definition of 6-7
    increases in 48
    for joints 10
    recommendation for 27
fluid balance 280
folate 267, 283
Food and Nutrition Board 265, 283
food guide pyramid 271, 273
food safety 271
fractures, osteoporosis-related 194f
frailty 134-138
Framingham Heart Study 8
free play 62
free weights 297-298
frequency
    definition 44
    of exercise for pregnant women 108
fruit 277
fruit and vegetables 268-269
fuel sources 276-277
functional independence 9-10

functional status 133-134, 220

G
gels, carbohydrate 276
gestational diabetes 117
Global Strategy on Diet (WHO) 38-40, 274
glucose levels 290, 290t
glucose levels, monitoring 208-209
glycogen storage 8
grain products 273
grains 268
gravity-reduced exercise 222
growth curves 264-265
Guidelines for Exercise Testing and Prescription 113, 127-128
gyms, home 300

H
Hagberg, J. 172-173
Harvard Alumni Study 6
head and neck cancers 163-164
health
    leading indicators 29
health and fitness outcomes for children and adolescents 78-79
Health Canada 34, 37-38, 273
health care costs 10-11
health status 133-134
Healthy Fitness Zone (HFZ) 81-82
Healthy People Consortium 30
Healthy People report
    history of 4-5
    mandatory physical education 81
    school and community health guidelines 84-85
    significant historical guidelines 29-30, 31f
heart disease
    physical inactivity as a risk factor for 15
    structural 11-12
heart rate calculations 45
heart rate monitors 299
heart rate reserve (HRR)
    ACSM recommendations 21
    definition 44
heart rate target zones 110
hematologic cancers 163
heme iron 269
hemodynamic response to exercise 260
high-carbohydrate bars 276-277
high-density lipoprotein
    effect of aerobic activity on 8
high-density loading forces 95
higher-intensity loading forces 95
high fasting blood triglycerides 151
high-risk behaviors 73

high-risk cardiac conditions 175
home gyms 300
hormonal changes associated with pregnancy 117
hormone levels 146-147
HRmax. *See* maximum heart rate
HRR. *See* heart rate reserve
hydration 274-276, 275*t*
hydrodensitometry 251
hyperglycemia 208
hypertension 9, 288
    ACSM guidelines for 169-171
    benefits of exercise for 167-169
    classification of blood pressure 168*t*
    international guidelines 171-172
    resting blood pressure 168*t*
hypoglycemia 208, 217

I
immune system 147
impaired fasting glucose 152
impaired glucose tolerance (IGT) 152
inactivity 56
infants
    AAP recommendations for 61
    healthy development 62
    muscular-strengthening exercises 118
    proper weight for 264-265
    sudden infant death syndrome 68
Institute of Medicine (IOM)
    Food and Nutrition Board 265
    new physical activity guidelines 5
    nutrient intake guidelines 32-34
    protein RDA 267
insulin bolus, premeal 216*t*
insulin resistance syndrome 151. *See also* diabetes
intensity
    of exercise for pregnant women 108
    measurement of 45
    talk test used for testing 46
intensity of physical activity 17, 19
    classified as 6.0 METs 20*t*
    FITT variable 44-45
International Association for the Study of Obesity 38-39, 93
International Consensus Conference 92
International Osteoporosis Foundation (IOF) 95
International Society of Hypertension 172
interval training 49
IOM. *See* Institute of Medicine
iron 269

isokinetic training 6*t*
isometric exercises during weight-lifting 106
isometric training 6*t*
isotonic training 6*t*

J
joint flexibility 10
Journal of Obstetrics and Gynaecology Canada 111
Journal of Pediatrics 78
Journal of the American Medical Association 24

K
Karvonen method 45
Kellogg Company 65
ketones 217
Kettles, M.A. 118
KidsHealth 110, 116
Kids in Action 65

L
Larsson, L. 107
LDL. *See* low-density lipoprotein
lean body mass 264
Leavitt, Mike 19
levadopa 226
lifestyle physical activities (LPAs) 56, 74
Lindqvist, P.G. 107
linoleic acid 267, 283
lung cancer 163
lymphedema 164

M
mandatory physical education 81
maximum heart rate (HRmax)
    ACSM recommendations 21
    calculating 45
    interindividual variation in 46
    older adults 122
maximum oxygen consumption. *See* $\dot{V}O_2$max
Mayo Clinic 255
meat or alternatives 273
medical disorders contributing to frailty 138
Medicine & Science in Sports & Exercise 22, 126, 169-170, 174
    cardiorespiratory fitness 27
    muscular strength 18
Mesa Petroleum 11
metabolic syndrome 150-152, 151*f*
MET-minute 44
METs 44, 258
    activities classified as 20*t*
    aerobic activity 17
microbial food-borne illness 271

milk and alternatives 273
milk and milk products 269
mode of activity 47
moderate intensity 19
mood-enhancing neurotransmitters 10
Morbidity and Mortality Weekly Report 75
morning stiffness 182
motor skills 74
multiple sclerosis 231-233
multivitamins 278
muscle fitness 56
muscular conditioning 119
muscular dystrophy
    Duchenne 228-229
    facioscapulohumeral 226-228
    guidelines for 226-228
muscular strength
    benefits of improvement in 48
    endurance and 18-19
musculoskeletal health 220
musculoskeletal injuries 11
MyActivity Pyramid 88
myocardial contractility 260-261
myocardial oxygen demand 12
MyPyramid 281-282
MyPyramid for Preschoolers 65

N
NASPE. *See* National Association for Sport and Physical Education
National Academies Press 82
National Academy of Sciences 34
National Association for Sport and Physical Education (NASPE)
    Active Start 63-64
    infants, toddlers, and children 15
    physical activity 75
National Association of State Boards of Education 82
National Center for Health Statistics 63
National Cholesterol Education Program (NCEP) 150-151
National Heart, Lung, and Blood Institute (NHLBI)
    guidelines on treatment for overweight and obesity 28-29
National Institute on Aging (NIA) 130
National Institutes of Health (NIH) 25
National Osteoporosis Foundation (NOF) 73
National Physical Activity Guidelines for Australians 39

National Sporting Goods Association (NSGA) 13
National Strength and Conditioning Association (NSCA) 32, 97
Nelson, M.E. 73
nephropathy 209
neurological benefits 123-124
neuromuscular training 140, 186
neuropathy 209
neurotransmitters, mood-enhancing 10
NHLBI. *See* National Heart, Lung, and Blood Institute
NIH. *See* National Institutes of Health
No Child Left Behind Act 81
noncommunicable diseases 39-40
NSCA. *See* National Strength and Conditioning Association
NSGA. *See* National Sporting Goods Association
nuclear imaging 260-261
Nurses' Health Study 123
nutrition. *See also* diet
    for athletes 279, 281-282
    connection between food and fitness 69
    dietary supplements 277-278
    food groups 270
    food guide pyramid 271, 273
    fruit and vegetables 268-269
    goals of care for cancer survivors 287-288
    grain products 273
    grains 268
    meat and alternatives 273
    milk and alternatives 273
    milk and milk products 269
    MyPyramid 281-282
    reducing the incidence of cancer 286-287
    vegetables and fruit 273
Nutrition and Physical Activity (PCPFS) 277

O
obesity
    BMI 264
    in children 67-69, 72
    International Association for the Study of Obesity 38-39
    metabolic syndrome 150-151
    NHLBI guidelines on treatment 28-29
    prevention 94-95
    risk of diabetes 152-153
    risk of health complications 72
Obesity Reviews 93
objective measures 45

Obstetrics and Gynecology 111
Office of the Surgeon General. *See* U.S. Office of the Surgeon General
older adults
    benefits of physical activity 121-124
    benefits of strength training 121-122
    flexibility training 126-127
    PAGA recommendations 124
omega-3 polyunsaturated fatty acids 287-288
1RM (1-repetition maximum) 44
osteoarthritis (OA) 183-185, 183*f*, 187-193
osteopenia 50
osteoporosis
    benefits of activity 194*f*
    bone density 194, 194*f*
    guidelines for 201-202
    overview 193
    prevention of 195-200
    role of diet in 288-289
    weight-bearing activities 195
Ottawa Panel 190-192
overload
    definition 7, 44
    progressive 33
overtraining 100
overuse injuries 100
oxidants 278
oxygen consumption 45

P
PAGA. *See Physical Activity Guidelines for Americans*
PALs 38-39
pancreatic cancer 150
Pangrazi, R.P. 74
parents as role models 62
Parkinson's disease
    guidelines for 224-226
PAR-Q (Physical Activity Readiness Questionnaire) 13
patterns in physical activity participation 12-13
PCPFS. *See* President's Council on Physical Fitness and Sports
Pediatric Diabetes 291
Pediatric Exercise Science 92
Pediatrics 62, 68, 77
pedometers 36, 299
perceived barriers to activity 13
perfusion imaging 261
periodization 7, 44
peripheral neuropathy 209
peripheral resistance 8
perishable foods 271

personal trainers 304
pharmacological stress testing 261
physical activity
    Australia's recommendations for 66
    changes and benefits resulting from 7-11
    definition of 5
    effect on health care costs 10-11
    versus exercise 6
    for functional independence 9-10
    health benefits of 8-9, 20*t*
    lack of, health problems related to 3-4
    perceived barriers to 13
    percentage of high school students participating in 72-73
    recommendations 269-270
Physical Activity and Health
    history of 4
    justification for guidelines 19
    significant historical guidelines 26
Physical Activity for Total Health Study 146
Physical Activity Guidelines for Americans (PAGA)
    activity for children under 6 y 65
    health benefits of activity 19, 20*t*
    lowering risk factors 21
    physical activities for health 4-5
    recent national guidelines 17
    recommendations for pregnant and postpartum women 103
Physical Activity Guide to Healthy Active Living (Canada) 38, 90
physical activity levels (PALs)
    definition 44
    total energy expenditure 33
physical activity pyramid 55-56, 56*f*, 87-88
Physical Activity Readiness Questionnaire (PAR-Q) 13
Physical Activity Recommendations for 5-12 Year Olds 91
physical education 81-82
playgrounds 69
pleasurability 73-74
PNF. *See* proprioceptive neuromuscular facilitation
postexercise nocturnal hypoglycemia 217
post-myocardial infarction testing 258
postpartum women
    aerobic exercise 119

breast-feeding 117-118
depression 10
muscular conditioning 119
need for physical activity 103-104
weight loss 285
potassium 270
Powell, K.E. 9
PowerBar 276
precipitation of labor 107
predicting target heart rate ranges 45*t*
preeclampsia 104
preexercise clearance 13
pregnant women. *See also* breast-feeding women; postpartum women
aerobic exercise 113
anatomical changes during pregnancy 105
benefits and risks of physical activity 104-106
Canadian guidelines for 109
concerns for competitive athletes 107-108
considerations for 112, 114
contraindications to exercise 106
data 104
effects of training and competition on 107-108
gestational diabetes 117
heart rate target zones 110
hormonal changes 117
iron requirements 269
maternal fitness 114
meeting metabolic needs of 285
need for physical activity 103-104
resistance training 113
premeal insulin bolus 216*t*
preparticipation exam
for high school students 99-101
Presidential Active Lifestyle Award 34, 89
President's Challenge 34-36, 89-90
President's Council on Physical Fitness and Sports (PCPFS) 34, 50-54, 74
President's Council on Youth Fitness 4
progression 7, 44
progressive overload 33
progressive resistance training 139
promoting activity through awareness 90
Promoting Physical Activity in Children and Youth 86
proprioceptive neuromuscular facilitation (PNF) 50

Prospective Diabetes Study (UKPDS) 207
prostate cancer 163
protein 267, 279, 283
Public Health Agency of Canada 38, 90-91

Q
quality of life 30
quantifying physical activity 36

R
Radcliffe, Paula 117
range-of-motion testing 227
rating of perceived exertion (RPE) 46
recommendations from the United Kingdom 90-92
repetitive stepping 296
resistance training
adaptations to 8
for adolescents 99
benefits of 31-32
for children 96-99
for CVD individuals 176-180
definition of 6
exercise prescription 51
guidelines for youth 99
for healthy adults 22
for people with arthritis 185
for pregnant women 113
recommendation for 27
variables in 49-50
respiratory tract infections 8
resting heart rate 45
retinopathy 209
revascularization of the heart 176
rheumatoid arthritis (RA) 181-183, 183*f*
rheumatoid nodules 182
risks of physical activity 11-12
risk stratification 261
Robert Wood Johnson Foundation 128, 131-132
rowing machines 296
Royal College of Obstetricians and Gynaecologists (RCOG) 110
RPE. *See* rating of perceived exertion

S
safety, of postpartum activity 117-118
SAT-9. *See* Stanford Achievement Test
school and community health guidelines
American Heart Association 85-87

general guidelines on the role of schools 82-83
Healthy People 84-85
overview 81
role of physical education in academic performance 81-82
schools
guidelines focusing on the role of 82-83
leadership role for 86
physical activity as part of dietary guidelines 88-89
physical activity pyramid 87-88
Schwarzenegger, Arnold 37, 67
sedentary individuals, guidelines for 50-51, 54
sedentary lifestyle 3-4
semirecumbent bicycle 295
sensitivity of exercise tests 257
serum rheumatoid factor 182
Shephard, R.J. 11
Shorter, Frank 31
skeletal bone growth and density 95
skin conditions 164
skinfolds 251
sleep quality 10
social influences 13
Society of Obstetricians and Gynaecologists of Canada (SOGC) 108-109
sodium 270
sodium, dietary 289
specificity
definition 7, 44
in resistance training programs 33
spinal cord injury and disability 233-235
sports drinks 276-277
Sports Research Digest 74
stability 123
stair steppers 296-297
Stanford Achievement Test (SAT-9) 81
The State of Aging and Health in America 121
static flexibility 6-7
static flexibility training 27
static stretching 188
stationary bicycles 294-295
strength
age-related loss of 10
benefits of training 122-123
strength exercises 133
benefits of 121-122
Canadian guidelines for 38
for people with RA 182
stroke and brain injury 235-236
stroke-related fatigue 237

structural heart disease 11-12
sudden cardiac death 12, 12*f*
sudden infant death syndrome
(SIDS) 63
sugar 267
supplements and hydration 280
symmetric arthritis 182
syndrome X 151

T
tachycardia 226
talk test 46
target heart rate range 45*t*
teens. *See* adolescents
testicular cancer 147
The President's Challenge 34-36
therapeutic lifestyle changes (TLC)
288
thermoregulation 106-107
*The State of Aging and Health in
America* 121
time 47
tobacco use 9
total energy expenditure (TEE)
doubly labeled water analyses 33
treadmill exercise testing 255, 259
treadmills 293-294, 295*t*
trends in physical activity 13
type 2 diabetes. *See* diabetes
type of activity 47

U
upper gastrointestinal cancer 163-
164
U.S. Department of Agriculture
(USDA) 34-35, 65
U.S. Department of Defense
(USDoD) 34
U.S. Department of Health
and Human Services
(USDHHS)
dietary guidelines 34-35
physical activity guidelines for
Americans 19, 20*tt*
U.S. Office of the Surgeon General
4, 121
USDA. *See* U.S. Department of
Agriculture
USDHHS. *See* U.S. Department
of Health and Human Ser-
vices
USDoD. *See* U.S. Department of
Defense

V
vasodilators 261
vegetables and fruit 273
very low-density lipoprotein
(VLDL) 8
videos 300
vigorous intensity 19
vitamin B 288-289
vitamin B6 285
vitamin D 267, 283
vitamins 278
VLDL. *See* very low-density lipoprotein
$\dot{V}O_2$max (maximum oxygen consumption)
definition 44
older adults 122
$\dot{V}O_2$R (oxygen consumption
reserve)
definition 44

W
Web sites 160
ACE Fit Facts 112
ACSM 277
Active Start 63
American Council on Exercise
56*f*
American Heart Association 32,
86
American Pregnancy Association
116
asthma 245-246
Australian Government Depart-
ment of Health and Ageing
39
balance and flexibility training
140
BMI-for-age calculators 265
bone health 199
California Center for Physical
Activity 137
California physical activity 66
Canada's food guide 274
Canadian Society for Exercise
Physiology 38
cancer organizations 159
Comprehensive School Physical
Activity Programs 84
Consensus Development Panel.
See National Institutes of
Health
Department of Health, United
Kingdom 135

Diet and Cancer Report 286
dietary guidelines 35, 89, 132
Guidelines for Americans 80
National Blueprint 131
National Heart, Lung, and Blood
Institute 29
NCPAD 140
NSCA publication 33
PAGA 125
Public Health Agency of Canada
38, 90
RCOG 112
rheumatology 187, 189
WHO nutrition publications 284
weight-bearing activities 50, 195,
197
weight control 9, 27-28
weight loss 28
weight machines 298
weight management 269
weights 297-298
Welk, G.J. 74
WHO. *See* World Health Organization
whole-grain products 270
women. *See also* postpartum women;
pregnant women
breast-feeding women 115
endometrial cancer 146-147
hormone levels 146-147
with osteopenia 50
workload 255
workout facilities 301-303
World Cancer Research Fund 149-
150, 286
World Health Organization (WHO)
dietary recommendations 38-40,
274
overweight adults 9
Physical Activity and Young
People 93
recommended amount of physical
activity 41
World Osteoporosis Day 202-203
worldwide noncommunicable diseases 39-40

X
X-linked recessive genetic disorders
226

Y
Youth Risk Behavior Survey 72

ABOUT THE AUTHOR

Riva L. Rahl, MD, is medical director of the Cooper Wellness Program at the Cooper Aerobics Center in Dallas, Texas, where she counsels patients and clients regarding appropriate physical activity programs.

Board certified in both internal and emergency medicine, Rahl is a staff physician at the Cooper Clinic in Dallas and was previously chief resident in emergency medicine at the University of Texas Southwestern (Parkland) Hospital. She is a member of the American College of Sports Medicine, the American College of Physicians, and USA Track and Field. She received her medical doctorate from the University of California at San Francisco in 1999.

Now a competitive marathoner, Rahl was a four-year varsity athlete in cross country and track and field at Rice University (Division I), where she studied biochemistry and exercise science. She has won several races, including the Dallas White Rock Marathon in 2000 and the Fort Worth Cowtown Marathon in 2000, 2002, and 2008.

Rahl and her husband, Brian, and two young sons reside in Dallas. Rahl enjoys training for race day, caring for her children, and traveling internationally.